Telecommunications Engineering

H. G. Brierley
MA, PhD, CEng, MIEE

School of Electrical Engineering and Science,
Royal Military College of Science—Cranfield,
Shrivenham, Swindon, UK

Edward Arnold

First published in Great Britain 1986 by Edward Arnold (Publishers) Ltd, 41 Bedford Square, London WC1B 3DQ

Edward Arnold (Australia) Pty Ltd, 80 Waverley Road, Caulfield East, Victoria 3145, Australia

Edward Arnold, 3 East Read Street, Baltimore, Maryland 21202, U.S.A.

British Library Cataloguing in Publication Data

Brierley, H. G.
　　Telecommunications engineering.
　　1. Telecommunication systems
　　I. Title
　　621.38　　　　TK5101

　　ISBN 0-7131-3558-1

Text set in 10/11 pt Times
by Macmillan India Ltd, Bangalore 25
Printed in Great Britain by J. W. Arrowsmith Ltd,
Bristol

Preface

Aims

This book has arisen out of the author's experience of teaching telecommunications systems engineering to undergraduates in the final year of an electrical-engineering degree course.

The primary aim is to assist undergraduates to improve their engineering skills by analysing a wide range of problems and, in so doing, to acquire a broad knowledge of techniques used in telecommunications systems.

A secondary aim is to provide a compact source of information for practising engineers from which they can derive insight into techniques used in telecommunications systems and obtain outlines of how the analysis of those techniques can be approached.

At first sight, to cover in a relatively small book the broad field suggested by the title *Telecommunications Engineering* would appear to necessitate a superficial coverage inconsistent wiitith the above aims. This has been overcome in part by assuming considerable prior knowledge of the basic principles of telecommunications, in part by adopting a concise style of writing and in part by limiting the number of topics covered. Details of the pre-requisite knowledge are given below and in each chapter references are given to help the reader find fuller explanations. As is the case with all books, the omission of a topic will occasionally disappoint the reader using this book for reference.

Pre-requisite knowledge

This book is not intended to be another introduction to the basic principles of telecommunications. There are now many excellent introductions and the reader of this book is assumed to be familiar with at least one of them.

Most of the basic telecommunications material which the reader is expected to have met can be found in the books entitled *Signals, Modulation, Noise, Wave transmission, Networks* and *Antennas* from the series *Introductory Topics in Electronics and Telecommunications* by F. R. Connor, published by Edward Arnold (Publishers) Ltd. This material may be summarised as follows.

Signals Time-domain analysis using impulse response and convolution; transfer functions and their relationship to impulse response; Fourier transforms and energy-spectra of finite-energy functions; Fourier series of periodic functions and the envelope of the spectrum as the transform of one period of the function; the Laplace transform; Shannon's sampling theorem; information content (entropy).

Modulation Basic properties of amplitude modulation and its variants (single-sideband, vestigial-sideband and suppressed-carrier double-sideband); coherent demodulation; quadrature amplitude modulation; the double-balanced modulator; the envelope detector; basic properties of angle modulation, instantaneous phase deviation and instantaneous frequency deviation; phasor diagrams to describe modulation; complex envelope and its relation to the spectra of modulated waveforms.

Noise Power spectra of random processes, the Wiener-Khintchine relations; noise temperatures of one-port and two-port networks; noise factor; noise bandwidth; descriptions and properties of narrow-band gaussian noise; comparison of the effects of noise in amplitude-modulation and angle-modulation systems, FM advantage; emphasis in frequency-modulation systems.

Wave transmission Propagation constant and characteristic impedance of transmission lines; reflexion coefficient.

Networks Analysis of linear networks; return loss; the prototype low-pass filter.

Antennas Gain and effective area; effective length; free-space loss.

In addition it is assumed that the reader is familiar with other topics usually covered in the first two years of an undergraduate course in electrical engineering. In particular he or she should have a little knowledge of the theory of electromagnetic waves (refraction, boundary conditions and group velocity), should understand the function of a shift register and should be familiar with first-order and second-order loops in classical control theory.

As far as possible, complicated mathematics has been avoided, but readers may well find some difficulty in finding the mathematical tools they require unless they have a thorough grounding in trigonometry, calculus, algebra and probability theory. The necessary algebra includes matrix partition and multiplication, the concept of a group and the manipulation of polynomials with binary coefficients. The probability theory used extends as far as generating functions, but the vast majority of its applications are either simple applications of discrete probability distributions (for example the binomial distribution) or derivations of means and variances from probability density functions.

Philosophy

Students of electrical engineering sometimes have considerable difficulty when confronted by a calculation in a system context, even when they have already performed very similar calculations as examples of previously expounded techniques. Since those techniques have presumably been taught because a competent electrical engineer should be able to apply them, it is disturbing to meet students who do not recognise situations in which they can be used.

The students' difficulty is often attributed to 'compartmentalisation', a bad habit all to often encouraged in degree courses. When related subjects are taught by different lecturers, they often appear to students as independent. In such situations, students are disinclined to compound their difficulties by looking for interactions between the subjects, preferring to isolate them in mental 'compartments' and even to resist lecturers' attempts to relate them. Indeed, compartmentalisation is a good strategy for passing traditional examinations, which are usually structured to reflect the independent teaching. By the time they reach the study of systems, some students have conditioned themselves not to consult more than one set of notes when attempting any one problem.

This difficulty is a reminder that a deeper understanding is required when choosing a technique appropriate to the solution of a given problem than when applying a prescribed technique. To find an appropriate technique requires a comprehensive perspective of the techniques available, their limitations and the situations in which they may be of use. A few gifted engineers may acquire such a perspective as they first study the techniques, but the majority achieve it gradually with experience. Most undergraduates need to work towards that deeper understanding by means of projects and tutorial work involving the application of techniques learned in earlier courses or discovered in books and journals.

A distinctive mark of the creative engineer is to be primarily an inventor and a problem-solver and secondarily an analyser and a technician. This does not mean that engineers do not need to be good analysers and technicians, but that those aspects of their work should support their primary task and not be the principal sources of their personal satisfaction.

In other words the engineers' attitude of mind should be different from that of the mathematician or that of the technician. Degree courses in engineering should be designed to encourage an engineering attitude of mind, and any assessment of the potential of an engineer should take into account his ability to solve problems and invent as well as his knowledge of techniques. The relative satisfaction obtained from the various aspects of an engineer's work can become very deeply rooted in an individual. It is therefore important to establish the right priorities during the education of an engineer and experience of applying previously learned material to the problems of realistic systems will help to do this.

The inclusion in an electrical-engineering degree course of systems-oriented components in which the emphasis is on the solving of problems rather than the learning of techniques is therefore desirable for at least these three reasons: they break down compartmentalism; they deepen understanding of earlier material; and they encourage an engineering attitude of mind.

Since not all kinds of electrical-engineering system can be covered by any one student during a degree course, considerable specialisation is necessary. Suitable systems range in magnitude from the detailed design of transistor amplifiers, which calls upon the students' experience of circuit analysis, feedback theory and electronic devices, to the outline design of a large telecommunications system, requiring knowledge of signal analysis, noise, modulation and the statistics of traffic. Undergraduates will usually prefer to study systems related to their intended fields of activity and thus as a by-product equip themselves with insights and knowledge relevant to their future careers.

This book is intended to be primarily a support for such a systems-oriented course on telecommunications. It can be used by students as a resource in the solution of assigned problems involving several aspects of system design or it can be used as a textbook which students work through. When using it in the former way, students will find many references to sources containing greater detail. Students using it in the latter way will almost certainly not have time to attempt all the exercises in the book and should select, or be directed to, the topics which interest them most.

Approximately half of the book is occupied by over two hundred exercises with solutions. This is a reflexion of the author's belief that learning a subject like telecommunications systems is not unlike learning to find one's way about a strange city, which is done more effectively and more quickly by driving oneself and occasionally getting lost, than by sitting in a taxi.

Since one of the objects of a systems-oriented course is to compel students to look at other courses for techniques necessary to solve the problems with which it presents them, this book has been designed not to be self-contained. It is recognised, however, that there are many demands on an undergraduate's time and that he or she cannot be expected to do an undue amount of looking-up or always to have other books to hand. This book therefore contains appendices which give many of the formulae and definitions the reader may need, but which do not give full explanations or derivations. The appendices also serve to establish the notation and terminology used throughout the book.

Structure

This book brings together a wide range of topics rarely found together in one book, particularly in a comparatively short textbook for students.

The wish to make the book suitable for students to choose topics from it made it undesirable to impose a rigorous logical development; the decision to keep the book reasonably short by omitting introductory material made such a rigorous development impossible. Nevertheless there are many interrelations between the topics and they have been ordered in such a way that the reader investigating the interrelations usually needs to refer backwards rather than forwards. Closely related topics have been grouped into chapters.

Broadly speaking the book may be regarded as comprising two main parts, Chapters 1–3 and Chapters 5–9, separated by a short chapter (Chapter 4) on switching systems and followed by an Appendix containing information useful for exercises throughout the book.

Chapters 1–3 are concerned primarily with the basic types of telecommunications signal, standards of quality, and techniques used in preparing the signals for transmission. Chapter 1 begins with an outline of standards of quality in telephony, continues with some techniques used in traditional line telephony, covers the digital encoding of speech and concludes with a section on programme sound. Chapter 2 deals with digital signals. It begins with an introduction to the terminology of telegraphy, continues with the detection of digital signals and error control, and concludes with a short discussion of data transmission over telephone channels. Chapter 3, on the transmission of pictorial information, concentrates on the video waveform for colour television and the digital encoding of pictures.

Chapter 4 gives a brief introduction to the basic principles of switching systems, with particular emphasis on the calculation of lost-call probability.

Chapters 5–9 are concerned with transmission. Chapter 5 deals with carrier telephony over coaxial cable and over FM radio and includes studies of the effects of noise, non-linearity and transmission deviations. The following chapter (Chapter 6) consists of topics important in the line transmission of digital signals, including intersymbol interference, spectrum shaping and the limitation of route length by noise, crosstalk and jitter. This is followed by three chapters (Chapters 7, 8 and 9) dealing with radio transmission. Chapter 7 concentrates on three techniques: threshold extension in FM receivers, diversity and spread-spectrum transmission. In Chapter 8 a selection of data is presented to enable estimates to be made of the effect of the atmosphere on propagation and of the noise introduced in radio paths. Chapter 9 consists of exercises relating to radio systems and is based on material covered in many different parts of the book.

The Appendix consists almost entirely of definitions and formulae useful for solving the exercises in Chapters 1–9. Areas covered include the Fourier transform, the response of a linear channel, parameters describing noise performance, antenna parameters and some basic concepts of information theory.

Acknowledgements

The author is indebted to colleagues and undergraduates at the Royal Military College of Science for the many conversations which have helped to clarify his thinking on many topics. Any errors remaining in this book are his sole responsibility and he would be grateful to any readers who draw his attention to them. He would also like to thank his wife, Joy, for her forbearance and for typing almost all the manuscript.

1984   H G B

Contents

Preface   iii

1   Telephony and programme sound   **1**

   1.1   Channel requirements for telephony   1
   1.2   Line transmission of telephony   4
   1.3   Digital encoding of speech   8
   1.4   Transmission of programme sound   14
   1.5   References   17
   Exercise topics   19
   Exercises   20

2   Digital communications   **37**

   2.1   Telegraphy   37
   2.2   Detection of digital signals in noise   39
   2.3   Error control   42
   2.4   Digital communications over telephone channels   51
   2.5   References   53
   Exercise topics   54
   Exercises   55

3   Transmission of pictorial information   **73**

   3.1   Facsimile   73
   3.2   Monochrome television   73
   3.3   Colour television   75
   3.4   Digital encoding of pictorial information   77
   3.5   References   79
   Exercise topics   80
   Exercises   81

4   Exchanges   **93**

   4.1   Telephone switching systems   93
   4.2   Store-and-forward systems   95
   4.3   Telephone-traffic theory   96
   4.4   References   101
   Exercise topics   102
   Exercises   103

5   Carrier telephony   **113**

   5.1   The carrier-telephony signal   113
   5.2   Signal degradations   115
   5.3   Line transmission   117

5.4 Radio transmission 122
5.5 Use of carrier telephony systems for signals other than speech 126
5.6 References 128
Exercise topics 130
Exercises 131

6 Topics in digital line transmission **143**

6.1 Intersymbol interference 143
6.2 Spectra of digital signals 145
6.3 Digital routes with repeaters 147
6.4 Digital multiplexing 150
6.5 References 152
Exercise topics 153
Exercises 154

7 Radio techniques **183**

7.1 The superhet receiver 183
7.2 Threshold extension 183
7.3 Diversity 186
7.4 Spread-spectrum techniques 188
7.5 References 193
Exercise topics 194
Exercises 195

8 Radio paths **215**

8.1 Propagation in the troposphere 215
8.2 Propagation in the ionosphere 221
8.3 Noise in radio paths 223
8.4 References 225
Exercise topics 226
Exercises 227

9 Miscellaneous radio systems—exercises **243**

Appendix **259**

A.1 The Fourier transform 259
A.2 The response of linear channels 263
A.3 rect(x) and sinc(x) 264
A.4 erfc(x) 264
A.5 Insertion gain, insertion loss and return loss 265
A.6 Noise temperature, noise bandwidth and noise factor 265
A.7 FM advantage 267
A.8 Systems aspects of antennas 268
A.9 Information theory 269

Bibliography **271**
Index **273**

1
Telephony and programme sound

1.1 Channel requirements for telephony

1.1.1 Subjective assessment

Assessment of the performance of a telephone connexion is ultimately a matter of opinion and the standards to which a system is designed are derived from experiments involving large numbers of subjects representative of the potential users of that system. In the case of a public telephone system, it is important that the subjects should have had no previous training.

A convenient form of experiment is the listening test, in which each subject listens to pre-recorded speech under varying conditions. Often the pre-recorded speech consists of groups of five short unrelated sentences, the conditions of the experiment remaining unchanged throughout each group. Each subject may be asked to indicate whether or not he has understood each sentence or to grade each group according to scales such as the following.

Loudness preference scale
- A   Much louder than preferred
- B   Louder than preferred
- C   Preferred
- D   Quieter than preferred
- E   Much quieter than preferred

Listening effort scale
- A   Complete relaxation possible; no effort required
- B   Attention necessary; no appreciable effort required
- C   Moderate effort required
- D   Considerable effort required
- E   No meaning understood with any feasible effort

Since the primary purpose of a telephone connexion is to enable a conversation to take place, more realistic assessments are obtained by arranging for pairs of subjects to engage in conversations with clear objectives. In a typical conversation test the caller and the responder are provided with maps which are identical except for the omission from the caller's map of the names of certain features. The caller is given the specific task of finding out those names from the responder. The time taken to perform the task is measured and the subjects may be required to grade the effort required on a scale similar to the listening-effort scale or to grade the quality of the connexion according to a scale such as the following.

- G   Good circuit; no appreciable effort required
- F   Fair circuit; some effort required in conversing
- P   Poor circuit; conversation possible but much difficulty on unfamiliar words
- B   Bad circuit; usable only with extreme difficulty

For a thorough discussion of channel assessment, see Richards[1].

1.1.2   Frequency response

Although the human ear is sensitive to frequencies extending from about 20 Hz to about 15 kHz, sufficient intelligibility and speaker recognition remain when a telephone channel restricts the transmitted bandwidth of speech to considerably less than this. Indeed, in some cases the removal of the higher frequencies improves the quality of the channel by increasing the signal-to-noise ratio at the receiver. The band of frequencies transmitted depends to some extent on the transmission system, but usually extends from about 300 Hz to about 3 kHz. There are international agreements applicable to various transmission systems.

Ohm's law of hearing, according to which the perceived quality of a sound depends only on its power spectrum, suggests that variation of phase shift with frequency should be of no significance in a telephone channel. However, variations of phase shift may accumulate in a long route to give variations of group delay (Appendix 2) comparable to the duration of syllables and therefore audible. Excessive group delay at high frequencies gives rise to a ringing effect and at low frequencies to a blurring of the speech. Typical specifications of group delay in a long route are that it should not exceed its minimum value by more than 60 ms at the upper frequency limit or by more than 30 ms at the lower. Group delay due to filtering of the speech signal is usually greatest near the frequency limits, but that due to inadequate stability margins (Section 1.2.3) may occur well inside the frequency band.

1.1.3   Noise and the psophometer

In common with all other electrical signals, the speech waveform in a telephone receiver is accompanied by electrical noise of various kinds and too low a signal-to-noise ratio will render a telephone channel unacceptable.

Since the combined subjective frequency response of a telephone receiver and a human ear is far from uniform, the annoyance caused by noise is not a function of noise power alone. A measure of the annoyance value of electrical noise in a telephone channel may be obtained by the use of a psophometer. A psophometer is basically an r.m.s. voltmeter with sensitivity varying with frequency according to an internationally agreed[2] characteristic and having a reduced response to bursts of noise lasting less than 200 ms. It is calibrated so that a continuous 800 Hz tone at a level of 1 mW gives a scale reading of 1 mWp (0 dBmp), the letter p indicating a psophometric measurement. The characteristic is such that 1 mW of gaussian noise spread uniformly over the frequency band from 300 Hz to 3400 Hz gives a reading of -2.5 dBmp.

In a commercial telephone channel the ratio of the mean power of continuous speech to the psophometrically measured noise power is usually not less than about 26 dB.

1.1.4   Dynamic range, mean power and volume

If a telephone channel cannot handle the largest excursions of the speech waveform without peak-clipping, distortion is introduced. Experience suggests that the distortion is not objectionable provided the mean power of continuous speech is at least 10 dB below that of a 1 kHz 'test-tone' which can be handled without peak-clipping. Hence, if the mean continuous speech power can be controlled, the level of the test-tone should in commercial telephony be at least 36 dB above that of the noise. If the mean continuous speech power is not controlled, it may vary from talker to talker by 30 dB or more and the ratio of test-tone power to noise power should be at least about 66 dB.

The mean continuous power of a speech waveform may be measured by a speech voltmeter, which is basically an r.m.s. voltmeter with a voice-operated switch to interrupt the averaging during pauses in the speech. Averaging times of a few seconds are used.

Use is sometimes made of the concept of 'volume'. Volume is an electrical quantity measured according to specific rules on a volume meter. The basic rule is that the operator should mentally average the peak swings of the meter, ignoring occasional high peaks. A volume meter is basically an r.m.s. voltmeter with a uniform frequency response and time constant small enough for syllabic variations to be followed (about 120 ms). It has a logarithmic scale identical to a decibel scale but marked in volume units (vu), 0 vu corresponding to the voltage which dissipates 1 mW in a 600 Ω load. The relationship between volume and mean continuous speech power varies from talker to talker, but typically 0 vu corresponds to a mean power of − 1.4 dBm.

Experiments show that the volumes of different talkers using a telephone channel are normally distributed with a standard deviation of 5.8 dB[3,4]. It follows that allowing for talker-to-talker variations of 30 dB is sufficient to accommodate 99 % of the population (Appendix 4). Note that when the mean powers expressed in dBm are normally distributed, the mean powers expressed in watts are said to be log-normally distributed.

1.1.5   Crosstalk

Crosstalk is unwanted coupling from one telephone channel to another.

It is usually most important for crosstalk to be unintelligible to the listener. Whether or not it is intelligible depends on the articulation of the speech overheard, the amount of noise accompanying it and the abilities of the listener. For there to be a less than a 1 % probability of intelligibility, crosstalk should be about 65 dB lower in level than the wanted speech. For measurement purposes this requirement means that crosstalk from a test-tone in one channel should be at least 65 dB below the test-tone in other channels.

Unintelligible crosstalk can usually be regarded as contributing to the noise in a telephone channel (Section 1.1.3). An exception to that is the adjacent-channel crosstalk occurring in carrier systems (Section 5.1.1); it is inverted in spectrum and therefore unintelligible, but its speech-like qualities make it annoying and it is usually required to be at least 60 dB below the level of the wanted speech[5].

In cables containing a number of pairs of conductors, crosstalk can arise from inductive or capacitive coupling between pairs. To assist system designers, specimen cables are subjected to tests in which a tone is injected at one end of a selected pair and measurements are made at the ends of other pairs. Measurements at the injection end give near-end crosstalk (NEXT). Measurements at the other end give far-end crosstalk (FEXT), which is usually expressed as a level relative to the level of the tone at the far end of the pair into which it is injected. The accumulation of crosstalk in long lengths of cable is minimised by carefully prescribed twisting of the pairs.

1.1.6   Sidetone and echoes

Sidetone is a talker's signal appearing with negligible delay in his own receiver. The level of the sidetone should approximate to that heard by direct propagation through the air when the talker is not using a telephone; otherwise the talker will instinctively adjust his volume to an unnatural level, which may be outside the range of volumes the channel is designed to handle. Sidetone may be produced by mismatching at several points in a telephone circuit and is usually controlled by arranging for it to be provided almost entirely by a network in the instrument based on a hybrid transformer (Section 1.2.2).

Mismatching at a distance from the transmitter may cause a talker to hear an echo of his own voice. Echoes delayed by less than 30 ms are perceived by most users as contributing to sidetone; echoes with a greater delay are perceived as echoes, thus causing a talker to slow down and disrupting conversation in other ways. For echoes at a given level, the greater the delay, the greater the disruption. In long routes, echo suppressors are used, in which the presence of a speech signal travelling in one direction inserts typically 50 dB of attenuation in the return channel. Interlocking is necessary to prevent speech from being suppressed in both directions simultaneously. The Bell System uses suppressors when there are echoes delayed by more than 45 ms.

Even in the absence of echoes, conversation can be disrupted by delay in a telephone channel. A delay of 400 ms in each direction is considered to be the maximum acceptable[6].

1.2   Line transmission of telephony

1.2.1   Lumped loading

The standard theory of transmission lines[7] shows that the complex amplitude V of a voltage wave at angular frequency ω moving along a transmission line in the direction of increasing x is given by

$$V = V_0 \exp(-\gamma x),$$

where the propagation constant γ is given in terms of the primary constants L, C, R and G by

$$\gamma = \sqrt{(R + j\omega L)(G + j\omega C)}. \tag{1.1}$$

L, C, R and G are respectively the inductance, capacitance, series resistance and shunt conductance of a unit length of the transmission line. γ may be separated into real and imaginary parts as $\gamma = \alpha + j\beta$, where α is the attenuation in nepers and β the phase lag in radians of a unit length of the line.

The parameters of balanced structures used for telephony usually satisfy the condition

$$\frac{G}{C} \ll \frac{R}{L}.$$

Insight into the variation of α with frequency may be obtained by considering three ranges of frequency. It may be shown from Eq. 1.1 that

(a)   when $\omega \ll \dfrac{G}{C}$, $\alpha \simeq \sqrt{RG}$,

(b)   when $\dfrac{G}{C} \ll \omega \ll \dfrac{R}{L}$, $\alpha \simeq \sqrt{\dfrac{\omega CR}{2}}$,

(c)   when $\dfrac{R}{L} \ll \omega$, $\alpha \simeq \dfrac{R}{2}\sqrt{\dfrac{C}{L}} + \dfrac{G}{2}\sqrt{\dfrac{L}{C}}$.

In practical telephone cables the audio frequencies transmitted usually roughly satisfy (b), so that α is approximately proportional to the square root of frequency.

It may be shown[8] that increasing L in a practical cable decreases α at all frequencies until L reaches the value satisfying

$$\frac{R}{L} = \frac{G}{C}. \tag{1.2}$$

Further increase of L increases α. Equation 1.2 is known as Heaviside's distortionless condition. When it is satisfied, $\alpha = \sqrt{RG}$ at all frequencies, the delay β/ω is independent of ω and the characteristic impedance is real and independent of ω.

L may be increased by loading a cable with material of high permeability; but that is usually considered too expensive and a much more common method is to insert inductors at intervals along the cable. Assuming that a transmission line has been divided into sections of length l by inductors of inductance L_1, one section of the loaded line may be represented at frequencies for which the wavelength exceeds about $12l$ by the circuit of Fig. 1.1.

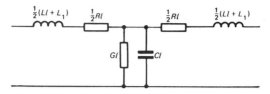

Fig. 1.1 Equivalent circuit of a section of line with lumped loading

Each section of a loaded line behaves therefore as a low-pass filter with cut-off frequency f_c given approximately by[8]

$$f_c = \frac{1}{\pi l \sqrt{C\left(L + \dfrac{L_1}{l}\right)}} \text{ Hz.}$$

The necessity for f_c to be above the range of frequencies transmitted imposes an upper limit on l for a given increase (L_1/l) in the inductance per unit length. The phase shift β of the section is given by[8]

$$\beta = \cos^{-1}\left\{1 - 2\frac{f^2}{f_c^2}\right\} \text{ rad}$$

for $f < f_c$.

1.2.2 The hybrid transformer

The hybrid transformer is a 4-port network widely used in telephony. It is usually represented by the symbol of Fig. 1.2(a) and if simultaneously matched at all four ports has the following properties.

(i) There is no transmission between Port A and Port C, nor between Port B and Port D.
(ii) Transmission between Port A and Port B is unaffected by changes either in the termination at Port C or in the termination at Port D (but not in general by changes in both). Corresponding results hold for other transmission paths.
(iii) A change in the termination at Port A results in transmission from Port B to Port D proportional to the reflexion coefficient of the termination at Port A. Corresponding results hold for other ports and transmission paths.

Figure 1.2(b) shows a realisation of a symmetric hybrid transformer with load resistance values which simultaneously match at all four ports. The description 'symmetric' indicates that the tapped winding is centre-tapped and that the matching loads at Port A and Port C are equal. Under matched conditions a signal introduced into one port of a symmetric hybrid is divided equally between two of the other ports. The matching resistance at Port D can be modified to any convenient value by modifying the turns-ratio of the transformer. It can be verified by straightforward circuit analysis that the circuit of Fig. 1.2(b) has properties (i), (ii) and (iii).

A simple application of a hybrid transformer is to suppress sidetone when the microphone and the receiver of a telephone instrument are both connected to the two wires

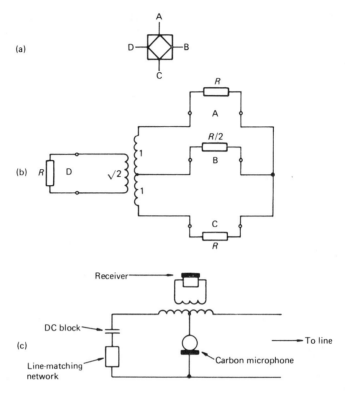

Fig. 1.2 The hybrid transformer (a) symbol (b) symmetric hybrid transformer with matching terminations (c) hybrid used to suppress sidetone

leading to the exchange. A possible circuit is shown in Fig. 1.2(c), but in practice resistors are added to allow a controlled amount of sidetone (Section 1.1.6). Other applications of the hybrid transformer form the basis of Exercises 1.5 to 1.11.

1.2.3 Repeaters

In a telephone circuit using traditional carbon microphones and moving-iron receivers the electrical signal at the receiver input should be approximately 14 dB below that at the microphone output. Therefore short connexions do not usually require the provision of any amplification.

When the length of a connexion makes it necessary to incorporate repeater amplifiers along the route, the most straightforward arrangement is to use four-wire operation, in which one pair of conductors carries signals in the one direction and another pair carries signals in the return direction (Fig. 1.3(a)). Unless there is a serious crosstalk problem between the pairs of conductors, the only echo path to be considered with four-wire operation is that involving acoustic transmission from receiver to microphone at the far terminal. Although acoustic transmission at the two terminals completes a signal loop, there is usually a net loss in the electrical path in each direction and instability of that loop is unlikely. Consequently repeater gains are limited by crosstalk rather than by stability considerations and repeaters can be separated by up to 60 km.

It is, however, usually uneconomic to provide two pairs of conductors for one telephone connexion and most telephone connexions operate over one pair of conductors (two-wire

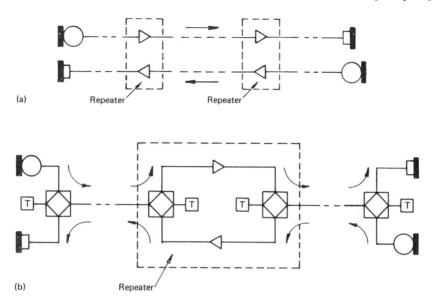

Fig. 1.3 Four-wire and two-wire operation (a) four-wire repeatered route (b) two-wire route with four-wire repeater (T denotes line-balancing network)

operation) for some or all of their length. Connexion of the microphones and receivers to the two wires necessitates hybrid transformers at the terminals (Section 1.2.2). Provision of simultaneous amplification in both directions makes the design of repeaters for two-wire operation more involved than in the four-wire case.

There are two approaches[9] to the design of a two-wire repeater. The first approach is to transform to four-wire operation for a short length and to insert unilateral amplifiers into that short length as shown in Fig. 1.3(b). Mismatching between repeater and lines creates a signal loop, causing ripples in the frequency response or even instability if the mismatching is severe (Exercise 1.11). A gain of 20 dB is typical of repeaters of this 'four-wire' or 'hybrid' type.

The second approach is the negative-impedance repeater of Exercise 1.10. Repeaters of the negative-impedance type are usually less tolerant of cable impedance than are repeaters of the four-wire type and are therefore restricted to lower values of gain. Every repeater in a two-wire connexion introduces a potential echo path for each direction of transmission.

1.2.4 Companding

In some transmission systems it is prohibitively expensive to provide the large dynamic range recommended in Section 1.1.4 for telephony. Hence it is useful to compress the dynamic range of speech for transmission, expanding it again at the receiver to restore its full natural variations. The combined process of compressing and expanding is called 'companding'.

A block diagram of a conventional compandor is shown in Fig. 1.4. The gains of the amplifiers are controlled by the speech power so that the power gain of the compressor amplifier is inversely proportional to its output power P_2 and the power gain of the expander amplifier is proportional to its input power P_3. A 2 dB increase in P_1 produces a 1 dB increase in P_2 and P_3 and a 2 dB increase in P_4 (Exercise 1.12). In the absence of speech the gain of the expander is low; thus noise is suppressed during quiet intervals.

The controlling voltages representing P_2 and P_3 are derived from the speech signals by

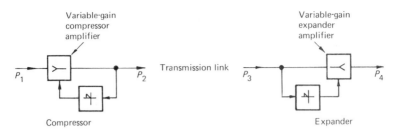

Fig. 1.4   Syllabic compandor for telephony

rectification and smoothing. The time constant associated with the smoothing is critical. If it is too long, loud syllables will suppress quiet syllables following immediately after them: if it is too short, the non-linear nature of the process will distort the fine structure of the speech waveform and change the character of its vowels (Section 1.3.3). A typical compromise[10] is an attack time a little less than 5 ms and a recovery time a little less than 22.5 ms. Since compressor and expander gains vary from syllable to syllable (the length of a syllable is usually between 250 ms and 600 ms[1]), the process is sometimes called 'syllabic companding' to distinguish it from the instantaneous companding used in PCM (pulse-code modulation—Section 1.3.1).

1.3   Digital encoding of speech

1.3.1   Pulse-code modulation (PCM)

It is increasingly common in telephone systems for speech to be transmitted as a discrete or 'digital' signal rather than as a continuous or 'analogue' waveform. The reasons for this include improved tolerance of noise, the possibility of regeneration (Section 6.3.1) and convenience of electronic implementation[9, 11, 12, 13]. The most straightforward method of encoding speech in digital form is pulse-code modulation (PCM), in which successive samples of the speech waveform are rounded to the nearest of $N(= 2^n)$ quantum levels and expressed as n-digit binary numbers.

It is well known that a waveform can in principle be reconstructed without distortion from accurate samples taken at a frequency at least twice the highest frequency present in the spectrum of the waveform. The reconstruction can be effected by passing the samples through a low-pass filter cutting off at half the sampling frequency, but if the samples are of short duration the filter output is weak. It is therefore often convenient to extend the duration of the samples before applying them to the filter (Exercise 1.13). It is standard practice to use a sampling frequency of 8 kHz when encoding speech of telephone quality. This allows a margin between the highest speech frequency to be transmitted and half the sampling frequency, thus rendering the reconstructing filter practicable.

Representation of each sample by a binary number with a finite number of digits introduces rounding errors which give rise to distortion in the reconstructed waveform. This distortion is referred to as 'quantisation distortion' or, because of its noise-like properties, 'quantisation noise'.

The effect of noise and distortion on the transmitted encoded signal is that some of the digits are received incorrectly. Such errors add further distortion to the reconstructed waveform.

Because of the wide variations in power between the syllables of speech, a separation of quantum levels sufficiently small for acceptable reconstruction of quiet syllables would give unnecessarily low distortion of loud syllables and therefore be wasteful of the capacity of the digital transmission link. Hence, in practice, a non-linear arrangement of quantum levels is used, in which the levels are close together at small signal voltages and widely spaced at large signal voltages. This may be implemented by preceding a uniform encoder (i.e. one having equally spaced quantum levels) by a non-linear network which compresses large signal voltages. The (uniform) decoder is followed by a complementary non-linear network which expands large signal voltages. This compressing and expanding process is called 'instantaneous companding'. Alternatively, instantaneous companding may be implemented by processing a uniformly encoded digital signal.

A simple formula for the mean-square quantisation distortion introduced by uniform encoding may be obtained by assuming that the sample voltages represented by each n-digit binary number are uniformly distributed over a voltage range equal to the separation δ between adjacent quantum levels. The probability density function of the quantisation distortion has a value $1/\delta$ and hence the mean-square quantisation distortion N_q is given by

$$N_q = \int_{-\delta/2}^{\delta/2} \frac{1}{\delta} v^2 \mathrm{d}v = \frac{1}{12}\delta^2. \tag{1.3}$$

In order to calculate the mean-square quantisation distortion when instantaneous companding is incorporated, first of all assume that the non-linear compressing network has a characteristic $y = y(x)$ such as that shown in Fig. 1.5. The variable x ranges from -1 to $+1$ and is proportional to the sample voltage to be encoded; the variable y is proportional to the sample voltage applied to the uniform encoder, which is assumed to have N quantum levels uniformly distributed between -1 and $+1$. $y(x)$ is assumed to be an odd function. The width δ_i of the quantum step centred on $x = x_i$ is given approximately by

$$\delta_i = \frac{2}{N}\left[\frac{\mathrm{d}y}{\mathrm{d}x}\bigg|_{x=x_i}\right]^{-1}.$$

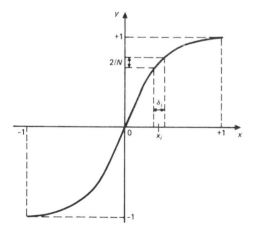

Fig. 1.5 Compression characteristic for PCM

Now assume that the signal to be encoded is distributed over the range $-1 \leqslant x \leqslant 1$ with probability density function $p(x)$. The signal power S is therefore given by

$$S = \int_{-1}^{1} p(x)x^2 \mathrm{d}x.$$

For large N we may assume that $p(x)$ does not vary within a quantum step and we may use

Eq. 1.3 to show that the mean-square quantisation distortion N_q is given by

$$N_q = \int_{-1}^{1} \frac{1}{12} \left(\frac{2}{N} \right)^2 p(x) \left(\frac{dx}{dy} \right)^2 dx.$$

The factor by which the companding reduces the mean-square quantisation distortion when the signal is in the vicinity of $x = 0$ is

$$\left[\frac{dy}{dx} \bigg|_{x=0} \right]^2$$

and is called the 'companding advantage'.

Now the ratio N_q/S is independent of $p(x)$ if

$$\frac{dy}{dx} = kx \qquad (k \text{ a constant}),$$

which is satisfied if y is proportional to $\ln x$. Hence compression characteristics are usually based on a logarithmic curve.

Since $\ln x \to -\infty$ as $x \to 0$, it is impractical for a compression characteristic to be logarithmic over its entire range. Practical characteristics are usually linear for small values of x. Examples of recommended[14] characteristics are the following.

(a) The A compression law[12]
$y(x)$ is an odd function, defined for positive x by:

$$y = \begin{cases} \dfrac{Ax}{1+\ln A} & \left(0 \leqslant x \leqslant \dfrac{1}{A} \right) \\[2ex] \dfrac{1+\ln (Ax)}{1+\ln A} & \left(\dfrac{1}{A} \leqslant x \leqslant 1 \right). \end{cases}$$

Thus $y(x)$ is exactly linear for $|x| \leqslant 1/A$. Both $y(x)$ and dy/dx are continuous at $x = \pm 1/A$.

(b) The μ compression law[15]
$y(x)$ is an odd function, defined for positive x by:

$$y = \frac{\ln (1+\mu x)}{\ln (1+\mu)} \qquad (0 \leqslant x \leqslant 1).$$

In this case $y(x)$ approaches linearity as $x \to 0$.

Although it depends in general on $p(x)$, bounds to the ratio N_q/S can sometimes be established. For example, for the A compression law N_q/S has been shown[12] to lie within the bounds shown in Fig. 1.6. For speech signals $p(x)$ is roughly exponential and an approximate expression in that case is derived in Exercise 1.17.

A suitable value of the parameter A for A-law compression of speech may be arrived at in the following way. Assume that the quantisation is uniform for the quietest talker (for whom the r.m.s. value of x is $x_{\text{r.m.s.}}$) and take the generally accepted criterion for the acceptability of uniformly quantised continuous speech, which is that each quantum step should be at least 16 dB below the r.m.s. voltage of the speech. Hence

$$\frac{2}{N} \frac{dx}{dy} = \frac{2}{N} \left(\frac{1+\ln A}{A} \right) < x_{\text{r.m.s.}} \times 10^{-0.8}.$$

It is also generally accepted that when the quantisation is logarithmic the quantum steps

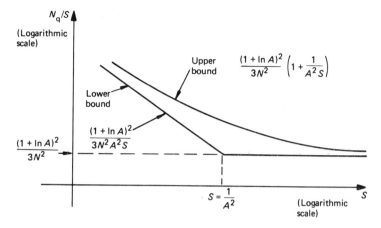

Fig. 1.6   Bounds on quantisation-distortion/signal ratio for A-law compression

should not exceed 1.4 dB. For the logarithmic region of the A-law characteristic this requirement may be written

$$\frac{1}{x}\left(\frac{2}{N}\frac{dx}{dy}\right) = \frac{2}{N}(1 + \ln A) < 10^{0.07} - 1.$$

Hence if A is chosen to be approximately equal to $1/x_{\text{r.m.s.}}$ the same value of N corresponds to the threshold of acceptability in both linear and logarithmic regions.

Eight digit PCM ($N = 256$) with instantaneous companding is in general use for civilian telephony. Fewer digits per sample are needed if the range of speaker volumes is compressed.

1.3.2   Delta modulation

In delta modulation each transmitted binary digit indicates either a rise or a fall in the speech waveform encoded. Since it encodes differences between samples, it is a form of differential encoding.

Delta modulation is of interest to designers of telephone systems for several reasons. It is simpler to implement than PCM; in some situations it gives better quality than PCM for a given rate of transmission of digits; and it is more tolerant than PCM of errors introduced during transmission.

The essential features of the basic linear delta modulator are shown in Fig. 1.7. The

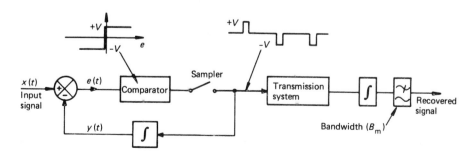

Fig. 1.7   Linear delta modulator

sampler is closed for short intervals at the sampling frequency $f_s (= 1/T)$. The feedback in the encoder is arranged so that $y(t)$ changes in steps so as to follow $x(t)$. The polar ($\pm V$) pulse train may be converted to a different format for transmission. In the absence of an input signal an accurately balanced encoder generates an idling pattern of alternate positive and negative pulses. If the change in $x(t)$ during a sampling period (T) is greater than the step change in $y(t)$, $y(t)$ cannot keep up with $x(t)$ and 'slope-clipping' distortion occurs. The ratio of the amplitude of a sinewave input on the threshold of slope-clipping to the amplitude of a sinewave input at the same frequency which just disturbs the idling pattern is sometimes called the 'amplitude range' of the delta modulator at that frequency.

If no errors are introduced during transmission, the waveform at the output of the integrator in the decoder is a replica of $y(t)$. This is not an exact replica of $x(t)$, but is corrupted by a form of quantisation distortion. The added distortion approximates to a sequence of triangles of duration T; hence its power spectrum is approximately uniform at low frequencies and extends to frequencies above f_s. Since in a satisfactory encoder f_s is considerably higher than frequencies encoded (see Exercise 1.19), most of the added distortion power can be removed by a low-pass filter in the decoder.

Provided that there is no slope-clipping, the mean-square quantisation distortion N_q may be shown by elementary arguments to be given by the expression:

$$N_q = K_q \frac{B_m \sigma^2}{f_s}$$

where σ is the height of the steps by which $y(t)$ changes, K_q is a constant depending on the nature of the signal, and B_m is the filter bandwidth. Typically K_q is about $1/3$.

In a modified form of delta modulation (called 'delta-sigma modulation'), $x(t)$ is integrated before encoding, this integration being compensated for by removal of the integrator from the decoder. The two integrators now required in the encoder may be replaced by one integrator situated at the output of the subtractor. Since there is no integrator in the decoder, the effect on the output of a transmission error does not persist as it does in basic linear delta modulation.

In many practical delta modulators the integrators are simple R–C circuits and their inputs are full-length pulses. Thus $y(t)$ changes exponentially and hence they are called 'exponential delta modulators'. It may be shown[16, 17] that for an input $E_m \sin 2\pi f_m t$ slope-clipping will not occur in an exponential delta modulator if

$$E_m < \frac{V}{\sqrt{1 + 4\pi^2 f_m^2 C^2 R^2}},$$

where the voltage applied to the integrator is $+V$ or $-V$ and where C and R are the capacitance and resistance used in the integrator. When encoding speech it is usual to arrange for the product CR to be about 1 ms so that the threshold of slope-clipping falls with increasing f_m in roughly the same way as the spectrum of speech. When the integrator output is zero, the step size is V/CRf_s (the 'central delta step'). Hence the amplitude range of an exponential delta modulator is

$$\frac{2CRf_s}{\sqrt{1 + 4\pi^2 f_m^2 C^2 R^2}}.$$

The range of signal amplitudes which can be encoded satisfactorily with a given sampling frequency can be extended by making the delta modulator adaptive. This is done by adjusting the amplitude (V) of the pulses fed to the integrator in accordance with the amplitude of the signal being encoded. The detector controlling the pulse amplitude has a time constant of the order of 10 ms, so that syllabic variations are followed. The modified delta modulator is therefore said to be syllabically companded. Signal amplitude is usually estimated from the output of the encoder so that an identical estimate can be made in the decoder to restore accurately the original variations of signal amplitude. One method is to measure the nearness to slope-overload by detecting unbroken sequences of say three 1s or three 0s in the transmitted sequence. Syllabically companded delta modulation with a sampling frequency of 32 kHz is adequate for most telephony purposes.

1.3.3 Vocoders

If it is assumed that talking consists of uttering a sequence of basic sounds (phonemes) chosen from an 'alphabet' of about 40 phonemes, and that no more than 10 phonemes can be pronounced per second, then elementary information theory (Appendix 9) indicates that the information content of speech cannot exceed $10 \log_2 40$ bit/s, i.e. about 53 bit/s. Obviously this description of speech is oversimplified, but it does suggest that PCM and delta modulation are very inefficient in that they require the transmission of binary digits at rates about a thousand times greater. Speech may be encoded more efficiently by a vocoder, which operates not by encoding the speech waveform itself but by transmitting sufficient information to enable a recognisably similar waveform to be synthesized in the receiver.

The synthesizer is therefore required to be a simplified electrical model of the mechanism of speech production. It usually contains two waveform generators to reproduce the two modes of operation of the larynx. One is a multivibrator to produce the periodic sawtooth waveform associated with voiced sounds; the other is a noise generator for the unvoiced sounds. The noise generator also serves to reproduce the noise-like vibrations sometimes produced by tongue, lips or teeth. The output of whichever waveform generator is in use is then passed through filters, the characteristics of which are varied to correspond to the filtering characteristics of the vocal tract. It is the spectral shaping produced by the cavities (throat, mouth and nose) comprising the vocal tract which gives to the various vowels their distinctive recognisable qualities. Realistic synthesis requires transient excitation of the filters for blockages (p, b, k, g, t, d) introduced by lips and tongue, but this refinement is usually omitted from vocoders.

The encoding part of a vocoder (the analyser) must extract the following information from the speech waveform if the synthesizer is to be effectively controlled.

(i) Voiced/unvoiced decision (to select which waveform generator should be used). This information may be obtained by comparing the speech power above about 5 kHz with that at lower frequencies, perhaps about 400 Hz.

(ii) Pitch (to control the frequency of the multivibrator). It is difficult to determine pitch reliably and errors can be very disturbing to a listener. Simple filtering is inadequate, since the fundamental component is often absent. Several methods of pitch extraction have been used or proposed, but none is entirely satisfactory. They include taking the discrete Fourier transform of the logarithm of the speech spectrum (cepstral analysis), autocorrelation of a processed waveform and the use of banks of comb filters.

(iii) Spectral envelope (to control the filters in the synthesiser). In the conventional channel vocoder (Fig. 1.8) this information is obtained by means of a bank of filters and detectors. More recently, interest has concentrated on linear-predictive-coding (LPC)—vocoders[18] in which the analyser derives weighting coefficients relating samples to previous samples. Digital filters in the synthesiser are controlled by these coefficients.

A brief survey of vocoders (and other methods of encoding speech) has been given by Gold[19].

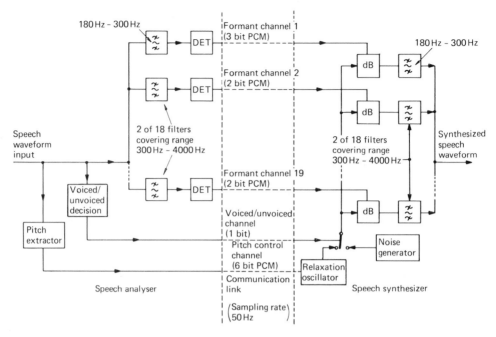

Fig. 1.8   A typical channel vocoder

1.4   Transmission of programme sound

1.4.1   Channel requirements for programme sound

Telecommunication systems make provision for programme sound of various degrees of quality, but the discussion here is confined to the transmission of high-fidelity sound, such as that broadcast by frequency-modulation radio.

As in telephony, assessment of channel quality is based on subjective tests, but for programme sound it is recommended[20] that the listeners include 'experts' with recent and extensive experience of such assessment. The following five-point scales are recommended[20].

	Quality		*Impairment*
5	Excellent	5	Imperceptible
4	Good	4	Perceptible but not annoying
3	Fair	3	Slightly annoying
2	Poor	2	Annoying
1	Bad	1	Very annoying

The transmitted bandwidth should cover as much as is practicable of the bandwidth of the human ear and is specified[21] to extend from 40 Hz to 15 kHz. The amplitude response is required to lie within 0.5 dB limits throughout the frequency range from 125 Hz to 10 kHz, with some relaxation outside that range. Less variation of group delay is permitted than in a telephone channel.

A reasonable degree of linearity is important if harmonics and intermodulation products are not to constitute an annoyance. It is recommended[21] that the total harmonic distortion due to a single tone at the overload level (a tone with amplitude equal to the largest excursions of the signal voltage) should not exceed 0.5%, although there are relaxations for tones at frequencies below 125 Hz. Intermodulation is specified separately in terms of the difference tone produced by two tones each 6 dB below the overload level.

The recommended[21] maximum level of noise power in a channel is such that noise having a flat spectrum should be at least 50 dB below the overload level (56 dB when measured on an appropriately weighted psophometer). This figure is, however, based on the performance of telephone systems (it is consistent with Section 5.2.2) rather than on what is desirable for programme sound.

As in telephony, intelligible crosstalk is very annoying. The permitted level varies with frequency and should be -74 dB relative to the overload level at frequencies at which the ear is most sensitive.

1.4.2   Noise reduction

The dynamic range of some programme-sound signals (for example those of symphony orchestras) may need to be reduced by up to about 20 dB for satisfactory transmission, even over a channel meeting CCITT recommendations (Section 1.4.1).

If the dynamic range is reduced by a compandor similar to the syllabic compandors used for telephony (Section 1.2.4), it is found that the attack time should be short (about 1 ms) to prevent distortion of sudden loud sounds and the recovery time considerably longer to prevent level fluctuations in sustained musical notes. In some systems the expander is controlled by an auxiliary signal (such as a frequency-modulated tone) transmitted by the compressor.

There is, however, a fundamental disadvantage inherent in the basic design of compandors, namely that the action of the variable-gain devices is sufficiently non-linear to produce unacceptable modulation effects associated with loud sounds. In particular, there is modulation of the noise by the signal, which is noticeable around loud sounds and sometimes during them. An alternative approach, used in what are usually called noise-reduction systems, is to arrange the compressor and expander so that their non-linear components affect the signal at low levels only.

The basic arrangement of a noise reduction system is shown in Fig. 1.9. The transfer functions $H_1(f)$ and $H_2(f)$ vary with the level of the signal. It may be shown (Exercise 1.25) that if $H_1(f) = H_2(f)$ at all times then the overall gain of the system is constant and independent of frequency. The non-linearity of the system resides in the devices represented by $H_1(f)$ and $H_2(f)$; these devices act as limiters and therefore have very little effect at high signal levels.

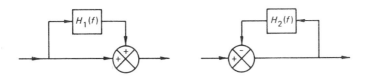

Fig. 1.9   Noise reduction system

The arrangement of Fig. 1.9 forms the basis of the well known Dolby noise-reduction systems[22]. In the Dolby 'A' system the signal is partitioned by filters into four frequency bands, the four components being separately limited. In the simpler Dolby 'B' system in widespread domestic use, $H_1(f)$ is a simple high-pass transfer function with its cut-off frequency controlled by signal level[23].

1.4.3   Digital encoding of programme sound

Digital encoding is being used increasingly by broadcasting authorities for the distribution of programme material to remote transmitters. The basic reason is that it renders the level of modulating signals at the transmitters independent of the gain of the transmission path, thus making it unnecessary for them to risk distortion in further automatic gain control at the transmitters. An additional reason is the ease with which pairs of channels can be aligned for stereo transmission (Section 1.4.4).

PCM is generally used, since it is considered that acceptable quality can be obtained at a lower digit rate with PCM than with delta modulation. A sampling frequency of 32 kHz enables the encoding of signal frequencies up to 15 kHz and is conveniently related to frequencies used in PCM telephony and carrier telephony (Sections 6.4.3 and 5.1.1).

Subjective tests show that for irregular signals considerably larger than the quantum step, the effects of quantisation noise are roughly the same as those of white gaussian noise of the same power, and that for either to be unobtrusive in quiet passages its mean power should be at least 72 dB below that of a sinewave on the threshold of overloading the encoder. Allowing for the possibility of the signal being encoded and decoded four times before modulating the transmitter, the mean quantisation noise for a single encoding should be at least 78 dB below a sinewave on the threshold of overloading. With uniform quantisation this can be achieved with a 2 dB margin by 13 digit PCM.

In order to make good use of available digital transmission systems, it is desirable to reduce the transmitted digit rate by some form of companding. For example, at ten digits per sample, six signals could be transmitted over a standard 2048 kbit/s link normally used for 30 PCM telephone channels. Distortion due to non-linearity in companding (Sections 1.2.4, 1.3.1 and 1.4.2) can be avoided by encoding uniformly with sufficient digits to give acceptable accuracy at low signal levels and carrying out the compression by digital processing of the PCM signal. In the BBC NICAM[24] (near-instantaneously companded audio multiplex) system, 10 digit PCM samples are transmitted, together with relatively infrequent indications of the coding range to which the samples relate. The coding range may be determined by the largest of a number of consecutive samples. A suitable number is 32, so that the coding range is up-dated every 1 ms.

Transmission errors in the more significant digits can be very annoying and it is often an advantage to add one or two parity checks (Section 2.3.1) to each transmitted sample. The simplest method of utilising the parity checks is error concealment, in which a parity violation causes the sample affected to be replaced by the mean of the adjacent samples, or more simply by a repeat of the previous sample. The less significant digits are usually not included in the parity checks, since errors in them are likely to introduce less distortion than the concealment process. The subjective effect of errors depends on several factors, including the method of error concealment used, and hence subjective tests are required for each system proposed. In a typical system using 13-digit uniform encoding, a parity check on the five most significant digits in each sample and error concealment by repeating the previous sample, the impairment due to a digit error rate of 1 in 2×10^5 has been found[25] to be imperceptible by 50% of listeners.

At low signal levels, PCM introduces a type of impairment known as granularity. Granularity occurs when the signal voltage is too small to reach the lowest quantum level but is nevertheless large enough for its absence to be perceptible. The imperceptibility of granularity as the signal level is reduced to the threshold of audibility requires 14 digit PCM if the signal is encoded directly, but it can be achieved with 13 digit PCM by the addition of dither before encoding. In this case, suitable dither is random noise at a level 4 dB below the quantisation-distortion. The dither increases the noise at the decoder output by about 1.5 dB, but this is within the 2 dB margin available with 13-digit PCM.

1.4.4 Stereophonic sound

The subjective effect of sound arriving from two sources is very sensitive to changes in the relative magnitude and relative time delay of the sources. Hence when two programme-sound channels are used as a stereophonic pair they must satisfy requirements additional to those in Section 1.4.1. For example, the gains of the channels should not differ by more than 0.8 dB over most of the frequency band and between 200 Hz and 4 kHz the phase shifts of the two channels should not differ by more than $15°$[21].

When stereophonic sound is broadcast by frequency-modulation radio, it is usually required that the transmitted signals should operate monophonic receivers satisfactorily and that little or no modification of radio-frequency transmitting equipment should be necessary. The usual method is to replace the monophonic modulating waveform with a multiplex stereophonic waveform of greater bandwidth, ensuring that the peak frequency deviation is not thereby increased. The lower frequencies of the multiplex waveform are occupied by a signal proportional to the sum of the two stereophonic signals and the higher frequencies (above 15 kHz) by a sub-carrier modulated by their difference. Thus a monophonic receiver gives an output proportional to the sum of the two signals and this is generally regarded as an acceptable alternative to a true monophonic signal.

Although a few stereophonic broadcasting systems use frequency-modulation of the sub-carrier, most use double-sideband amplitude modulation, which renders the difference signal more affected by noise but less affected by interference. There are two standard systems[26] using double-sideband amplitude modulation. In the (USSR) 'polar modulation' system the sub-carrier is at 31.250 kHz and is partially suppressed; in the 'pilot-tone' system the carrier is 38 kHz, but is fully suppressed and replaced by a pilot-tone at 19 kHz. Further details of the pilot-tone system are given in Exercise 1.27. Figure 1.10 shows one possible method of decoding the multiplex waveform. It should be noted that emphasis is applied to each of the two stereophonic signals before they are combined to form the multiplex waveform and no further emphasis is used. In Figure 1.10, A and B denote the instantaneous values of the two signals after emphasis.

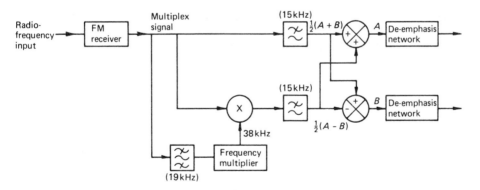

Fig. 1.10   Pilot-tone stereophonic decoder

1.5 References

1   Richards, D. L., *Telecommunication by Speech*, Butterworths, 1973
2   *CCITT Yellow Book*, Vol. V, Recommendation P53, Geneva, 1980
3   Panter, P. F., *Communication Systems Design: Line-of-sight and Tropo-scatter Systems*, McGraw-Hill, 1972

4   Holbrook, B. D., and Dixon, J. T., 'Load rating theory for multi-channel amplifiers,' *Bell System Technical Journal*, **18**, pp. 624–44, October 1939

5   *CCITT Yellow Book*, Vol. III, Recommendation G232, Geneva, 1980

6   *CCITT Yellow Book*, Vol. III, Recommendation G114, Geneva, 1980

7   Glazier, E. V. D., and Lamont, H. R. L., *The Services' Textbook of Radio, Vol. 5, Transmission and Propagation*, HMSO, 1958

8   *The Royal Signals Handbook of Line Communication*, HMSO, 1947

9   Flood, J. E., (Editor), *Telecommunication Networks*, IEE Telecommunications Series 1, Peter Peregrinus, 1975

10   *CCITT Yellow Book*, Vol. III, Recommendation G162, Geneva, 1980

11   Inose, H., *An Introduction to Digital Integrated Communication Systems*, Peter Peregrinus, 1979

12   Cattermole, K. W., *Principles of Pulse Code Modulation*, Iliffe, 1969

13   Bylanski, P., and Ingram, D. G. W., *Digital Transmission Systems*, IEE Telecommunications Series 4, Peter Peregrinus, 1976

14   *CCITT Yellow Book*, Vol. III, Recommendation G711, Geneva, 1980

15   Smith, B., 'Instantaneous companding of quantized signals,' *Bell System Technical Journal*, **36**, pp. 653–709, May 1957

16   Steele, R., *Delta modulation Systems*, Pentech Press, 1975

17   Betts, J. A., *Signal Processing, Modulation and Noise'*, The English Universities Press, 1970

18   Atal, B. S., and Hanauer, S. L., 'Speech analysis and synthesis by linear prediction of the speech wave,' *The Journal of the Acoustical Society of America*, **50**, pp. 637–55, 1971

19   Gold, B., 'Digital speech networks,' *Proc. IEEE*, **65**, pp. 1636–58, December 1977

20   *CCIR*, Vol. XII Transmission of Sound Broadcasting and Television Signals over Long Distances (CMTT) Report 623, Geneva, 1982

21   *CCITT Yellow Book*, Vol III, Recommendation J21, Geneva, 1980

22   Dolby, R. M., 'An audio noise reduction system', *Journal of the Audio Engineering Society*, October 1967

23   Berkovitz, R., and Gundry, K., 'Dolby B-Type noise reduction system, *Audio*, September 1973 and October 1973

24   Caine, C. R., English, A. R., and O'Clarey, J. W. H., 'Nicam 3: near-instantaneously companded digital transmission system for high-quality sound programmes,' *The Radio and Electronic Engineer*, **50**, pp. 519–30, October 1980

25   Shorter, D. E. L., and Chew, J. R., 'Application of pulse-code modulation to sound-signal distribution in a broadcasting network,' *Proc. IEE*, **119**, pp. 1442–8, October 1972

26   *CCIR*, Vol. X Broadcasting Service (Sound), Recommendation 450–1, Geneva, 1982

Exercise topics

1.1   Line attenuation at audio frequencies
1.2   Inductance loading for minimum attenuation
1.3   Inductance loading with prescribed spacing
1.4   Velocity of propagation in loaded cable
1.5   Amplifiers interconnected by hybrid transformers
1.6   Two-wire repeater using hybrid transformers
1.7   Effects of mismatching on hybrid transformer
1.8   Asymmetric hybrid transformer
1.9   Return–loss measurement using hybrid transformer
1.10   Negative-impedance two-wire repeater
1.11   Effect of mismatching on gain of two-wire repeater
1.12   Syllabic companding
1.13   Amplitude distortion due to pulse stretching
1.14   Noise due to errors in uniformly quantised PCM
1.15   Quantisation noise in μ-law companded PCM
1.16   Quantisation noise in A-law companded PCM
1.17   Exponentially distributed signal with A-law companding
1.18 A-law quantisation for acceptable telephony
1.19   Dynamic range of linear delta modulation
1.20   Linear delta modulation for acceptable telephony
1.21   Linear delta-sigma modulation
1.22   Exponential delta modulation
1.23   Syllabically companded exponential delta modulation
1.24   Noise due to errors in exponential delta modulation
1.25   Noise reduction for high-quality sound
1.26   PCM for programme-sound
1.27   'Pilot-tone' stereophonic transmission

Exercises

1.1   Line attenuation at audio frequencies

Show that at angular frequencies much greater than R/L and G/C, where L, C, R and G are the primary constants per unit length, the attenuation of a transmission line in nepers per unit length is approximately

$$\frac{R}{2}\sqrt{\frac{C}{L}} + \frac{G}{2}\sqrt{\frac{L}{C}}.$$

Solution outline
Write γ in the form

$$\gamma = \sqrt{j\omega L\left(1+\frac{R}{j\omega L}\right)j\omega C\left(1+\frac{G}{j\omega C}\right)}.$$

Expansion by the binomial theorem and discarding second and higher order terms in $R/\omega L$ and $G/\omega C$ gives

$$\gamma \simeq \frac{1}{2}\left(R\sqrt{\frac{C}{L}} + G\sqrt{\frac{L}{C}}\right) + j\omega\sqrt{LC}.$$

1.2   Inductance loading for minimum attenuation

A telephone cable which has the following primary constants

$$L = 0.5\ \mu\text{H/m} \qquad C = 50\ \text{pF/m}$$
$$R = 0.02\ \Omega/\text{m} \qquad G = 10^{-8}\ \text{S/m}$$

is to be loaded with coils of negligible resistance to minimise its attenuation. Calculate the maximum separation of the coils if the cut-off frequency of each section must not be less than 10 kHz.

Solution outline
Attenuation is minimised by increasing the average inductance per unit length to CR/G (Eq. 1.2), which in this case is 10^{-4} H/m. Substituting 10^{-4} for $L + (L_1/l)$ in the formula for cut-off frequency in Section 1.2.1 shows that the cut-off frequency is greater than 10 kHz provided that l is less than 450 m.

1.3   Inductance loading with prescribed spacing

The cable described in Exercise 1.2 is to be loaded with coils of negligible resistance at intervals of 5 km. Calculate the maximum inductance permissible for each coil if the cut-off frequency should not be less than 4 kHz. Is this inductance sufficient for the minimum-attenuation condition to be satisfied?

Solution outline
From the formula for cut-off frequency in Section 1.2.1, $L + (L_1/l)$ must not exceed $5.07\ \mu\text{H/m}$. Hence L_1 must not exceed 22.8 mH. The maximum allowable average inductance per unit length $(5.07\ \mu\text{H/m})$ is well below the value $100\ \mu\text{H/m}$ required to satisfy Eq. 1.2.

1.4 Velocity of propagation in loaded cable

The cable described in Exercise 1.2 is loaded with coils of inductance 50 mH at intervals of 2 km. Calculate the phase-shift per section at 1 kHz and the average velocity of propagation of a 1 kHz sinewave in the loaded cable.

Solution outline
Substitution in the formula for cut-off frequency (Section 1.2.1) gives $f_c = 4457$ Hz. Hence the following formula for phase-shift gives $\beta = 0.453$ rad at 1 kHz. The delay of each section at 1 kHz is therefore $(0.453) \div (2\pi \times 10^3)$ s, i.e 7.20×10^{-5} s, and therefore the average velocity of propagation is $2000 \div (7.20 \times 10^{-5})$ m/s, i.e 2.8×10^7 m/s.

1.5 Amplifiers interconnected by hybrid transformers

A 20 dB amplifier inserted in (and matching) a long length of 75 Ω coaxial cable is replaced by a unit containing two identical 20 dB amplifiers and two hybrid transformers as shown in Fig. 1.11.

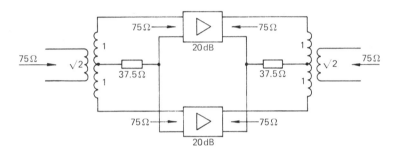

Fig. 1.11 Diagram of Exercises 1.5 and 1.6

(i)   Calculate the insertion gain of the replacement unit.
(ii)   Calculate the insertion gain of the unit when a power supply failure prevents all transmission through one of the amplifiers without affecting the driving-point impedances.
(iii)   Calculate the insertion gain of the unit when a fault in one of the amplifiers not only prevents all transmission through that amplifier but also short circuits its output terminals.
(iv)   Assuming that each of the replacement amplifiers is identical to the original amplifier, compare the effects of amplifier noise in the two arrangements.
(v)   One of the 37.5 Ω resistors becomes disconnected. How is the performance of the unit affected?

Solution outline
No detailed analysis is necessary.
(i)   There are two transmission paths, each of gain 14 dB. Their output voltages add and thus the gain of the unit is 20 dB.
(ii)   The output voltage is halved, reducing the gain to 14 dB.
(iii)   Provided that other ports of the output hybrid remain correctly terminated, the short circuit will not affect transmission through the unbroken path and the gain remains 14 dB.
(iv)   If the noise power at the output of one amplifier is N, the noise at the output of the unit due to one amplifier is $N/2$ and hence the total output noise is also N.

(v)   The effect of faults is changed. In particular the gain of one path is affected by changes in amplifier impedance in the other path.

1.6   Two-wire repeater using hybrid transformers

The direction of one of the amplifiers in Fig. 1.11 is reversed so that the unit forms a two-wire repeater. Calculate the gain of the repeater.

In what ways may departures of the cable impedance from 75 Ω impair the quality of a connexion using such a repeater?

Solution outline
As in Exercise 1.5 (ii), the gain in one path (and hence in one direction) is 14 dB.
   Possible impairments are
(i)   echoes due to reflections at the input and output of the repeater
(ii)   variation of gain with frequency (or even instability) since transmission through the hybrids completes a feedback loop.

1.7   Effects of mismatching on hybrid transformer

In the symmetric hybrid arrangement of Fig. 1.2(b), Port D is connected to the output port of an amplifier with output impedance 480 Ω, Port B to a 300 Ω resistor and Ports A and C to 600 Ω loads. Calculate the change (in dB) of signal power in one load when the other load is disconnected. Calculate also the output impedance seen by one load when the other is (a) connected and (b) disconnected.

Solution outline
Represent the amplifier by a voltage source e in series with 480 Ω. When both loads are connected, the voltage at Port A is $600e/\sqrt{2}(600+480) = e/\sqrt{2}(1.8)$. When the load at Port C is disconnected, the voltage at Port A is

$$e\left(\frac{1800}{1800+480}\right)\left(\frac{1}{\sqrt{2}}\right)\left(\frac{600}{600+300}\right) = \frac{e}{\sqrt{2}(1.9)}.$$

The required change is $20 \log (1.8/1.9)$ dB $= -0.47$ dB.
(a)   When the load is connected, one approach to calculating the impedance is to insert in the circuit diagram the voltage and current constraints of the ideal transformer as in Fig. 1.12.

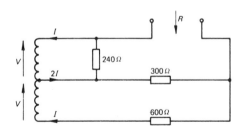

Fig. 1.12   Diagram of solution to Exercise 1.7

Writing the port voltage in two ways,

$$2V - 600I = V + 300(2I + V/240)$$
$$\therefore \quad V = -4800I.$$

Hence

$$R = \frac{2V - 600I}{I + V/240} = 537\ \Omega.$$

(b)   With the 600 Ω resistor removed, by inspection $R = 540\ \Omega$.

1.8   Asymmetric hybrid transformer

Figure 1.13 shows an ideal hybrid transformer used to connect a pilot receiver to a traffic path in such a way that the traffic path is unaffected by changes in the impedance of the pilot receiver. The ports carrying traffic have matched 75 Ω terminations and the insertion loss to traffic is 0.5 dB. Calculate the loss in the pilot path and the value of R.

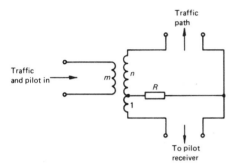

Fig. 1.13   Asymmetric hybrid of Exercise 1.8

Solution outline
If there is no transmission to R, the sum of the power gains of the two paths is unity. Hence the gain of the pilot path is $10 \log (1 - 10^{-0.05})$ dB, corresponding to a loss of 9.64 dB.
    From the ratio of the powers at the two outputs, $n = 8.20$ and the pilot receiver input impedance is 9.14 Ω. For the input impedance to be 75 Ω,

$$\frac{m^2}{(n+1)^2}\ (75 + 9.14) = 75.$$

Hence $m^2 = 75.4$. To calculate R, obtain an expression for the impedance seen at the traffic output port assuming no current flows into the pilot receiver and equate it to 75. Thus

$$75\frac{n^2}{m^2} + R = 75.$$

Hence $R = 8.16\ \Omega.$

1.9   Return-loss measurement using hybrid transformer

The symmetric hybrid transformer of Fig. 1.2(b) may be used to enable return loss (Appendix 5) to be measured by equipment intended for the measurement of insertion loss. Show that if Port D is terminated with a one-port network of impedance Z and Port B remains terminated with resistance $R/2$, the insertion loss from Port A to Port C measured between source and load resistances R is

$$20 \log \left[2 \left| \frac{Z + R}{Z - R} \right| \right]\ \text{dB}.$$

Solution outline
With the notation of Appendix 5, $R_1 = R_2 = R$ and $P_{20} = E^2/4R$. To obtain an expression for P_2, insert voltage and current constraints as in Fig. 1.12.

1.10   Negative-impedance two-wire repeater

Figure 1.14 shows in unbalanced form the basic circuit of a negative-impedance repeater. The coil is centre-tapped and there is perfect coupling between the halves. Show that if $Z_1 Z_2 = R^2$ the network will simultaneously match a resistance R at each port.

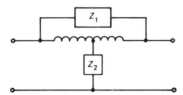

Fig. 1.14   Negative-impedance-repeater configuration of Exercise 1.10

Determine two pairs of real values for Z_1 and Z_2 for which the network has a gain of 20 dB when working between matched 600 Ω terminations.
Determine the driving point impedances 'seen' by Z_1 and Z_2 under matched conditions.

Solution outline
Assume one port to be terminated with R and obtain an expression for the driving-point impedance at the other port by inserting voltage and current constraints as in Fig. 1.12.
When $Z_1 Z_2 = R^2$, the gain of the network between matched terminations may be shown to be

$$20 \log \left| \frac{2R - Z_1}{2R + Z_1} \right| \text{ dB.}$$

Equating this to 20 gives $Z_1 = -1467\,\Omega$ (whence $Z_2 = -245\,\Omega$) or $Z_1 = -982\,\Omega$ (whence $Z_2 = -367\,\Omega$) when $R = 600\,\Omega$.
From the basic properties of the hybrid transformer, Z_1 'sees' 1200 Ω and Z_2 'sees' 300 Ω.

1.11   Effect of mismatching on gain of two-wire repeater

Show that the insertion gain G of the two-wire repeater of Fig. 1.15 is given by

$$G = 20 \log \left| \frac{1}{4}(1 - \rho_1 \rho_2) \frac{A_1}{1 - \dfrac{A_1 A_2 \rho_1 \rho_2}{16}} \right|,$$

where the reflexion coefficients ρ_1 and ρ_2 are given by

$$\rho_1 = \frac{Z_1 - R}{Z_1 + R} \quad \text{and} \quad \rho_2 = \frac{Z_2 - R}{Z_2 + R}.$$

Given that the insertion gain of each amplifier with a source of resistance R and a load of resistance R is 26 dB and that the return losses with respect to R of source and load are not less than 35 dB, estimate the maximum peak-to-peak variation of G which can occur due to variations in the arguments of A_1, A_2, ρ_1 and ρ_2.

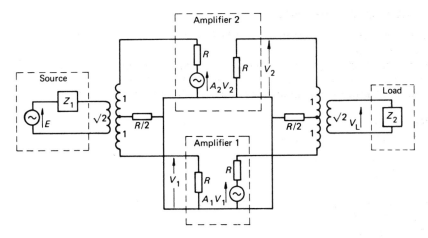

Fig. 1.15 Repeater configuration of Exercise 1.11

Solution outline

Apart from the source and load impedances, both hybrid transformers have matched terminations. Using the result derived in Exercise 1.9 and other properties of the hybrids,

$$V_1 = -\frac{ER}{(R+Z_1)\sqrt{2}} - \frac{A_2 V_2 \rho_1}{4},$$

$$V_L = -\frac{A_1 V_1 Z_2}{(R+Z_2)\sqrt{2}},$$

$$V_2 = -\frac{A_1 V_1 \rho_2}{4}.$$

Hence V_L/E may be derived. Without the repeater the load voltage V_{L0} is given by:

$$\frac{V_{L0}}{E} = \frac{Z_2}{Z_1 + Z_2}.$$

Hence V_L/V_{L0} may be derived. The required gain G is given by:

$$G = 20 \log \left| \frac{V_L}{V_{L0}} \right|.$$

Return losses greater than 35 dB imply:

$$|\rho_1|, |\rho_2| < 10^{-1.75} = 0.0178.$$

Amplifier insertion gain 26 dB implies $|A_1| = |A_2| = 40$. The principal contribution to the variation of G is the factor

$$\left| 1 - \frac{A_1 A_2 \rho_1 \rho_2}{16} \right|.$$

Now when r is real and positive, the ratio of the maximum and minimum values of $|1 - r \exp(j\theta)|$ as θ varies is $|1 + r|/|1 - r|$ and hence the peak-to-peak variation of G should not exceed

$$20 \log \frac{1 + (40)^2 (0.0178)^2/16}{1 - (40)^2 (0.0178)^2/16} \text{ dB} = 0.55 \text{ dB}.$$

1.12   Syllabic companding

Show that the syllabic compandor described in Section 1.2.4 halves the range of speaker volumes transmitted.

Before compression, speaker volumes are normally distributed with standard deviation 5.8 dB about a mean 20 dB below the volume for which the compressor has 0 dB gain. Determine the mean and standard deviation of the volumes after compression.

Solution outline
With the notation of Fig. 1.4, in the compressor

$$\frac{P_2}{P_1} = \frac{k_1}{P_2}$$

for some constant k_1. Hence $P_2 = \sqrt{k_1 P_1}$ and thus a 2 dB increase in P_1 gives rise to a 1 dB increase in P_2.

For the expander

$$\frac{P_4}{P_3} = k_2 P_3$$

for some constant k_2. Hence $P_4 = k_2 P_3^2$ and thus a 1 dB increase in P_3 gives rise to a 2 dB increase in P_4.

After compression the distribution of volumes remains normal and has standard deviation 2.9 dB. The mean volume after compression is 10 dB below the volume at which the compressor has 0 dB gain.

1.13   Amplitude distortion due to pulse stretching

Obtain an expression for the frequency response of a pulse-stretching network which lengthens pulses of duration τ_1 into pulses of duration τ_2.

A speech waveform with spectrum extending from 300 Hz to 3400 Hz is transmitted by amplitude modulation of an 8 kHz train of very short pulses. At the receiver the pulses are lengthened to a duration of 125 μs to form a 'box-car' waveform before being demodulated by a low-pass filter. Show that the lengthening of the pulses introduces about 2.7 dB of amplitude distortion over the frequency band of the speech waveform.

Solution outline
The Fourier transform of a pulse of unit amplitude extending from $t = 0$ to $t = \tau$ is (Appendix 1)

$$\int_0^\tau \exp(-j2\pi ft)\,dt = \frac{1}{\pi f}\exp(-j\pi f\tau)\sin(\pi f\tau).$$

Hence the required frequency response $H(f)$ is given by

$$H(f) = \exp\{j\pi f(\tau_1 - \tau_2)\}\frac{\sin(\pi f\tau_2)}{\sin(\pi f\tau_1)}.$$

If τ_1 is small,

$$|H(f)| \simeq \frac{\sin(\pi f\tau_2)}{\pi f\tau_1}.$$

Putting $\tau_2 = 125$ μs and evaluating $|H(f)|$ at 300 Hz and 3400 Hz give for the amplitude distortion:

$$20\log\{|H(300)|/|H(3400)|\} = 2.7 \text{ dB}.$$

1.14 Noise due to errors in uniformly quantised PCM

For a uniformly quantised 8 digit PCM system with a sampling rate of 8 kHz, estimate the ratio of the power in a 1 kHz sinewave at the threshold of overloading to the noise power due to random transmission errors with digit error rate 0.1 %. Assume that each error adds or subtracts a rectangular pulse of duration 125 μs to a box-car waveform before the final low-pass filter.

Estimate the digit error rate at which the noise power due to errors equals the quantisation noise power.

Solution outline
Assume that the amplitude of the sinewave is unity and that its power is 0.5.

An error in the most significant digit adds or subtracts a pulse of unit amplitude to the box-car waveform. Hence errors in the most significant digits contribute $0.001 \times (1)^2$ to the noise power.

Assuming that there is at most one error per sample, the total noise power due to errors in all eight digits is therefore

$$0.001 [(1)^2 + (1/2)^2 + (1/4)^2 + \ldots + (1/128)^2], \quad \text{i.e } 0.001\,33.$$

The power spectrum of the noise due to errors is of $\{\text{sinc}\,(f)\}^2$ form and from Appendix 4 it may be seen that a demodulating low-pass filter cutting off at half the sampling frequency removes about 20 % of this noise.

Hence the required ratio is $0.5 \div (0.001\,33 \times 0.8)$, which is approximately 27 dB.

The noise power due to errors equals the quantisation noise power when the digit error rate p_e is given by (Eq. 1.3):

$$1.33 p_e = \frac{1}{12} \left(\frac{2}{256} \right)^2.$$

Hence $p_e = 3.8 \times 10^{-6}$.

1.15 Quantisation noise in μ-law companded PCM

An 8 digit PCM encoder with μ-law companding has a companding advantage of 15 dB. Calculate the signal/quantisation-distortion ratio when the signal voltage is uniformly distributed over the range of the encoder.

Solution outline

$$20 \log \left\{ \frac{dy}{dx} \bigg|_{x=0} \right\} = 20 \log \left\{ \frac{\mu}{(1 + \mu x) \ln (1 + \mu)} \right\} = 15.$$

Hence $\mu \simeq 15.9$.

For a uniformly distributed voltage,

$$p(x) = \begin{cases} 0.5 & |x| \leq 1 \\ 0 & |x| > 1 \end{cases}.$$

Hence

$$N_q = 2 \int_0^1 \frac{1}{12} \left(\frac{2}{N} \right)^2 (0.5) \left\{ \frac{(1 + \mu x) \ln (1 + \mu)}{\mu} \right\}^2 dx,$$

$$S = 2 \int_0^1 (0.5) \, x^2 \, dx,$$

where $N = 2^8 = 256$. Evaluating the integrals gives

$$\frac{S}{N_q} = 20\,500 \qquad \text{or} \qquad 43.1 \text{ dB}.$$

1.16   Quantisation noise in A-law companded PCM

Derive expressions for the ratio of quantisation-distortion to signal power in an A-law PCM encoder when the signal is uniformly distributed over the range $-b \leqslant x \leqslant b$ in the two cases: (a) $b < 1/A$, and (b) $1/A < b < 1$.

Solution outline

(a) $\quad N_q = 2 \int_0^b \frac{1}{12} \left(\frac{2}{N}\right)^2 \frac{1}{2b} \left(\frac{1 + \ln A}{A}\right)^2 dx = \frac{1}{3N^2} \left(\frac{1 + \ln A}{A}\right)^2.$

$\quad\quad S = 2 \int_0^b \frac{1}{2b} x^2 \, dx = \frac{b^2}{3}.$

Hence

$$\frac{N_q}{S} = \frac{(1 + \ln A)^2}{N^2 \, b^2 \, A^2}.$$

(b) $\quad N_q = 2 \int_0^{1/A} \frac{1}{12} \left(\frac{2}{N}\right)^2 \frac{1}{2b} \left(\frac{1 + \ln A}{A}\right)^2 dx + 2 \int_{1/A}^b \frac{1}{12} \left(\frac{2}{N}\right)^2 \frac{1}{2b} x^2 (1 + \ln A)^2 \, dx.$

$\quad\quad S = 2 \int_0^b \frac{1}{2b} x^2 \, dx.$

Hence

$$\frac{N_q}{S} = \frac{(1 + \ln A)^2}{3N^2} \left(1 + \frac{2}{A^3 \, b^3}\right).$$

1.17   Exponentially distributed signal with A-law companding

Show that when the input signal is exponentially distributed [i.e. $p(x) = \frac{1}{2} k \exp(-k|x|)$] the ratio of quantisation-distortion to signal power S in an A-law PCM encoder is approximately

$$\frac{(1 + \ln A)^2}{3N^2} \left[\frac{1}{SA^2} + \exp\left(-\frac{\sqrt{2}}{A\sqrt{S}}\right) \left\{\frac{\sqrt{2}}{A\sqrt{S}} + 1\right\}\right].$$

[Neglect the contribution of peak-clipping ($|x| > 1$) to the quantisation-distortion.]

Solution outline

$$N_q = \frac{1}{12} \left(\frac{2}{N}\right)^2 \left[2 \int_0^{1/A} \frac{k}{2} \exp(-kx) \left(\frac{1 + \ln A}{A}\right)^2 dx \right.$$

$$+ 2 \int_{1/A}^1 \frac{k}{2} \exp(-kx) \, x^2 (1 + \ln A)^2 \, dx \Bigg]$$

$$= \frac{(1 + \ln A)^2}{3N^2} \left[\frac{1}{A^2} + \exp\left(-\frac{k}{A}\right) \left(\frac{2}{Ak} + \frac{2}{k^2}\right) - \exp(-k) \left(1 + \frac{2}{k} + \frac{2}{k^2}\right)\right].$$

$$\left[\text{Note that } \int kx^2 \exp{(-kx)} \, dx = -\exp{(-kx)} \left(x^2 + \frac{2x}{k} + \frac{2}{k^2} \right). \right]$$

$$S = 2 \int_0^\infty \frac{k}{2} \exp{(-kx)} \, x^2 \, dx = \frac{2}{k^2}.$$

Neglect the term $\exp{(-k)}(1 + 2/k + 2/k^2)$ in the expression for N_q and substitute $\sqrt{2}/\sqrt{S}$ for k.

1.18 *A*-law quantisation for acceptable telephony

Determine a suitable value of A and the minimum number of digits per sample for an *A*-law PCM encoder for telephony, taking into account the noise requirement of Section 1.1.3, the overload recommendations of Section 1.1.4 and the desirability of accommodating the volumes of 99 % of the population (Section 1.1.4).

Solution outline
The r.m.s. value of x for a sinewave on the threshold of overloading the encoder is $1/\sqrt{2}$. The r.m.s. value of x for continuous speech from the loudest talker should be at least 10 dB below $1/\sqrt{2}$, and from the quietest talker at least 40 dB below $1/\sqrt{2}$. Hence A should be chosen to be $100\sqrt{2}$ (Section 1.3.1).

Allowing for psophometric weighting, the quantisation-distortion should be at least 23.5 dB below the r.m.s. value of x for the quietest talker. Hence

$$\frac{1}{12} \left(\frac{2}{N} \right)^2 \left(\frac{1 + \ln A}{A} \right)^2 < \left(\frac{1}{100\sqrt{2}} \right)^2 10^{-2.35}.$$

Substituting for A gives $N > 51.4$. Hence six-digit PCM ($N = 64$) should suffice.

1.19 Dynamic range of linear delta modulation

In the linear delta modulator of Fig. 1.7, each output pulse causes a step change σ in $y(t)$. Assuming that

$$x(t) = E_m \sin 2\pi f_m t,$$

obtain expressions for
(i)   the value of E_m at which the idling pattern is just disturbed,
(ii) the value of E_m on the threshold of slope-clipping.

Solution outline
(i)   The idling pattern causes $y(t)$ to be a square wave with peak-to-peak voltage σ. Hence the idling pattern is disturbed if E_m exceeds $\sigma/2$.
(ii) The average rate of change of $y(t)$ cannot exceed σf_s. The rate of change of $x(t)$ is greatest at the zero crossings, where it is $2\pi f_m E_m$. Hence slope-clipping occurs if E_m exceeds $\sigma f_s / 2\pi f_m$.

1.20 Linear delta modulation for acceptable telephony

A linear delta modulator is to encode speech signals with spectra extending from 300 Hz to 3400 Hz. It is to encode without slope-clipping a 1 kHz sinusoidal test-tone of amplitude

5 V and the quantisation-distortion is to be at least 60 dB below the test-tone. Determine the minimum acceptable clock frequency and an appropriate value for the step height at the integrator output. Take $K_q = 1/3$ (Section 1.3.2).

Solution outline
Avoidance of slope-clipping requires (Exercise 1.19)

$$\sigma f_s > 2\pi \times 5000.$$

Acceptable distortion requires

$$\frac{1}{3} \times \frac{3100 \, \sigma^2}{f_s} < \frac{1}{2} (5)^2 \times 10^{-6}.$$

The minimum value of f_s for which both conditions can be simultaneously satisfied is 434 kHz. If $f_s = 434$ kHz, the two conditions require $\sigma = 0.0724$ V.

1.21　Linear delta-sigma modulation

Repeat Exercise 1.19 for a linear delta-sigma modulator and show that the ratio of the two values of E_m obtained is equal to the corresponding ratio for the linear delta modulator.

Solution outline
Regard the encoder as integrating $x(t)$ and the pulses in separate integrators before subtracting them.
(i)   After integration the amplitude of $x(t)$ is $E_m/2\pi f_m$. Hence the idling pattern is disturbed if E_m exceeds $\pi f_m \sigma$.
(ii)   After integration the greatest rate of change of $x(t)$ is E_m. Hence slope-clipping occurs if E_m exceeds σf_s.

1.22　Exponential delta modulation

Assume that in exponential delta modulation the quantisation-distortion at the input to the low-pass filter in the decoder has the same total power with a signal present as when the encoded waveform is a steady zero volts and has a power spectrum varying with frequency in the same manner as the energy spectrum of a triangular pulse of duration $2/f_s$. Show that the quantisation-distortion at the filter output is approximately $\sigma^2 B_m/4f_s$, where B_m is the filter bandwidth, f_s the sampling frequency and σ the central delta step.

Use the above result to show that when the encoded signal is a sinewave at frequency f_m well above the cut-off frequency of the integrator, the maximum signal/quantisation-distortion ratio obtainable is approximately $f_s^3/2\pi^2 f_m^2 B_m$.

Solution outline
The energy spectrum of a triangular pulse of unit height and duration $2/f_s$ has (two-sided) density $1/f_s^2$ at low frequencies (Appendix 1). The energy in the pulse is $2/3 f_s$, which may be regarded as the product of the low-frequency energy density $1/f_s^2$ and an equivalent (two-sided) bandwidth $2f_s/3$.

When the input signal is zero, the quantisation-distortion at the filter input is roughly a triangular wave with peak-to-peak voltage σ, the power in which may be shown to be $\sigma^2/12$.

From the foregoing, the low-frequency, two-sided, power spectral density of the quantisation-distortion at the filter input is $\sigma^2/12 \div 2f_s/3$ which should be multiplied by $2B_m$ to obtain the required result.

The maximum signal power with no slope-clipping is approximately

$$\frac{1}{2}\left(\frac{V}{2\pi f_{\mathrm{m}}CR}\right)^2 .$$

Since $\sigma = V/CRf_{\mathrm{s}}$, the required result follows.

1.23 Syllabically companded exponential delta modulation

A syllabically companded exponential delta modulator operates with voltages $+AV$ and $-AV$ applied to its integrator. The time-constant (CR) of the integrator is 1 ms and the sampling frequency is 50 kHz. The multiplier A is derived from the output bit stream in such a way that when the input is a 1 kHz sinewave of amplitude E,

$$A = 1 + 100\left(\frac{E}{E_{\max}}\right)^2 ,$$

where $E_{\max}$ is the value of E on the threshold of slope-clipping for that value of A. Calculate the value of A when a 1 kHz sinewave input at a constant amplitude is on the threshold of slope-clipping and estimate the amplitude range, assuming that A remains at that value. Calculate also the amplitude range assuming that the amplitude varies sufficiently slowly for A to follow.

The amplitude is reduced slowly to a steady value 20 dB below the steady value on the threshold of slope-clipping. Estimate the new value of A and the change in signal/quantisation-noise ratio.

The amplitude is then increased again sufficiently rapidly for A to remain unchanged. Determine by how much it may be increased before slope-clipping occurs.

Solution outline
On the threshold of slope-clipping, $E = E_{\max}$ and hence $A = 101$. From the formula in Section 1.3.2, the amplitude range when A remains unchanged is 23.9 dB. For a steady signal just disturbing the idling pattern, $E_{\max}/E$ is equal to the amplitude range and hence $A = 1.40$. Thus the amplitude range for slow variations of amplitude is

$$[23.9 + 20\log(101/1.40)]\ \mathrm{dB}, \quad \text{i.e. } 61.1\ \mathrm{dB}.$$

When the amplitude has been reduced slowly by 20 dB from the slope-clipping value, $E = 10.1\,V/\sqrt{1+4\pi^2}$. The new value of A may be determined by using that equation and the equations:

$$E_{\max} = \frac{AV}{\sqrt{1+4\pi^2}} \quad \text{and} \quad A = 1 + 100\left(\frac{E}{E_{\max}}\right)^2 .$$

Hence $A = 22.026$. The signal level has fallen by 20 dB and the quantisation-distortion by $20\log(101/22.026)$ dB, i.e. 13.2 dB. Hence the ratio has fallen by 6.8 dB.

Slope-clipping occurs when the amplitude is increased rapidly by $20\log(E_{\max}/E)$ dB. When $A = 22.026$, this is 6.8 dB.

1.24 Noise due to errors in exponential delta modulation

Show that the output noise power due to random transmission errors in an exponential delta modulation system is

$$\frac{8V^2 p_{\mathrm{e}}f_1}{f_{\mathrm{s}}}\tan^{-1}\left(\frac{B_{\mathrm{m}}}{f_1}\right),$$

where

p_e is the digit error rate,
f_s (Hz)   is the sampling frequency,
f_1 (Hz)   is the cut-off frequency $(1/2\pi CR)$ of the integrator,
B_m (Hz)   is the bandwidth of the low-pass filter at the decoder output, and
$\pm V$ are the voltage levels applied to the integrator in the decoder.

Assume $B_m \ll f_s$.

Solution outline
At the input to the integrator in the decoder, the effect of transmission errors is to add a random bipolar pulse stream in which $+2V$ pulses and $-2V$ pulses each occur with probability $p_e/2$. Since the autocorrelation of this pulse stream is triangular, its power spectrum is a (sinc)^2 function which may be assumed uniform at frequencies below B_m. Since the power in the pulse stream is $4p_e V^2$, the low-frequency value of its two-sided power spectrum is $4p_e V^2/f_s$.

The transfer function of the integrator is $(1 + \mathrm{j}f/f_1)^{-1}$ and hence the output noise power due to errors is

$$\int_{-B_m}^{B_m} \frac{4p_e V^2}{f_s} \left(1 + \frac{f^2}{f_1^2}\right)^{-1} \mathrm{d}f,$$

which integrates to give the required result.

1.25 Noise reduction for high-quality sound

Figure 1.16 shows a noise-reduction system for high-quality sound. f_1 is controlled by the level of the sound, varying from 1.5 kHz at low levels to 5 kHz at high levels. Show that the feedback in the receiver compensates for the feedforward in the transmitter and show also that the effect of channel noise is reduced at frequencies above f_1. Calculate the amount in dB by which the feedback reduces output noise power due to white noise at P when the level of the sound is low.

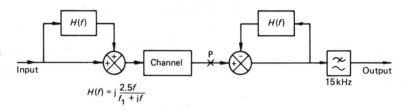

$$H(f) = \mathrm{j}\,\frac{2.5f}{f_1 + \mathrm{j}f}$$

Fig. 1.16 Noise reduction system of Exercise 1.25

Solution outline
The transfer function of the transmitter is $1 + H(f)$ and by writing $V_{out} = V_{in} - H(f)V_{out}$ that of the receiver may be shown to be $\{1 + H(f)\}^{-1}$.

The power gain of the receiver is $|1 + H(f)|^{-2}$, which is less than unity and decreases with increasing f, thus reducing noise at high frequencies.

The receiver reduces white noise at P by

$$10 \log \left\{\frac{1}{15} \int_0^{15} \left|1 + H(f)\right|^{-2} \mathrm{d}f\right\} \, \mathrm{dB}.$$

When $f_1 = 1.5$, this is about 9.1 dB.

1.26 PCM for programme-sound

A 14 digit PCM system for programme-sound uses uniform encoding and a sampling frequency of 32 kHz. Calculate the maximum acceptable digit error rate for the transmission link if the total distortion power due to random errors and quantisation should be at least 72 dB below the power in a sinewave signal at the threshold of overloading. Assume the decoder output to be in unfiltered box-car form.

Repeat the calculation with the system modified to use 13 digit uniform encoding and to transmit a parity check on the five most significant digits in each sample. Assume that errors in the protected digits contribute to the distortion only when there are double errors in the protected digits and their parity check.

Solution outline
Using Eq. 1.3 and arguments used in Exercise 1.14,

$$\frac{4}{3} p_e \{1 - (0.25)^{14}\} + \frac{1}{12} (2^{-13})^2 < 0.5 \times 10^{-7.2}$$

where p_e is the digit error rate. Hence $p_e < 2.27 \times 10^{-8}$.

By enumerating all possibilities or otherwise, the mean-square distortion due to double errors in the protected digits and their parity check may be shown to be $(4/3) 5 p_e^2 \{1 - (0.25)^5\}$. Hence in this case,

$$\frac{4}{3} 5 p_e^2 \{1 - (0.25)^5\} + \frac{4}{3} p_e \left(\frac{1}{32}\right)^2 \{1 - (0.25)^8\} + \frac{1}{12} (2^{-12})^2 < 0.5 \times 10^{-7.2}.$$

Hence $p_e < 1.86 \times 10^{-5}$.

1.27 'Pilot-tone' stereophonic transmission

In the pilot-tone stereophonic broadcasting system the instantaneous frequency deviation of the transmitted signal is proportional to the multiplex signal

$$0.9 \{0.5 (A + B) + 0.5 (A - B) \sin \omega_1 t\} + 0.09 \sin (0.5 \omega_1 t),$$

where A and B are the instantaneous values of the audio signals after emphasis and $\omega_1 = 2\pi \times 38 \times 10^3$ rad/s. The transmitter is set up so that provided $|A| < 1$ and $|B| < 1$ the permitted maximum deviation (75 kHz) is not exceeded.
(i)   Show that the magnitude of $0.5 (A + B) + 0.5 (A - B) \sin \omega_1 t$ cannot exceed the greater of $|A|$ and $|B|$.
(ii)   Calculate the crosstalk introduced in the decoder of Fig. 1.10 due to a $10°$ error in the 19 kHz pilot. Describe also the effects of $90°$ and $180°$ phase errors.
(iii)   The A and B signals may be partially separated by multiplying the multiplex signal by the two 38 kHz square waves shown in Fig. 1.17. Calculate the crosstalk and suggest refinements to reduce it.

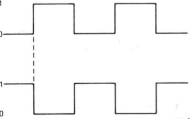

Fig. 1.17   38 kHz switching waveforms of Exercise 1.27 (iii)

(iv)   For a receiver containing the decoder of Fig. 1.10, show that if the FM signal is accompanied by white gaussian noise, the noise accompanying the $(A - B)$ signal is approximately 16 dB greater than that accompanying the $(A + B)$ signal.

(v)   Assuming emphasis equivalent to voltage gain $1 + j\omega/\omega_2$, where $\omega_2 = 20 \times 10^3$ rad/s, compare the noise accompanying the A (or B) signal after de-emphasis in the stereophonic receiver of Fig. 1.10 with the noise accompanying the $0.5 (A + B)$ signal in a monophonic receiver.

Solution outline

(i) $|0.5 (A + B) + 0.5 (A - B) \sin \omega_1 t|$ cannot exceed $0.5 (|A + B| + |A - B|)$, which may be shown not to exceed the greater of $|A|$ and $|B|$ by considering the different possible signs and relative magnitudes of A and B.

(ii)   If there is a phase error ϕ in the subcarrier used to demodulate the difference signal, A is replaced by $0.5 (A + B) + 0.5 (A - B) \cos \phi$ which may be written $0.5 A (1 + \cos \phi) + 0.5 B (1 - \cos \phi)$. The crosstalk is therefore $20 \log \{(1 + \cos \phi)/(1 - \cos \phi)\}$ dB below the level of the signal in the A channel. In this case $\phi = 20°$ (after frequency doubling) and hence the crosstalk is 30.1 dB down in the A channel. A similar argument holds for the B channel.

(iii)   After multiplication the following waveforms are available.

$$m_1 (t) = \{(A + B) + (A - B) \sin \omega_1 t\} \left\{ \frac{1}{2} + \frac{2}{\pi} \sin \omega_1 t + \dots \right\},$$

and $$m_2 (t) = \{(A + B) + (A - B) \sin \omega_1 t\} \left\{ \frac{1}{2} - \frac{2}{\pi} \sin \omega_1 t + \dots \right\}.$$

After low-pass filtering,

$$m_1' (t) = A \left(\frac{1}{2} + \frac{1}{\pi} \right) + B \left(\frac{1}{2} - \frac{1}{\pi} \right),$$

and $$m_2' (t) = A \left(\frac{1}{2} - \frac{1}{\pi} \right) + B \left(\frac{1}{2} + \frac{1}{\pi} \right).$$

The crosstalk ratio is therefore $20 \log \{(\frac{1}{2} + \frac{1}{\pi})/(\frac{1}{2} - \frac{1}{\pi})\}$ dB, i.e. 13.1 dB. It may be improved by forming appropriate linear combinations of $m_1' (t)$ and $m_2' (t)$.

(iv)   At the input to the product demodulator in Fig. 1.10 the multiplex signal must be $0.5 (A + B) + 0.5 (A - B) \sin \omega_1 t + 0.1 \sin (0.5 \omega_1 t)$ and the product demodulator must multiply by $2 \sin \omega_1 t$. It follows (from the modulation theorem, Appendix 1, or otherwise) that the noise accompanying the $0.5 (A - B)$ signal after product demodulation equals that between 23 kHz and 53 kHz accompanying the multiplex signal. Since the power spectral density of noise accompanying the multiplex signal is proportional to the square of frequency, the required ratio is

$$10 \log \left\{ \frac{\left| \int_{23}^{53} f^2 \, df \right|}{\left| \int_0^{15} f^2 \, df \right|} \right\} \text{ dB} = 16.1 \text{ dB}.$$

(v)   If the power spectral density of noise accompanying $0.5 (A + B)$ is kf^2, that of noise accompanying $0.5 (A - B)$ is $k \{ (38 + f)^2 + (38 - f)^2 \}$. Hence the noise power accompanying A (or B) after de-emphasis is

$$k \int_0^{15} \{f^2 + (38 + f)^2 + (38 - f)^2 \} \left(1 + \frac{f^2}{f_2^2} \right)^{-1} df = 128 \, 40 \, k.$$

The noise power accompanying $0.5\,(A+B)$ in a monophonic receiver is (after de-emphasis)

$$k\int_0^{15} f^2\left(1+\frac{f^2}{f_2^2}\right)^{-1}\,\mathrm{d}f = 108\,k.$$

The ratio of these two noise powers is about 21 dB.

2
Digital communications

2.1 Telegraphy

2.1.1 Telegraph codes

Telegraphy is the transmission of characters (such as letters and numbers) in coded form and has a history extending over more than 2000 years[1]. Modern telegraphy usually takes place between teleprinters, which provide a printed copy of messages and can be operated 'off-line'; i.e. messages can be typed into a store and transmitted automatically when required. The traditional telegram, in which printed messages are delivered to addressees, has declined in recent years, but there is increasing demand for the Telex service, in which subscribers' equipment automatically prints telegraph messages received over a dedicated network.

Each of the keyboard characters of a teleprinter is usually encoded as five binary digits, together with 'start' and 'stop' signals separating the characters. The five binary digits representing each character are agreed internationally and the code is known as the 'CCITT International Telegraph Alphabet No. 2' or the '5-unit code'[2]. The number of characters which can be encoded is extended beyond 32 by the use of 'shift' characters, analogous to the shift keys of a typewriter. In fact the teleprinter keyboard resembles the keyboard of a typewriter, the principal difference being that it has no lower-case letters and provides numerals by shifting the top row of letters. Conventionally the binary digits are referred to as Z (mark) and A (space). The start signal is an A; the stop signal resembles a Z but is half as long again.

An alternative alphabet[2] (the CCITT International Telegraph Alphabet No. 3 or Van Duuren code) represents each character by seven binary digits, using only the 35 permutations of four As and three Zs. Since transmission errors usually give rise to incorrect proportions of As and Zs, this code is useful when error detection is required (Section 2.3.1).

The well known Morse code[2] (1832) proved inconvenient for use with electro-mechanical teleprinters because its characters were of different lengths, but it continues to be used when teleprinters are not available.

2.1.2 The baud

The unit of telegraphic signalling speed is called the 'baud' (after Baudot). The speed of signalling in bauds is defined as the reciprocal of the duration in seconds of the shortest signalling element. In the case of the 5-unit code, the shortest element is an A or a Z; in the case of the Morse code it is a dot. The bandwidth required for the transmission of telegraphy is largely determined by the baud rate.

From the foregoing it is evident that in a 75 baud telegraph system using the 5-unit code up to ten characters may be transmitted in one second. This corresponds to the transmission of about 100 words per minute and is typical of conventional telegraph systems. Much higher signalling speeds are used in data transmission systems.

It should be noted that baud is not synonymous with bit/s, the unit of the speed of transmission of information.

2.1.3 Error rates

The quality of a telegraph channel is usually expressed in terms of the rate at which errors occur in a received message. The usual definitions of error rate are

$$\text{Character error rate} = \frac{\text{Number of characters in message which are received incorrectly}}{\text{Total number of characters in message}}$$

$$\text{Digit error rate} = \frac{\text{Number of binary digits in message which are received incorrectly}}{\text{Total number of binary digits in message}}$$

If errors occur at random, the above quantities are respectively the character error probability and the digit error probability.

Traditionally a digit error rate of 10^{-4} has been considered acceptable in telegraphy, but data transmission systems are usually required to have much lower error rates.

2.1.4 Synchronous and start-stop operation

The transmission of digital signals is said to be 'synchronous' when the digits are transmitted in a fixed time pattern of which the receiver has prior knowledge. In synchronous transmission the received waveform can be examined at those instants most likely to give a correct indication of the transmitted message. When the transmitted signal consists of rectangular pulses, synchronous transmission enables the use of an integrate-and-dump receiver (Section 2.2.1), which gives an error rate close to the theoretical minimum when the signal is accompanied by white gaussian noise.

When there is no fixed time pattern, the transmission is said to be 'asynchronous'. Asynchronous transmission is less efficient than synchronous transmission, since it requires either an additional symbol to separate message symbols or an encoding scheme in which no two successive transmitted symbols are the same.

Start-stop operation of teleprinters (Section 2.1.1) is both synchronous and asynchronous. Characters are transmitted asynchronously to permit direct transmission from a keyboard as the operator types ('on-line' operation), but within each character the digits are transmitted synchronously. The timing mechanisms within the teleprinters must be sufficiently accurate for two teleprinters to remain in step for the duration of one character.

2.1.5 Telegraph distortion

The two-level voltage waveform representing a teleprinter character becomes distorted during transmission and subsequently has its edges sharpened in the receiver to restore the rapid transitions necessary to operate a teleprinter. In the process, changes in the relative times of the transitions may occur. Such changes constitute telegraph distortion.

The basic measure of telegraph distortion is 'individual distortion' (δ_{IND}) which is defined by

$$\delta_{\text{IND}} = \frac{(\text{actual instant of transition}) - (\text{nominal instant of transition})}{\text{theoretical pulse length}}$$

In situations where it is impossible to assign an absolute value to the nominal instant of a transition an alternative measure is 'isochronous distortion' (δ_{IS}), which is defined as the difference between the largest and the smallest values of δ_{IND}. In start-stop telegraphy the nominal instant is assumed to be relative to the initial transition of the start pulse and with that assumption the 'start-stop distortion' (δ_{ST}) is defined to be $|\delta_{IND}|$. Start-stop distortion is used in CCITT recommendations; for example, after a single link δ_{ST} should not exceed 8% at 50 baud[2].

2.2 Detection of digital signals in noise

2.2.1 The matched filter

At some point in a receiver for binary digital signals decisions are made about the digit transmitted by comparing samples derived from the received signal with a threshold voltage. If noise accompanying the signal causes a sample to be on the wrong side of the threshold, an error is introduced. In many situations[3] the sample voltages have a gaussian probability density function and the probability that a given deviation is exceeded may be calculated as outlined in Appendix 4.

In order to minimise the probability of error, the comparator should be preceded by a network designed to distort each pulse so that at the instant of sampling its voltage is as large as possible relative to the root-mean-square noise voltage. Such a network is said to be a 'matched filter' for the received pulse.

Assuming that the noise is white and gaussian, it may be shown[4,5,6,7] that the transfer function of a matched filter is the complex conjugate of the Fourier transform of the pulse to which it is matched. Such a filter renders all the spectral components of the pulse in phase, thus ensuring that large samples are available, and ensures a large signal-to-noise ratio by matching its gain to the pulse spectrum. It may also be shown[5,6,7] that a pulse of energy E accompanied by white gaussian noise with one-sided power spectral density N_0 produces at the output of a matched filter a pulse with signal-to-noise ratio $2E/N_0$ at its peak.

It follows that an equivalent definition of a matched filter is that its impulse response should have the shape of the pulse reversed in time. Hence the response of a matched filter to the pulse to which it is matched is the autocorrelation of that pulse. Obviously matched filters can be used only in systems in which the pulse shape is known.

The nearest practical approach to a matched filter is the integrate-and-dump receiver for synchronously transmitted rectangular pulses (Exercise 2.5). It may be shown that the time-constant of the integrator should be as long as possible.

2.2.2 The decision threshold

In general the threshold voltage in a binary receiver should be chosen to minimise the total probability of error. In other words it should maximise the probability that the symbol transmitted was the symbol chosen by the decision made in the receiver.

In Fig. 2.1(a), v_0, (or v_1) is the value of v, the received voltage at a sampling instant, if the symbol 0 (or 1) is transmitted and is received unaccompanied by noise. The probability density functions $p(v|0)$ and $p(v|1)$ describe the distributions of v due to the addition of noise when the transmitted symbol is 0 and 1 respectively. If the symbols 0 and 1 are transmitted with probabilities p_0 and p_1 respectively (the a priori probabilities) and the threshold voltage is set at $v = V$, the total error probability P_e is given by

$$P_e = \int_{-\infty}^{V} p_1 p(v|1)\mathrm{d}v + \int_{V}^{\infty} p_0 p(v|0)\mathrm{d}v$$

and is represented by the sum of the shaded areas of Fig. 2.1(b). Thus P_e is minimised if V

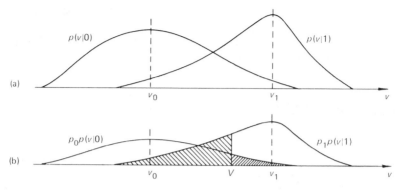

Fig. 2.1

is chosen to be the abscissa at which the curves of Fig. 2.1(b) intersect. Hence V should be chosen so that

$$p_0 p(V|0) = p_1 p(V|1).$$

This result can be derived more formally as the choice of V by Bayes' test to minimise an appropriate cost function[8,9]. The probability that a given symbol was transmitted when it is known what decision the receiver has made is called an 'a posteriori' probability and therefore a decision based on this choice of V is called a 'maximum-a-posteriori-probability decision'.

If p_0 and p_1 are not known at the receiver, V can be chosen so that

$$p(V|0) = p(V|1).$$

Decisions based on this choice of V are called 'maximum-likelihood' decisions because they are based on choosing the largest of the 'likelihood functions' $p(v|0)$ and $p(v|1)$[3,9].

The amount (in dB) by which noise power may be increased before a specified error probability is reached is sometimes called the 'noise margin' of the system under consideration.

2.2.3 Digitally modulated carriers

Many transmission systems (particularly radio systems) require the spectra of digital signals to be shifted. The shifting is most straightforwardly achieved by modulating a sinewave carrier with the digital signal. The basic methods of modulation are amplitude-shift keying (ASK), frequency-shift keying (FSK) and phase-shift keying (PSK). This section contains a brief description of the basic methods and some of their properties. A fuller treatment may be found in many modern books on modulation[3,4,10,11].

In ASK, the information is contained in the amplitude of the carrier. The simplest form of ASK is on-off keying, in which binary digits are represented by the presence or absence of the carrier. Since amplitude variations due to fading (Section 7.3.1) cannot always be distinguished from amplitude modulation, ASK is used only under non-fading conditions (see for example Section 2.4.2). An example of the use of refined ASK is in the high-speed vestigial-sideband transmission of data (Section 5.5.3) in which the carrier component is removed to reduce the required transmitted power and one sideband is restricted to reduce bandwidth.

In FSK, the information is contained in the frequency of the sinewave carrier. In binary systems the carrier frequency is switched between two values. In order to avoid unnecessary spreading of the spectrum of an FSK waveform, continuity of carrier phase should be maintained at the switching instants[1]. FSK offers no advantage over ASK in the presence of added noise, but an FSK receiver is largely unaffected by amplitude variations and FSK is therefore useful in systems subject to fading.

In PSK, the information is contained in the phase of the sinewave carrier. In its simplest form PSK consists of the transmission of a binary message sequence by reversing the phase of the carrier. This may be regarded as on-off keying with the carrier component removed and consequently may be shown to have a 3 dB advantage over on-off keying in the presence of added noise. The rate of transmission of information can be doubled without any increase in bandwidth if four 'levels' of phase are used (quaternary PSK or QPSK), each level representing two binary message digits (one di-bit). In QPSK the rate of transmission of binary message digits is equal to twice the signalling speed in baud. If adjacent levels differ by 90°, the QPSK signal may be regarded as being made up of two binary PSK signals with their carriers in quadrature.

The need for a reference phase in a PSK receiver can be avoided by using differential encoding, in which the message is represented by the difference in phase between successive transmitted signalling elements. An application of differential QPSK (DQPSK) is mentioned in Section 2.4.3.

Distinction may be made between coherent demodulation, in which knowledge of the phase and frequency of the carrier is utilised in the receiver, and incoherent demodulation, in which it is not. In general, coherent demodulation is the more difficult to implement, but gives the lower probability of error, particularly at low signal-to-noise ratios. Demodulation of differentially encoded PSK by comparison of the phase of successive signalling elements is said to be 'differentially coherent'. Differentially coherent demodulation is based on the comparison of two noisy signals and therefore gives a higher probability of error than fully coherent demodulation, in which one noisy signal is compared with an almost noise-free reference. There is also a tendency for errors to occur in pairs when differentially coherent demodulation is used.

The minimum obtainable probability of error in a signalling element for basic binary FSK and PSK systems is shown in Table 2.1, in which E represents the received energy per binary message digit and N_0 the one-sided power spectral density of white gaussian noise accompanying the signal.

Table 2.1 Probability of error in basic binary modulation schemes

	Probability of error in a signalling element
Coherent FSK	$\frac{1}{2}\,\mathrm{erfc}\left(\sqrt{\dfrac{E}{2N_0}}\right)$
Incoherent FSK*	$\frac{1}{2}\exp\left(-\dfrac{E}{2N_0}\right)$
Coherent PSK	$\frac{1}{2}\,\mathrm{erfc}\left(\sqrt{\dfrac{E}{N_0}}\right)$
Differentially coherent PSK	$\frac{1}{2}\exp\left(-\dfrac{E}{N_0}\right)$

* The receiver is assumed to be based on two matched filters followed by envelope detectors, the outputs of which are compared.

2.3   Error control

2.3.1   Error-detection coding

Error-detection coding is based on the introduction of redundancy into the transmitted digital sequence. In other words interrelations exist between the transmitted digits. The receiver looks for these interrelations in the received sequence and the violation of any of them indicates the presence of errors.

A very simple form of error-detection coding for binary signals is to separate the sequence of message digits (the data) into blocks and to add one parity check to each block so that all transmitted blocks have the same parity. A block has even (or odd) parity if it contains an even (or odd) number of 1s. The presence of an odd number of errors in a block can then be detected by calculating the parity of each block in the receiver. In this simple scheme the presence of an even number of errors in a block goes undetected, but it is possible to reduce the probability that a block contains undetected errors by incorporating more than one parity check in each block. More general parity-check codes are introduced in Section 2.3.4. Another simple error-detection code is the Van Duuren code (Section 2.1.1).

Some systems incorporating error-detection coding (for example the transmission of digitally encoded programme sound described in Section 1.4.3) can be designed to give acceptable performance if all the data in blocks showing parity violations is disregarded; but in more stringent applications error-detection coding is usually associated with the existence of a return channel used to request the retransmission of doubtful data. When the request and the retransmission take place automatically in a system, that system is said to be an 'automatic request repeat' (ARQ) system.

The usual arrangement in an ARQ system is for an indication of parity violation to cause the transmitter to repeat the most recently transmitted blocks, beginning with the one associated with the parity violation. The number of blocks repeated for each parity violation depends upon the delays of the forward and return transmission channels, and the system becomes very inefficient if the delays are long. The inefficiency is largely due to the discarding of blocks received correctly after a parity violation but before the repeat has reached the receiver, and suggestions[12] have been made for improving the efficiency of ARQ satellite systems by not retransmitting blocks already received correctly.

An ARQ system is an example of a decision-feedback system in that the signals fed back to the transmitter are decisions made at the receiver about whether or not to request a repeat. If the return channel has sufficient capacity, more reliable communication can be obtained by the use of information feedback, in which decisions to repeat are made at the transmitter with the aid of information (perhaps in the form of parity checks) fed back from the receiver.

2.3.2   Forward error correction

Errors in a binary digital sequence can be corrected if their locations are known. Hence when no return channel is available it is sometimes advantageous to introduce so much redundancy into the transmitted sequence that the receiver not only detects errors but also locates and corrects them. This method of reducing error probability is known as 'forward error correction'.

A simple code used commercially for forward error correction of telegraphy is the Marconi Autospec code. In this code the five data digits for each character are followed by five check digits, which are either a repeat of the data digits or their complement, the choice being determined by the parity of the data digits.

A code proposed by Hamming[8] uses parity checks within blocks of data (Section 2.3.1) to correct one error in each block. In this code a transmitted block contains $2^m - 1$ binary digits in locations numbered from 1 to $2^m - 1$; the m digits in locations 1, 2, 4, 8 etc. are parity-check digits and the remaining locations are occupied by data. If the parity checks are calculated as shown in Table 2.2 for the case $m = 4$, the location of an error is equal to the sum of the locations of the check digits showing parity violations. Hamming's code is efficient, since m binary digits are just sufficient to indicate whether or not there is an error and, if there is, in which of $2^m - 1$ locations it occurs.

Table 2.2 Hamming's single error correcting code

Location of check digit	Locations of digits contributing to parity check
1	1, 3, 5, 7, 9, 11, 13, 15
2	2, 3, 6, 7, 10, 11, 14, 15
4	4, 5, 6, 7, 12, 13, 14, 15
8	8, 9, 10, 11, 12, 13, 14, 15

Codes such as Hamming's code in which the data sequence is divided into blocks within which the parity checks are applied are known as 'block codes'. Significant improvements in error probability can be obtained with block codes provided that the errors occur at random. Unfortunately there is a tendency in many transmission systems for errors to be introduced in bursts by fading or interference and when protection against bursts is required block codes are inefficient. In some systems the bursts are dispersed by scrambling the order of the digits before transmission and restoring it at the receiver.

Another approach to error-correction coding is provided by convolutional codes, in which transmitted digits are derived from data digits in a continuous process. An early example is the Hagelbarger code[1, 8] in which the transmitted sequence contains data digits 1, 2, 3, ... alternated with check digits in the following way:

$$C_{47}, 1, C_{58}, 2, C_{69}, 3, \ldots ,$$

where C_{ij} is a parity check on data digits i and j. This code was intended to deal with errors occurring in widely separated bursts and can correct all the errors in a burst of length six provided that it is preceded by at least 19 correct digits. More recent examples are described in Section 2.3.7. A useful survey of the applicability and costs of error-correction codes has been given by Berlekamp[13].

2.3.3 Hamming distance

A block code in which blocks of k binary data digits are encoded for transmission as blocks of n binary digits $(n > k)$ is described as an (n, k) block code. Its codewords are 2^k of the 2^n possible n-tuples. The Hamming distance between two n-tuples is the number of digits in which they differ. The minimum distance (d) of a block code is the minimum Hamming distance between two codewords of that code.

If a block code is to detect e errors in a block, its minimum distance must not be less than $e + 1$; if it is to correct t errors in a block, its minimum distance must not be less than $2t + 1$.

It is sometimes helpful to illustrate the error-detecting and error-correcting capabilities of a block code by a diagram such as Fig. 2.2. Distance along the line represents Hamming distance. The seven points on the line represent two codewords separated by the minimum distance of the code (in this case six) and the five n-tuples occurring in a digit-at-a-time transition between them. An n-tuple adjacent to a codeword is a codeword with one error, an n-tuple next-but-one to a codeword is a codeword with two errors etc. The diagram

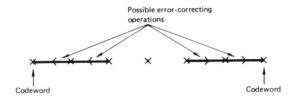

Fig. 2.2   Error-correcting capability of block code with minimum distance six

suggests that a code with minimum distance six could be used to correct double errors and detect triple errors or alternatively to detect up to five errors in any codeword.

2.3.4   Parity-check codes

The codewords of an (n, k) parity-check code consist of k data digits and $(n-k)$ check digits, each check digit being a parity check on some of the data digits. If the data digits are assumed to form the first k digits of a codeword, the $(1 \times n)$ vector c describing the codeword may be derived from the $(1 \times k)$ vector d describing the data by the matrix multiplication:

$$c = dG$$

where the 'generator matrix' G is a $(k \times n)$ matrix. The elements of the vectors and matrices used in this Section are 0 or 1 and addition is modulo-2.

It follows from the above that the sum of two codewords is a codeword, that $c = 0$ is a codeword and hence that the codewords form a group under modulo-2 addition. The code is sometimes referred to as a 'linear' or 'group' code.

The generator matrix G may be partitioned:

$$G = (I \,|\, P),$$

where I is the $(k \times k)$ unit matrix. The 'parity-check matrix' H is the $(n-k) \times n$ matrix defined by:

$$H = (P' \,|\, I_{n-k}),$$

where I_{n-k} is the $(n-k) \times (n-k)$ unit matrix and P' is the transpose of P. It may be shown that

$$GH' = 0.$$

If c is a codeword, $c = dG$ for some data d and hence

$$cH' = dGH' = 0.$$

Conversely, if $eH' = 0$ for an n-tuple e, the parity checks are satisfied and therefore e is a codeword.

The 'syndrome' s of an n-tuple e is the $1 \times (n-k)$ vector defined by

$$s = eH'.$$

The presence of a 1 in the i-th position of s indicates that the parity check specified by the i-th column of H' (or of P) fails when applied to e. s may be obtained by calculating the check digits appropriate to the first k digits (the data digits) of e and adding them to the last $(n-k)$ digits of e.

It may be shown that for any (n, k) linear parity-check block code the 2^n possible n-tuples may be arranged in a rectangular array called a 'standard array'. The array has 2^{n-k} rows, each containing all the 2^k n-tuples with a given syndrome, and 2^k columns, each headed by a codeword. The 2^k n-tuples with the same syndrome are said to form a 'coset'. A given code does not have a unique standard array, but the grouping of all n-tuples into cosets is unique for that code. A standard array can be used as a decoding array for error correction if each received n-tuple is associated with the codeword heading the column in which it occurs.

A standard array may be constructed as follows. Write down all the codewords to form the first row, beginning with the all-zero n-tuple at the left-hand end. Form the second row by adding the same n-tuple (not a codeword) to each codeword in turn. Form the third row by adding another n-tuple (neither a codeword nor an n-tuple appearing in the second row) to each codeword in turn and continue in the same way. As a very simple illustration, consider the code generated by

$$G = \begin{pmatrix} 1 & 0 & 1 \\ 0 & 1 & 1 \end{pmatrix}$$

One possible standard array is the following.

	Coset leaders			
Coset with syndrome 0 (codewords)	000	101	011	110
Coset with syndrome 1	001	100	010	111

In the above array the second row was derived from the first by choosing 001 as coset leader and adding it to each codeword in turn. An example of a standard array for a (5, 3) code occurs in Exercise 2.12.

The minimum distance of a linear parity-check block code is the Hamming weight (i.e the number of 1s) of the non-zero codeword with minimum Hamming weight. Since 0 is a codeword, the minimum distance cannot exceed the minimum weight and since the sum of two codewords is also a codeword it cannot be less.

For a more rigorous exposition of the material of this Section see Lucky et al[14].

2.3.5 Cyclic codes

Theoretical study of parity-check block codes has concentrated on a class within them called cyclic codes. This Section outlines their basic properties but makes no attempt to derive them. More explanation can be found in Lucky et al[14] or in more specialist texts.

A cyclic code is defined to be a block code having the property that every cyclic shift of a codeword is also a codeword. For example, if 101001 is a codeword, so is 001101.

The 'code polynomial' $c(X)$ representing the codeword

$$c = \{c_{n-1}, c_{n-2}, \ldots, c_1, c_0\}$$

is defined by

$$c(X) = c_{n-1} X^{n-1} + c_{n-2} X^{n-2} + \ldots + c_1 X + c_0.$$

Under the rules that $X^n = 1$, $X^{n+1} = X$ etc. and that the coefficients c_i add modulo-2, multiplication of $c(X)$ by X corresponds to a cyclic shift of the codeword by one digit.

Write the generator matrix G of a parity-check code in the form

$$G = \begin{pmatrix} & \begin{matrix} h_0 \\ h_1 \\ h_2 \\ \vdots \\ h_{k-1} & g_{n-k-2} & \cdots & g_1 & g_0 \\ (h_{k-1} = g_{n-k-1}) \end{matrix} \end{pmatrix}$$

If the code is cyclic, $h_0 = g_0 = 1$, since otherwise a cyclic shift would lead to an n-tuple not a codeword. When the code is cyclic, the polynomial

$$g(X) = X^{n-k} + g_{n-k-1} X^{n-k-1} + \ldots + g_1 X + g_0$$

is called the 'generator polynomial' of the code and the polynomial

$$h(X) = X^k + h_{k-1} X^{k-1} + \ldots + h_1 X + 1$$

is called the 'parity-check polynomial' of the code. It may be proved that

$$g(X)h(X) = X^n + 1$$

and thus that $g(X)$ is a divisor of $X^n + 1$. Conversely it may be shown that every divisor of $X^n + 1$ generates a cyclic (n, k) code.

The generator matrix of a cyclic code may be constructed from the generator polynomial $g(X)$ in the following way.

(i)   Write down the k-th row directly from the coefficients of $g(X)$.

(ii)   To obtain the $(k-1)$-th row, shift the k-th row one place to the left. If the k-th element is 0, this is the required $(k-1)$-th row; if it is 1, add the k-th row to it.

(iii)   To obtain further rows, shift the previous row one place to the left and if necessary add the k-th row to make the k-th element 0.

For the example, the generator matrix of the $(7, 4)$ cyclic code generated by $X^3 + X + 1$ may be shown by these rules to be

$$\begin{pmatrix} 1 & 0 & 0 & 0 & 1 & 0 & 1 \\ 0 & 1 & 0 & 0 & 1 & 1 & 1 \\ 0 & 0 & 1 & 0 & 1 & 1 & 0 \\ 0 & 0 & 0 & 1 & 0 & 1 & 1 \end{pmatrix}$$

The 2^k codewords of a cyclic code are the 2^k possible sums of the rows of its generator matrix and their code polynomials are the 2^k possible multiples of its generator polynomial which have degree not greater than $n - 1$. The code has no code polynomial with degree less than or equal to $n - k$ other than its generator polynomial.

The code polynomial of the codeword containing the k-bit data block with polynomial $d(X)$ may be shown to be

$$X^{n-k}d(X) + r(X),$$

where $r(X)$ is the remainder after division of $X^{n-k} d(X)$ by $g(X)$. Also the polynomial of the syndrome of an n-tuple with polynomial $e(X)$ is the remainder after division of $e(X)$ by $g(X)$.

Division of one polynomial by another can be implemented by a feedback shift-register, which can therefore form the basis of convenient methods of encoding data or calculating syndromes. The basic forms of two types of encoder are shown in Fig. 2.3.

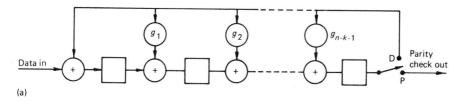

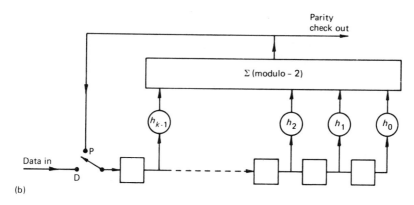

Fig 2.3 Basic encoders for cyclic codes (a) $(n-k)$-stage encoder (b) k-stage encoder

Before encoding with the $(n-k)$-stage encoder (Fig. 2.3(a)), the bi-stable circuits are cleared to 0 and the switch set to D. The register is then clocked k times as the data is fed in and a further $n-k$ times with 0s fed in. It then contains the $n-k$ check digits, which can be shifted out with the switch set to P. By re-arrangement of the input and output the number of shifts necessary to calculate the check digits can be reduced from n to $n-k$ (see Exercise 2.17).

Before encoding with the k-stage encoder (Fig. 2.3(b)), the bi-stable circuits are cleared to 0 and the switch set to D. The register is then clocked k times as the data is fed in and a further $n-k$ shifts with the switch set to P generate the parity checks at the output.

2.3.6 BCH (Bose–Chauduri–Hocquenghem) codes

BCH codes form a class of cyclic codes and provide efficient correction of multiple errors. A rigorous exposition[15] of BCH codes is mathematically involved and will not be attempted here.

The coefficients of the polynomials introduced in Section 2.3.5 are the elements of the simple Galois field $GF(2)$, which contains just the two elements 0 and 1 and is defined for the operations of multiplication and modulo-2 addition. However, the zeros of those polynomials will not in general be elements of $GF(2)$ and can be described only by regarding them as elements of some wider field, in a way analogous to the use of complex numbers to describe the zeros of polynomials with real coefficients.

Regarding the zeros of $X^n + 1$ as elements of that wider field, they may be listed:

$$\alpha^0 (= 1), \alpha, \alpha^2, \alpha^3, \ldots, \alpha^{n-1}.$$

In other words they are the powers of one of them, which is called a 'primitive element'. Note that they form a group under cyclic multiplication ($\alpha^{n+i} = \alpha^i$). Consecutive zeros in the above list (e.g. α^i and α^{i+1}) are said to be 'consecutive zeros' of $X^n + 1$. Since the generator polynomial of any (n, k) cyclic code divides $X^n + 1$ (Section 2.3.5), its zeros are a subset of the above list.

The construction of BCH codes is based on the BCH bound, which states[14] that the minimum distance of an (n, k) cyclic code is strictly greater than the length of the longest unbroken sequence of consecutive zeros of $X^n + 1$ which are also zeros of the generator polynomial of the code. In this context α^{n-1} and α^0 are regarded as consecutive. Hence an (n, k) cyclic code can be designed to have a minimum distance not less than d by choosing as factors of its generator polynomial a set of factors of $X^n + 1$ containing at least $d - 1$ consecutive zeros.

A BCH code is an (n, k) cyclic code with its generator polynomial chosen so as to include the longest possible sequence of consecutive zeros of $X^n + 1$ for given values of n and k. Thus for a given number of parity checks a BCH code is a code which maximises the minimum distance guaranteed by the BCH bound.

When $n = 2^m - 1$ (m an integer), it can be shown[14,15] that there exists an $(n, n - mt)$ BCH code which can correct up to t errors in every block. A BCH code with block length $n = 2^m - 1$ is said to be 'primitive'.

Two simple primitive BCH codes are explored in Exercises 2.18 and 2.19.

2.3.7   Convolutional codes

Convolutional codes are claimed to give better performance than block codes of similar complexity and are becoming widely used for protection against errors.

A simple example of a convolutional code is that generated by the encoder of Fig. 2.4(a), which operates as follows. After each data digit has been fed into the shift register from the left, each modulo-2 adder in turn contributes a digit to the output sequence. In this case the output digit rate is twice the data digit rate and the code is said to have a code rate of $\frac{1}{2}$.

The encoder of Fig. 2.4(a) may be regarded as a system with four possible states, its state being defined by the contents of the first two stages of the register. The state after the arrival of a data digit is determined partly by that digit and partly by the state before its arrival. All possible transitions between states are shown in the de Bruijn diagram of Fig. 2.4(b), in which the input X and the output YZ associated with each transition are shown as $(X : YZ)$. Figure 2.4(c) shows a representation of possible transitions as a trellis diagram. In these diagrams the first digit of a pair describing the state is the content of the first stage of the register: some authors[16,17] write the digits in the reverse order.

The decoding process for a convolutional code is based on finding the path through the trellis diagram most likely to have been that followed by the encoder. If each transition is assigned a measure (or 'metric') related to its likelihood in view of the digits actually received, the problem becomes analogous to that of finding the shortest path through a network made up of branches of different lengths. An appropriate metric for a transition is the Hamming distance between the encoder output digits for that transition and the digits received. The metric for a path through the trellis diagram is the sum of the metrics for the transitions making up the path. Finding the path with the lowest metric corresponds to finding the sequence of states for which the received sequence would have the fewest disagreements with the encoder output associated with it. There are several methods of finding this path. We shall consider only Viterbi's method[18], which has been shown to yield maximum-likelihood decisions (Section 2.2.2).

Viterbi's method is to find the path of least metric (Hamming distance) leading from an assumed initial state at level 0 to every node at each level in turn, noting the metric of that path and the data sequence (the 'survivor') associated with it. The calculation is carried out level by level. At each level two metrics and survivors are obtained for each node (since there are always two branches leading to it) and the lower metric and associated survivor are retained for use at later levels. The process is terminated after a suitable number of

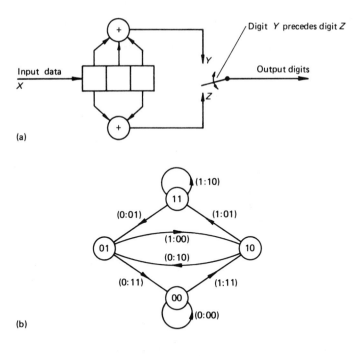

(a)

(b)

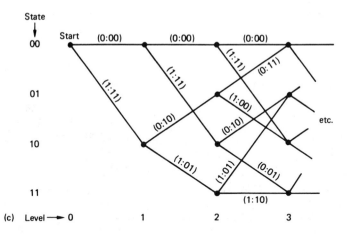

(c) Level ⟶ 0   1   2   3

Fig. 2.4 A convolution encoder and its state diagrams (a) a simple convolutional encoder (b) de Bruijn diagram for encoder in (a) (c) trellis diagram for encoder in (a) initially in state 00

levels (typically five times the encoder register length) and a decision is made on a single data digit by selecting the first digit in the survivor at the node with the lowest metric at the final level. The process is then repeated, beginning one level later. An example of the calculation of metrics and survivors is given in Fig. 2.5 for the encoder of Fig. 2.4(a) and the received sequence 01000000. Termination at level 4 suggests that the first digit of the survivor 0000 should be accepted as a data digit. This is consistent with the correction of a single error when the encoder output is 00000000.

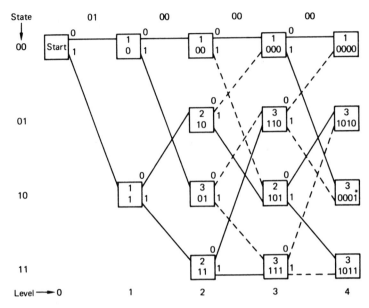

Fig. 2.5 Example of calculation of Hamming distance and survivors by Viterbi's method for convolution code of Fig. 2.4
Digits by each transition indicate encoder input associated with that transition
Windows at nodes show minimum metric above survivor
Rejected transitions are shown – – –
* indicates that in arriving at this survivor an arbitrary tie-break rule has been used (selection of the upper of the two transitions)

    Some of the error-correcting properties of the code of Fig. 2.4 can be derived by inspection of the de Bruijn diagram, as may be seen from the following example. Assume that the correct sequence of states is 00 00 00 00 00. (There is no loss of generality in making this assumption[17].) There is an alternative path from state 00 to state 00 by way of states 10 and 01. The received sequence corresponding to this alternative path is 111011 and therefore has a Hamming weight of five. There is no other incorrect path from 00 to 00 with lower Hamming weight. Hence it is reasonable to conclude that the code can correct two errors in a sequence of six transmitted digits.

    A systematic approach to the determination of the error-correcting capabilities of a convolutional code is to derive a generating function in the form of a power series in the following way. Ignoring the self-loop at the node corresponding to the all-zero state, split that node and redraw the de Bruijn diagram as a flow graph with the all-zero node as source and sink nodes. Assign to each transition the branch gain D^k, where k is the Hamming weight of that transition and express the gain of the flow graph as a power series in D. That power series is the required generating function, the presence of a term $a_i D^i$ indicating that there are a_i possible paths with Hamming weight i. As an example, the flow graph for the encoder of Fig. 2.4 is shown in Fig. 2.6. Its gain a'/a may be obtained by simplifying the flow graph or by eliminating b, c and d from the equations

$$b = D^2 a + c$$
$$c = Db + Dd$$
$$d = Db + Dd$$
$$a' = D^2 c.$$

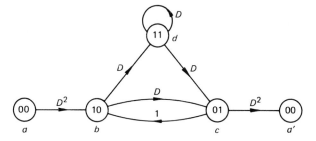

Fig. 2.6 Flow graph showing Hamming weights for encoder of Fig. 2.4

Hence

$$\frac{a'}{a} = \frac{D^5}{1 - 2D} = D^5 + 2D^6 + 4D^7 + \ldots .$$

Hence there is one path with weight five, two paths with weight six etc.

2.4 Digital communications over telephone channels

2.4.1 Channel limitations

It is obviously very attractive to use existing telephone facilities for the transmission of digital signals, but channels which transmit speech satisfactorily may have serious limitations when used for digital signals. In particular, there are three aspects of channel performance which can seriously impair digital transmissions, namely delay distortion, echoes and frequency offset.

Digital signals are transmitted over telephone channels as modulation of an audio-frequency carrier, since telephone channels do not usually transmit frequencies below about 300 Hz (Section 1.1.2). Hence in order to ensure that the digital signal suffers no delay distortion it is sufficient to ensure no variation of group delay over the frequency band of the modulated signal (Appendix 2). As a rough guide, delay distortion is unacceptable if the variation exceeds the reciprocal of the signalling speed in baud and, therefore, a variation of 60 ms, although it might be acceptable for telephony (Section 1.1.2), would impose severe restrictions on signalling speed. Fortunately most of the variation occurs near the edges of the frequency band[19] and experience has shown that transmission at 1200 baud is usually possible over a dialled connexion if the spectrum of the transmitted signal is limited to a range from about 900 Hz to about 2600 Hz. If the variation is reduced to less than 0.5 ms over that range by transversal-filter equalisation (Section 6.1.3), transmission at 1200 baud can be virtually guaranteed.

While echoes delayed by as much as 30 ms are rarely of importance in telephony, an echo delayed by as little as 0.5 ms may cause sufficient intersymbol interference in a 1200 baud digital signal (Section 6.1) to increase its susceptibility to errors. Severe echoes can be reduced by a transversal filter.

The inclusion of a carrier-telephony link in a telephone connexion usually results in a constant frequency offset of all the spectral components of a signal. This causes very severe distortion of digital signals, unless they are in the form of a modulated sinewave, in which case carrier and sidebands are offset together and the digital signal can be recovered without distortion.

Many telephone systems use in-band signalling in the form of tones within the frequency band of the channel. Since it is important that digital signals should not cause false operation of signalling receivers, there is sometimes a further restriction on the frequency band which they can occupy, particularly at frequencies below 900 Hz.

2.4.2   Voice-frequency telegraphy (VFT)

For many years telephone channels have been used for the transmission of signals in the form of modulated voice-frequency sinewaves. Several telegraph signals modulating carriers at different frequencies can share one telephone channel. The principle of sharing a communication link in this way by division of the frequency band transmitted is known as frequency-division multiplex (FDM). By international agreement[20] carriers separated by 120 Hz are used to transmit 24 50 baud telegraph signals simultaneously.

Early systems used ASK, but FSK is now favoured because of its tolerance of rapid level changes. The frequency deviation recommended is ± 30 Hz or ± 35 Hz and a 3 dB bandwidth of 80 Hz is typical for the band-pass separation filters.

2.4.3   Modems for data transmission

A modem is a device which converts digital data to a modulated form suitable for transmission and demodulates data received in modulated form.

Over dialled connexions, the rate of transmission of data is restricted by international agreement[21] to be 300, 600, 1200, 2400 or 4800 bit/s. Over leased connexions other rates are allowed, provided that they are expressible as $600n$ bit/s where n is a positive integer not exceeding 18[22].

The standard 300 baud system is capable of simultaneous two-way operation (duplex operation) over a telephone circuit. Transmission is by FSK, using 980 Hz and 1180 Hz in the forward direction and 1650 Hz and 1850 Hz in the return direction. The return channel can be used either for data or for error-control feedback.

The standard 600/1200 baud system transmits in one direction by FSK (1300 Hz and 1700 Hz for 600 baud, 1300 Hz and 2100 Hz for 1200 baud); in the other direction there is a 75 baud FSK channel (390 Hz and 450 Hz). To prevent undue spreading of the spectrum by phase discontinuities in the FSK waveform, modulation is usually effected by switching the aiming voltage of a multivibrator. Since one signalling element lasts for little more than one cycle of the lower frequency (1300 Hz), discriminators are impractical and demodulation is usually carried out by a zero-crossing detector[23].

Signalling speeds in excess of 1200 bauds are considered too high for use over telephone channels and therefore data transmission at 2400 bit/s uses DQPSK (Section 2.2.3). In some systems the four possible di-bits are represented by phase shifts of $0°$, $90°$, $180°$ and $270°$. In other systems the phase shifts are $45°$, $135°$, $225°$ and $315°$, thus ensuring that signal-element synchronisation can be extracted whatever the information pattern. Demodulation may be coherent or differentially coherent (Section 2.2.3). The carrier frequency is 1800 Hz and a 75 baud return channel is provided. When intended for use on dialled connexions, the modems should be capable of operating at 1200 bit/s by (binary) DPSK when the circuit quality is poor.

Transmission at 4800 bit/s is achieved by 8 level DPSK at 1200 baud and at 9600 bit/s by simultaneous 8 level DPSK and binary ASK. In either case equalisation of the channel is necessary. Scrambling (re-ordering of the message digits for transmission) is required to avoid prominent tones due to repeated patterns. The standard scrambler uses seven-stage shift registers. A simplified scrambler with three-stage registers is the subject of Exercise 2.26.

2.5 References

1 Bennett, W. R., and Davey, J. R., *Data Transmission*, McGraw-Hill, 1965
2 Renton, R. N., *Telegraphy*, Pitman, 1976
3 Roden, M. S., *Digital and Data Communication Systems*, Prentice Hall, 1982
4 Haykin, S., *Communication Systems*, Wiley, Second Edition 1983
5 Betts, J. A., *Signal Processing, Modulation and Noise*, English Universities Press, 1970
6 Dunlop, J., and Smith, D. G., *Telecommunications Engineering*, Van Nostrand Reinhold (UK), 1984
7 Clark, A. P., *Principles of Digital Data Transmission*, Pentech Press, Second Edition 1983
8 Rosie, A. M., *Information and Communication Theory*, Van Nostrand Reinhold, Second Edition 1973
9 Stiffler, J. J., *Theory of Synchronous Communications*, Prentice-Hall, 1971
10 Shanmugan, K. S., *Digital and Analog Communication Systems*, Wiley, 1979
11 Carlson, A. B., *Communication Systems*, McGraw-Hill-Kogakusha, Second Edition 1975
12 Lin, S., and Yu, P. S., 'An effective error control scheme for satellite communications', *IEEE Trans.*, COM-28, pp. 395–401, 1980
13 Berlekamp, E. R., 'The technology of error-correcting codes', *Proc. IEEE*, **68**, pp. 564–93, 1980
14 Lucky, R. W., Salz, J., and Weldon, E. J., *Principles of Data Communication*, McGraw-Hill, 1968
15 Peterson, W. W., and Weldon, E. J., *Error-correcting Codes*, MIT Press, Second Edition 1972
16 Spilker, J. J., *Digital Communication by Satellite*, Prentice-Hall, 1977
17 Viterbi, A. J., Convolutional codes and their performance in communication systems', *IEEE Trans.*, COM-19, pp. 751–72, 1971
18 Forney, G. D., 'The Viterbi algorithm', *Proc. IEEE*, **61**, pp. 268–78, 1973
19 Ridout, P. N., and Rolfe, P., 'Transmission measurements of connexions in the switched telephone network', *Post Office Electrical Engineers Journal*, **63**, pp. 97–104, July 1970
20 *CCITT Yellow Book*, Vol. III, Recommendation H23, Geneva, 1980
21 *CCITT Yellow Book*, Vol. VIII, Recommendations V5, V21, V23, Geneva, 1980
22 *CCITT Yellow Book*, Vol. VIII, Recommendation V6, Geneva 1980
23 Renton, R. N., *Data Telecommunication*, Pitman, 1973
24 *CCITT Yellow Book*, Vol. VIII, Recommendation V27, Geneva, 1980

Exercise topics

2.1 Error probability for polar signalling
2.2 Optimum threshold for on-off signalling
2.3 Optimum threshold for ternary signalling
2.4 Matched-filter response to sinewave burst
2.5 Baseband integrate-and-dump receiver
2.6 Non-coherent ASK receiver
2.7 Non-coherent FSK receiver
2.8 Probability of error in di-bit in QPSK
2.9 ARQ using single parity check
2.10 Hamming single-error-correcting code
2.11 Minimum distance of Van Duuren, Autospec and Hamming codes
2.12 Block code (using matrices)
2.13 Hamming bound
2.14 Cyclic code (using polynomial manipulation)
2.15 Cyclic codes of word-length six
2.16 A (7, 4) cyclic code
2.17 Encoders for cyclic codes
2.18 Construction of (7, 3) cyclic code with minimum distance four
2.19 Construction of (15, 5) triple-error-correcting cyclic code
2.20 Comparison of ARQ with forward error correction
2.21 Decision feedback
2.22 Trellis diagram for convolutional code
2.23 de Bruijn diagram for convolutional code
2.24 Error-correcting capability of convolutional code
2.25 Generating function of convolutional code
2.26 Scrambler

Exercises

2.1 Error probability for polar signalling

At a decision device with a threshold at 0 V a synchronous binary signal switching rapidly between $+2.5$ V and -2.5 V is accompanied by gaussian noise. The signal-to-noise ratio is 13 dB. Calculate the digit error probability.

Assuming that the two signalling elements are transmitted with equal probabilities, recalculate the digit error probability if the threshold is offset by 0.5 V.

Solution outline
The signal power (in 1 Ω) is $(2.5)^2$ W. The noise power (σ^2) is 13 dB lower. Hence $\sigma = 0.56$ V. Hence the digit error probability is (Appendix 4)

$$\frac{1}{2}\text{erfc}\left(\frac{2.5}{0.56\sqrt{2}}\right) = \frac{1}{2}\text{erfc}\,(3.16) \simeq 4 \times 10^{-6}.$$

With the threshold offset, the digit error probability becomes

$$\frac{1}{2}\left[\frac{1}{2}\text{erfc}\left(\frac{3}{0.56\sqrt{2}}\right) + \frac{1}{2}\text{erfc}\left(\frac{2}{0.56\sqrt{2}}\right)\right] \simeq 9 \times 10^{-6}.$$

2.2 Optimum threshold for on-off signalling

Marks and spaces are transmitted with probabilities p_1 and p_0 ($p_1 + p_0 = 1$) and are presented to a decision circuit as samples of voltages V and zero respectively. Show that when there is accompanying gaussian noise with mean-square voltage N the decision threshold should be at

$$\frac{V}{2} + \frac{N}{V}\ln\left(\frac{p_0}{p_1}\right).$$

Solution outline
The probability density function $p(v\,|\,0)$ for the voltage v presented to the decision circuit when a space is transmitted is given by

$$p(v\,|\,0) = \frac{1}{\sqrt{2\pi N}}\exp\left(\frac{-v^2}{2N}\right)$$

and when a mark is transmitted the probability density function $p(v\,|\,1)$ is given by

$$p(v\,|\,1) = \frac{1}{\sqrt{2\pi N}}\exp\left\{\frac{-(v-V)^2}{2N}\right\}.$$

Equating $p_0 p(v\,|\,0)$ to $p_1 p(v\,|\,1)$ and solving for v gives the required result.

2.3 Optimum threshold for ternary signalling

At the decision device in a synchronous ternary digital transmission system the digits are either -1 V, 0 V, or $+1$ V at the sampling instants and are accompanied by gaussian noise with root-mean-square voltage 0.2 V. Assuming that the symbols $-, 0, +$ are transmitted with probabilities $\frac{1}{4}, \frac{1}{2}, \frac{1}{4}$ respectively, calculate suitable threshold voltages.

Solution outline
Extending the argument of Section 2.2.2 to ternary signalling, the upper threshold should

be that value for v which satisfies

$$\frac{1}{4}\exp\left\{\frac{-(v-1)^2}{2N}\right\} = \frac{1}{2}\exp\left\{\frac{-v^2}{2N}\right\} + \frac{1}{4}\exp\left\{\frac{-(v+1)^2}{2N}\right\}.$$

Solving for v,

$$v = N\sinh^{-1}\{\exp(1/2N)\}.$$

In this case $N = 0.04$ and hence $v = 0.53$ V. By a similar argument the lower threshold should be at -0.53 V.

In this example the probabilities of a $+$ being mistaken for a $-$ and vice versa are so small that the result of Exercise 2.2 could have been applied directly.

2.4 Matched-filter response to sinewave burst

A burst of the sinewave $A\sin(2\pi t/T)$ extending from $t = 0$ to $t = 2T$ is applied to a matched filter. Obtain an expression for the output waveform.

Solution outline
The impulse response of the matched filter is of the form $A\sin\{2\pi(t_0 - t)/T\}$ extending from $t = t_0 - 2T$ to $t = t_0$. For convenience assume $t_0 = 0$, although the filter is then not realisable unless additional delay is incorporated. Hence the output waveform is (by convolution)

(a) for $-2T < t < 0$

$$\int_0^{t+2T}\left\{A\sin\left(\frac{2\pi\tau}{T}\right)\right\}\left\{A\sin\frac{2\pi(\tau-t)}{T}\right\}d\tau = \frac{A^2}{2}\left\{(2T+t)\cos\left(\frac{2\pi t}{T}\right) - \frac{T}{2\pi}\sin\left(\frac{2\pi t}{T}\right)\right\}$$

(b) for $0 < t < 2T$

$$\int_t^{2T}\left\{A\sin\left(\frac{2\pi\tau}{T}\right)\right\}\left\{A\sin\frac{2\pi(\tau-t)}{T}\right\}d\tau = \frac{A^2}{2}\left\{(2T-t)\cos\left(\frac{2\pi t}{T}\right) + \frac{T}{2\pi}\sin\left(\frac{2\pi t}{T}\right)\right\}$$

(c) zero elsewhere.

2.5 Baseband integrate-and-dump receiver

Figure 2.7 shows a baseband integrate-and-dump receiver. At intervals of T seconds, $v_2(t)$ is sampled by closing S1 for a short period. Immediately after each sample is taken, any charge on the capacitor is dumped (i.e. $v_2(t)$ is returned to zero) by closing S2 for a short period. When $v_1(t)$ consists only of white gaussian noise with (one-sided) power spectral density N_0, it may be shown that at time t after dumping $v_2(t)$ is gaussian-distributed with zero mean and variance

$$\frac{N_0}{4CR}\left\{1 - \exp\left(\frac{-2t}{CR}\right)\right\}.$$

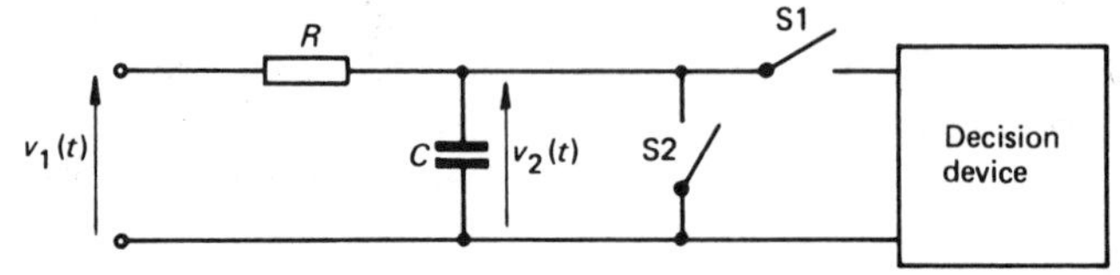

Fig. 2.7 Baseband integrate-and-dump receiver of Exercise 2.5

Show that when the signal component of $v_1(t)$ is a rectangular pulse extending from one sampling instant to the next, the signal-to-noise ratio for the samples approaches its matched-filter value if CR is made large.

Solution outline

If the amplitude of the rectangular pulse is V, its energy (E) is V^2T and hence the matched-filter value of the signal-to-noise ratio for the samples $(2E/N_0)$ is $2V^2T/N_0$.

The signal component of $v_2(t)$ is $V\{1 - \exp(-t/CR)\}$. Hence the signal-to-noise ratio for $v_2(t)$ at $t = T$ is

$$\frac{4CRV^2\{1 - \exp(-T/CR)\}^2}{N_0\{1 - \exp(-2T/CR)\}} = \frac{4CRV^2}{N_0} \tanh\left(\frac{T}{2CR}\right),$$

which approaches the value $2V^2T/N_0$ as $CR \to \infty$.

2.6 Non-coherent ASK receiver

A synchronous 100 baud binary ASK signal consists of an equiprobable random sequence of marks and spaces (the marks being 10 ms bursts of amplitude 1 V) and is accompanied by gaussian noise with one-sided power spectral density (N_0) 0.0002 V^2/Hz. It is received by a receiver consisting of a band-pass filter (with unity gain at the carrier frequency and noise bandwidth 100 Hz), an envelope detector, a sampler and a decision device. Each mark has just reached 1 V and each space 0 V when sampled at the detector output. Assuming that at the decision-device noise accompanying marks is gaussian and noise accompanying spaces is Rayleigh-distributed (Section 7.3.1), calculate
(i)   the optimum voltage for the decision threshold,
(ii) the digit error rate with that threshold voltage.
Calculate also the minimum digit error rate obtainable with a coherent matched-filter receiver.

[Hint: the mean-square value of the Rayleigh-distributed noise voltage is twice that of the gaussian-distributed noise voltage.]

Solution outline

(i)   Let the probability density function for the sample voltage be $p(v\,|\,1)$ when the transmitted digit is a mark and $p(v\,|\,0)$ when it is a space. Then

$$p(v\,|\,1) = \frac{1}{\sigma\sqrt{2\pi}} \exp\left\{\frac{-(v-1)^2}{2\sigma^2}\right\}$$

and $$p(v\,|\,0) = \frac{v}{\sigma^2} \exp\left\{\frac{-v^2}{2\sigma^2}\right\},$$

where $\sigma^2 = 0.0002 \times 100$. Since marks and spaces are equally probable, the optimum threshold is V_T given by

$$p(V_T\,|\,0) = p(V_T\,|\,1).$$

Hence $V_T = 0.545$ V.
(ii)   The digit error rate is

$$\frac{1}{2}\int_{0.545}^{\infty} \frac{v}{\sigma^2} \exp\left\{\frac{-v^2}{2\sigma^2}\right\} dv + \frac{1}{4}\operatorname{erfc}\left\{\frac{0.455}{\sigma\sqrt{2}}\right\} \simeq 6.2 \times 10^{-4}.$$

For a coherent matched-filter receiver the noise is gaussian during both marks and spaces.

Denote the mean-square noise voltage at the threshold by σ_1^2. If a mark produces a peak voltage V_1, the threshold should be set at $V_1/2$ and the digit error rate is then

$$\frac{1}{2}\operatorname{erfc}\left\{\frac{V_1}{2\sigma_1\sqrt{2}}\right\},$$

which may be shown equal to 2.0×10^{-4} by using the fact (Section 2.2.1) that

$$\frac{V_1^2}{\sigma_1^2} = \frac{2E}{N_0} = 2 \times \tfrac{1}{2}(1)^2(10^{-2}) \div 0.0002.$$

2.7   Non-coherent FSK receiver

An equiprobable binary data sequence is transmitted at 200 bit/s by FSK, using frequencies 980 Hz and 1180 Hz. It is received, accompanied by white gaussian noise, by a receiver consisting of a discriminator, a rectangular low-pass filter (cutting off at 200 Hz), a sampler and a decision device. Assuming that the signal component of the filter output reaches a steady value before each sample is taken and that the noise component is gaussian, compare the E/N_0 ratio required at the receiver input for a given error rate with that required by a coherent matched-filter receiver.

Solution outline

The error probability for a coherent matched-filter receiver is $0.5\operatorname{erfc}(\sqrt{E/2N_0})$ (Section 2.2.3).

With the notation of Appendix 7, the noise power σ^2 at the filter output is given by

$$\sigma^2 = \int_0^{200} \frac{8\pi^2 K^2 N_0 f^2}{A^2}\,\mathrm{d}f = \frac{4\pi^2 K^2 (2N_0)(200)^3}{3A^2}\,\mathrm{V}^2.$$

The signal levels at the filter output are separated by $400\pi K$ V. Assuming that the decision threshold is midway between them, the error probability is $0.5\operatorname{erfc}(200\pi K/\sigma\sqrt{2})$. This may be written $0.5\operatorname{erfc}(\sqrt{3E/8E_0})$, since E, the energy per bit, is $A^2/400$. Thus the E/N_0 ratio required is $10\log(4/3)$ dB ($\simeq 1.2$ dB) greater than in a coherent matched-filter receiver.

2.8   Probability of error in di-bit in QPSK

Show that the probability that a di-bit is in error in a coherent matched-filter QPSK receiver is

$$\operatorname{erfc}\left(\sqrt{\frac{E}{N_0}}\right) - \frac{1}{4}\left\{\operatorname{erfc}\left(\sqrt{\frac{E}{N_0}}\right)\right\}^2,$$

where E is the energy per binary digit and N_0 the one-sided power spectral density of the added gaussian noise.

Solution outline

Regard the QPSK waveform as the sum of two binary PSK waveforms with their carriers in quadrature, each demodulated in a coherent demodulator.

Each binary waveform has energy E per di-bit and hence for each demodulator the probability of error in a di-bit is $0.5\operatorname{erfc}(\sqrt{E/N_0})$ (Section 2.2.3).

The probability of an error in the decoded di-bit is the probability that there is an error in one or both of the demodulator outputs and is therefore

$$1 - \{1 - 0.5\operatorname{erfc}(\sqrt{E/N_0})\}^2.$$

2.9 ARQ using single parity check

In a certain telegraph system, each character is transmitted as five binary data digits followed by one parity-check digit. During transmission random errors occur with digit-error probability 0.1. Calculate

(i) the probability that a character is received incorrectly but without parity violation,
(ii) the probability that a received character satisfying the parity check is correct,
(iii) the average number of times a character must be transmitted before it is received without parity violation.

Solution outline
(i) The required probability is the probability that there are 2, 4 or 6 errors in the received character and is therefore

$$\binom{6}{2}(0.1)^2(0.9)^4 + \binom{6}{4}(0.1)^4(0.9)^2 + (0.1)^6 = 0.0996.$$

(ii) The probability that a received character satisfies the parity check and is correct is $(0.9)^6 = 0.5314$. Hence the required probability is

$$\frac{0.5314}{0.0996 + 0.5314} = 0.8421.$$

(iii) Let p be the probability that a character is received with parity violation. [$p = 1 - 0.0996 - 0.5314 = 0.3689$.] The average number of transmissions required is

$$(1 - p) + 2p(1 - p) + 3p^2(1 - p) + \ldots = (1 - p)^{-1} = 1.585.$$

[Alternatively argue that a proportion $1 - p$ of the received characters satisfy the parity check.]

2.10 Hamming single-error-correcting code

In a single-error-correcting code of the type proposed by Hamming (Section 2.3.2), blocks of 31 transmitted digits include parity checks in positions 1, 2, 4, 8 and 16. Locate the single error in the following received block.

1 1 0 1 1 1 0 1 0 1 0 0 0 1 1 1 1 0 1 0 1 1 0 0 1 1 0 1 0 1 0

Solution outline
If it is assumed that the parity checks should show even parity, the check digit in position 8 is correct, but those in positions 1, 2, 4 and 16 are incorrect. Hence there could be a single error in position $1 + 2 + 4 + 16 = 23$.

If the parity checks should show odd parity, there could be a single error in position 8.

2.11 Minimum distance of Van Duuren, Autospec and Hamming codes

Determine the minimum distance of
(i) the Van Duuren seven-unit code (Section 2.1.1.),
(ii) the Marconi Autospec code (Section 2.3.2),
(iii) the Hamming code of Section 2.3.2.

Solution outline
(i) The Hamming distance between any two codewords is an even number and hence the minimum distance is two.
(ii) Codewords differing in one data digit also differ in four check digits and hence are separated by Hamming distance five; codewords differing in two data digits also differ in

two check digits and hence are separated by Hamming distance four; etc. The minimum distance is four.

(iii)   Codewords differing in one data digit also differ in at least two check digits; codewords differing in two data digits also differ in at least one check digit. Hence the minimum distance is three.

2.12   Block code (using matrices)

The generator matrix G for a (5, 3) parity-check block code is

$$\begin{pmatrix} 1 & 0 & 0 & 1 & 0 \\ 0 & 1 & 0 & 1 & 1 \\ 0 & 0 & 1 & 0 & 1 \end{pmatrix}.$$

(i)   Derive the parity-check matrix H and verify that $GH' = 0$.
(ii)   Derive the codeword which has data digits 110.
(iii)   Derive the syndrome of 11011.
(iv)   Construct a standard array having 00000, 10000, 01000, 00100 as coset leaders and verify that the code could be used to correct single errors in the data digits.

Solution outline

(i) $H = \begin{pmatrix} 1 & 1 & 0 & 1 & 0 \\ 0 & 1 & 1 & 0 & 1 \end{pmatrix}.$

(ii) $c = d\,G = (1 \quad 1 \quad 0) \begin{pmatrix} 1 & 0 & 0 & 1 & 0 \\ 0 & 1 & 0 & 1 & 1 \\ 0 & 0 & 1 & 0 & 1 \end{pmatrix} = (1 \quad 1 \quad 0 \quad 0 \quad 1).$

(iii) $s = e\,H' = (1 \quad 1 \quad 0 \quad 1 \quad 1) \begin{pmatrix} 1 & 0 \\ 1 & 1 \\ 0 & 1 \\ 1 & 0 \\ 0 & 1 \end{pmatrix} = (1 \quad 0).$

(iv)

Syndromes	Coset leaders							
00	00000	10010	01011	00101	11001	01110	10111	11100
10	10000	00010	11011	10101	01001	11110	00111	01100
11	01000	11010	00011	01101	10001	00110	11111	10100
01	00100	10110	01111	00001	11101	01010	10011	11000

2.13   Hamming bound

By considering the column of coset leaders in a standard array show that an upper bound on the number t of errors which can be corrected in one block in an (n, k) parity-check code is given by

$$2^{n-k} \geqslant \sum_{i=0}^{t} \binom{n}{i}.$$

Solution outline

In the n-tuples in the column of coset leaders the 1s represent the patterns of correctable errors. For a code to be able to correct up to t errors that column must include:

the all-zero n-tuple,

$\binom{n}{1}$ n-tuples with one 1,

$\binom{n}{2}$ n-tuples with two 1s,

continuing as far as

$\binom{n}{t}$ n-tuples with t 1s.

Since there are only 2^{n-k} n-tuples in the column, the result follows.

2.14 Cyclic code (using polynomial manipulation)

A (6, 4) parity-check block code has generator polynomial

$$g(X) = X^2 + X + 1.$$

(i) Construct the generator matrix.

(ii) List all the codewords and verify that the code is cyclic.

(iii) Construct the parity-check polynomial $h(X)$ and verify that

$$g(X)h(X) = X^6 + 1.$$

(iv) Determine the minimum distance of the code and comment on its ability to correct or detect errors.

Solution outline

(i) Using the rules in Section 2.3.5,

$$G = \begin{pmatrix} 1 & 0 & 0 & 0 & 1 & 1 \\ 0 & 1 & 0 & 0 & 1 & 0 \\ 0 & 0 & 1 & 0 & 0 & 1 \\ 0 & 0 & 0 & 1 & 1 & 1 \end{pmatrix}.$$

(ii) There are $2^4 = 16$ codewords. Trying all possible data digits shows the codewords to be:

0 0 0 0 0 0		(1)
0 0 0 1 1 1	and cyclic shifts	(6)
0 0 1 0 0 1	and cyclic shifts	(3)
0 1 1 0 1 1	and cyclic shifts	(3)
0 1 0 1 0 1	and cyclic shifts	(2)
1 1 1 1 1 1		(1)
	Total	16

(iii) From G,

$$h(X) = X^4 + X^3 + X + 1.$$

(iv) From the list of codewords the minimum weight is 2 and therefore the **minimum distance is 2**. Hence the code can detect single errors, but no error correction is possible.

2.15 Cyclic codes of word-length six

List all the codewords in each of the cyclic codes of length six and determine the error-correcting capability of each code.

Solution outline

The codes are generated by factors of $X^6 + 1$, which may be written

$$X^6 + 1 = (X + 1)^2 (X^2 + X + 1)^2.$$

There are seven codes. Construct the generator matrix in each case and determine the codewords by inspection of it. The results are shown in Table 2.3.

Table 2.3

Generator polynomial	Sample codewords	No. in cycle	No. in code	Minimum distance	Error-correcting capability
$X+1$	100001	6			
	010001	6			
	001001	3			
	000000	1	32	2	0
	111111	1			
	110110	3			
	110011	6			
	110101	6			
$(X+1)^2$	100010	6			
	110011	6	16	2	0
	011011	3			
	000000	1			
X^2+X+1	100011	6			
	010010	3			
	000111	3	16	2	0
	010101	2			
	111111	1			
	000000	1			
$(X+1)(X^2+X+1)$	100100	3			
	110110	3	8	2	0
	111111	1			
	000000	1			
$(X+1)^2(X^2+X+1)$	101101	3			
	000000	1	4	4	1
$(X^2+X+1)^2$	101010	2			
	111111	1	4	3	1
	000000	1			
$(X+1)(X^2+X+1)^2$	111111	1			
	000000	1	2	6	2

2.16   A (7,4) cyclic code

For the $(7, 4)$ cyclic code generated by $X^3 + X^2 + 1$,
   (i)   obtain the generator matrix,
   (ii)   list the codewords,
   (iii)   determine the error-correcting capability,
   (iv)   obtain the parity-check matrix,
   (v)   use the parity-check matrix to calculate the syndrome of 0111100 and verify your result by polynomial division,
   (vi)   list all the 7-tuples with the same syndrome as 0111100,
   (vii)   design a 3 stage encoder,
   (viii)   design a 4 stage encoder.

Solution outline

(i)
$$G = \begin{pmatrix} 1 & 0 & 0 & 0 & 1 & 1 & 0 \\ 0 & 1 & 0 & 0 & 0 & 1 & 1 \\ 0 & 0 & 1 & 0 & 1 & 1 & 1 \\ 0 & 0 & 0 & 1 & 1 & 0 & 1 \end{pmatrix}.$$

(ii) Codewords are cyclic shifts of 1000110, 0010111, 1111111 and 0000000.

(iii) The minimum distance is three and hence single errors can be corrected.

(iv)
$$H = \begin{pmatrix} 1 & 0 & 1 & 1 & 1 & 0 & 0 \\ 1 & 1 & 1 & 0 & 0 & 1 & 0 \\ 0 & 1 & 1 & 1 & 0 & 0 & 1 \end{pmatrix}.$$

(v) Multiplying 0111100 by H' gives 101. Polynomial division gives

$$(X^5 + X^4 + X^3 + X^2) \div (X^3 + X^2 + 1) = (X^2 + 1) \text{ with remainder } (X^2 + 1).$$

(vi) 0111100 may be obtained from the codeword 0110100 by adding 0001000.
7-tuples with the same syndrome are obtained by adding 0001000 to each of the other codewords.

(vii) As shown in Fig. 2.3(a) with three register stages, $g_1 = 0$, $g_2 = 1$.

(viii) As shown in Fig. 2.3(b) with four register stages, $h_1 = 0$, $h_0 = h_2 = h_3 = 1$.

2.17 Encoders for cyclic codes

Show that Fig. 2.8(a) is an alternative form of the $(n - k)$-stage encoder for the $(7, 4)$ cyclic code generated by $X^3 + X + 1$ and requires just four shifts to load the data and calculate the parity checks.

Solution outline
Argue that Fig. 2.8(a) is equivalent to Fig. 2.8(b), which is equivalent to Fig. 2.3(a) with $n = 7$, $k = 4$, $g_1 = 1$ and $g_2 = 0$.

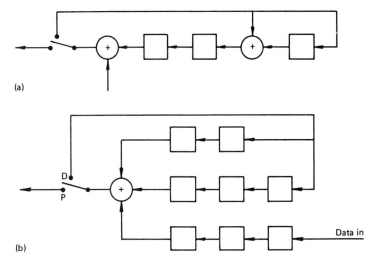

Fig. 2.8 Diagrams of Exercise 2.17

2.18   Construction of (7,3) cyclic code with minimum distance four

Verify that for coefficients in $GF(2)$

$$X^7 + 1 = (X + 1)(X^3 + X + 1)(X^3 + X^2 + 1).$$

Given that

$$\alpha^3 + \alpha + 1 = 0,$$

show that α, α^2, α^4 are roots of $X^3 + X + 1 = 0$ and that

$$\alpha^3, \alpha^5, \alpha^6 \text{ are roots of } X^3 + X^2 + 1 = 0.$$

Obtain the generator matrix of a (7,3) cyclic code with minimum distance not less than four and determine the minimum distance by listing the codewords. Construct sufficient of a standard array to show that the code can correct all single errors, all double errors occurring in consecutive digits and inversion of all seven digits.

Solution outline
Since $\alpha^3 + \alpha + 1 = 0$, α is a root of $X^3 + X + 1 = 0$. Since

$$(\alpha^3 + \alpha + 1)^2 = \alpha^6 + \alpha^2 + 1 = (\alpha^2)^3 + \alpha^2 + 1,$$

α^2 is also a root of $X^3 + X + 1 = 0$, and since

$$(\alpha^3 + \alpha + 1)^4 = (\alpha^6 + \alpha^2 + 1)^2 = (\alpha^4)^3 + \alpha^4 + 1,$$

α^4 is also a root of $X^3 + X + 1 = 0$.

Since $\alpha^9 = \alpha^2$

$$\alpha^6 + \alpha^2 + 1 = \alpha^9 + \alpha^6 + 1 = (\alpha^3)^3 + (\alpha^3)^2 + 1$$

and therefore α^3 is a root of $X^3 + X^2 + 1 = 0$. Since

$$(\alpha^9 + \alpha^6 + 1)^2 = \alpha^{18} + \alpha^{12} + 1 = (\alpha^6)^3 + (\alpha^6)^2 + 1,$$

α^6 is also a root of $X^3 + X^2 + 1 = 0$, and since

$$\{(\alpha^6)^3 + (\alpha^6)^2 + 1\}^2 = \alpha^{36} + \alpha^{24} + 1 = \alpha^{15} + \alpha^{10} + 1,$$

so also is α^5.

Choosing $g(X) = (X + 1)(X^3 + X + 1) = X^4 + X^3 + X^2 + 1$ gives as required three consecutive roots $(\alpha^0, \alpha^1, \alpha^2)$. Hence

$$G = \begin{pmatrix} 1 & 0 & 0 & 1 & 1 & 1 & 0 \\ 0 & 1 & 0 & 0 & 1 & 1 & 1 \\ 0 & 0 & 1 & 1 & 1 & 0 & 1 \end{pmatrix}.$$

The eight codewords are 0000000 and the seven cyclic shifts of 1001110. The minimum weight (and hence the minimum distance) is therefore four.

Since they all have different syndromes, the following sixteen n-tuples may be chosen as coset leaders:

0000000,   1000000,   0100000,   0010000,   0001000,   0000100,
0000010,   0000001,   1100000,   0110000,   0011000,   0001100,
0000110,   0000011,   1000001,   1111111.

These 7-tuples represent the error patterns described.

2.19 Construction of (15,5) triple-error-correcting cyclic code

Derive the generator matrix of a BCH code of block-length 15 which is capable of correcting three errors in each block.

[Hint:

$$X^{15} + 1 = (X+1)(X^2+X+1)(X^4+X+1)(X^4+X^3+1)(X^4+X^3+X^2+X+1)$$

Assume the primitive element α to satisfy $\alpha^4 + \alpha + 1 = 0$.]

Solution outline
By arguments similar to those used in Exercise 2.18, powers of α are associated with the factors of $X^{15} + 1$ as follows:

$X+1$	α^0
X^2+X+1	α^5, α^{10}
X^4+X+1	$\alpha, \alpha^2, \alpha^4, \alpha^8$
X^4+X^3+1	$\alpha^7, \alpha^{11}, \alpha^{13}, \alpha^{14}$
$X^4+X^3+X^2+X+1$	$\alpha^3, \alpha^6, \alpha^9, \alpha^{12}$

The code must have a minimum distance of at least seven and hence its generator polynomial should have at least six consecutive zeros. Choosing

$$g(X) = (X^2+X+1)(X^4+X+1)(X^4+X^3+X^2+X+1)$$

ensures that zeros include $\alpha, \alpha^2, \alpha^3, \alpha^4, \alpha^5, \alpha^6$. Hence

$$g(X) = X^{10} + X^8 + X^5 + X^4 + X^2 + X + 1.$$

Thus $n - k = 10$, the code is a (15,5) cyclic code and has generator matrix:

$$\begin{pmatrix}
1 & 0 & 0 & 0 & 0 & 1 & 0 & 1 & 0 & 0 & 1 & 1 & 0 & 1 & 1 \\
0 & 1 & 0 & 0 & 0 & 1 & 1 & 1 & 1 & 0 & 1 & 0 & 1 & 1 & 0 \\
0 & 0 & 1 & 0 & 0 & 0 & 1 & 1 & 1 & 1 & 0 & 1 & 0 & 1 & 1 \\
0 & 0 & 0 & 1 & 0 & 1 & 0 & 0 & 1 & 1 & 0 & 1 & 1 & 1 & 0 \\
0 & 0 & 0 & 0 & 1 & 0 & 1 & 0 & 0 & 1 & 1 & 0 & 1 & 1 & 1
\end{pmatrix}$$

Alternatively, choose

$$g(X) = (X^2+X+1)(X^4+X^3+1)(X^4+X^3+X^2+X+1),$$

giving a (15,5) cyclic code with codewords time inverses of those of the above matrix.

2.20 Comparison of ARQ with forward error-correction

Teleprinter characters, each represented by five binary signalling elements, are transmitted synchronously by FSK at 50 baud (i.e. at 10 characters per second). White gaussian noise accompanying the received signal gives rise to random digit errors, which produce an error probability of 0.1 in the printed characters. Investigate whether characters printed at an average rate of 10 per second would have a lower probability of error for the same transmitted power if either of the following error-reduction methods were used. Assume that the spectral density of the accompanying noise is unaltered and that coherent matched-filter receivers are used.

(i) Forward error correction using a (15,5) BCH code.
(ii) ARQ using the Van Duuren seven-unit code. Assume that the effect of each request for a repeat causes the transmitter to go back four characters and that the receiver ignores the first three characters received after requesting a repeat. Assume also that the return channel is error-free.

Solution outline

Denote digit error probability by p_d and probability of error in a printed character by p_c. For a coherent matched-filter receiver

$$p_d = \tfrac{1}{2} \operatorname{erfc} (\rho),$$

where ρ^2 is proportional to the energy per transmitted digit (Section 2.2.3).

In the original system, $p_c = 0.1$ and

$$p_c = 1 - (1 - p_d)^5.$$

Hence $p_d = 0.021$ and $\rho = 1.44$ (Appendix 4).

(i)   In this case $\rho = 1.44 \div \sqrt{3}$ and hence $p_d = 0.12$. The code can correct upto three errors (Exercise 2.19). Hence p_c is the probability that there are four or more errors and is

$$1 - [(1 - p_d)^{15} + \tbinom{15}{1} p_d (1 - p_d)^{14} + \tbinom{15}{2} p_d^2 (1 - p_d)^{13} + \tbinom{15}{3} p_d^3 (1 - p_d)^{12}],$$

which is approximately 0.095, a very slight improvement.

(ii)   In this case, the probability p_v that a received character is valid is given by

$$p_v = (1 - p_d)^7 + 12 p_d^2 (1 - p_d)^5 + 18 p_d^4 (1 - p_d)^3 + 4 p_d^6 (1 - p_d)$$

(since a transmitted character is received as a valid character if as many errors occur among its As as among its Zs).

Now argue that the proportion of the time occupied by valid characters is $p_v / \{p_v + 4(1 - p_v)\}$ and that therefore the transmission rate should be increased by the factor $1 + 4(1 - p_v)/p_v$. (Alternatively sum a series as in Exercise 2.9.) Hence

$$p_d = \tfrac{1}{2} \operatorname{erfc} \left\{ 1.44 \times \sqrt{\frac{5}{7}} \div \sqrt{1 + \frac{4(1 - p_v)}{p_v}} \right\}.$$

We thus have two equations relating p_d and p_v and they can be solved by numerical methods to give $p_d = 0.288$ and $p_v = 0.321$.

The probability that a valid character is in error is

$$\{ p_v - (1 - p_d)^7 \} / p_v.$$

With the above values for p_d and p_v this is approximately 0.71, a considerable degradation.

2.21   Decision feedback

In a synchronous binary data-transmission system, the sampled signal voltage is $+1$ V or -1 V and is compared with a threshold voltage of 0 V. The message symbols are equiprobable and there is no intersymbol interference. The digit error probability due to gaussian noise accompanying the signal is 10^{-3}, which is unacceptably high. Two schemes are proposed for reducing it to 10^{-6}.

Scheme A   Modify the decision device to have two thresholds at $-a$ V and $+a$ V and add a return channel. A sample falling between the thresholds requests the transmitter to repeat the last n digits transmitted. The receiver ignores that sample and the next $n - 1$ samples.

Scheme B   Reduction of the signalling speed. Calculate
(i)   the minimum value of a necessary in Scheme A, and
(ii)   the maximum value of n for which the rate at which samples are accepted in Scheme A exceeds the rate of transmission of data in Scheme B.

Solution outline
Denoting the r.m.s. noise voltage accompanying the samples by σ,

$$\tfrac{1}{2} \operatorname{erfc} \left(\frac{1}{\sigma \sqrt{2}} \right) = 10^{-3}.$$

Hence $1/\sigma \sqrt{2} = 2.185$.

(i) In Scheme A, the probability p_e that a sample is accepted but in error is given by

$$p_e = \tfrac{1}{2} \operatorname{erfc} \left(\frac{1+a}{\sigma \sqrt{2}} \right)$$

and the probability p_c that a sample is accepted and correct is given by

$$p_c = 1 - \tfrac{1}{2} \operatorname{erfc} \left(\frac{1-a}{\sigma \sqrt{2}} \right).$$

The probability that an accepted sample is in error is $p_e/(p_e + p_c)$, which may be shown to equal 10^{-6} when a is approximately 0.543.

(ii) The probability p_a that a sample in Scheme A falls outside the thresholds and is therefore acceptable is given by

$$p_a = p_e + p_c = 0.921.$$

By arguing as in Exercise 2.20 that accepted characters occupy a proportion $p_a/\{p_a+n(1-p_a)\}$ of the time, Scheme A may be shown to divide the rate of transmission of data by the factor $1+n(1-p_a)/p_a$.
   In Scheme B the signalling speed should be divided by the factor k given by

$$\tfrac{1}{2} \operatorname{erfc} (2.185 \sqrt{k}) = 10^{-6}.$$

Hence $k = 2.37$. The inequality

$$1 + n(1-p_a)/p_a < k$$

is satisfied if $n \leqslant 15$.

2.22   Trellis diagram for convolutional code

The data sequence $00000000 \ldots$ is encoded by the encoder of Fig. 2.4(a) and transmitted. It is received as the sequence $111000000 \ldots$ and is decoded by a Viterbi decoder starting in state 00 and operating as illustrated in Fig. 2.5. Calculate the metrics and survivors at each level until it is clear that a correct decision for the first digit will never be obtained

Solution outline
At level 3, reading from top to bottom, the metrics are $2, 3, 0, 3$, and the survivors are 101, $110, 101, 111$. All these survivors begin with 1 and therefore no subsequent survivors can begin with 0.

2.23   de Bruijn diagram for convolutional code

Construct the de Bruijn diagram for the encoder shown in Fig. 2.9. Show that the diagram contains a loop of zero Hamming weight which does not include the state 00 and comment on the undesirability of that feature.

Solution outline
The de Bruijn diagram is shown in Fig. 2.10.

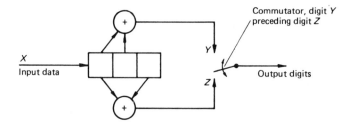

Fig. 2.9   Convolution encoder of Exercise 2.23

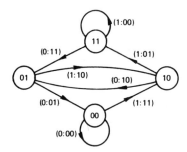

Fig. 2.10   Solution to Exercise 2.23

The self-loop containing the state 11 has zero weight. Assume an all-zero sequence transmitted. If errors have caused the decoder to go into state 11, it remains in that state when there are no errors and is said to display 'catastrophic error propagation'.

2.24   Error-correcting capability of convolutional code

Construct the de Bruijn diagram for the convolution encoder of Fig. 2.11. Determine by inspection of the diagram the error-correcting capability of the code.

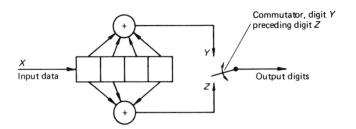

Fig. 2.11   Convolution encoder of Exercise 2.24

Solution outline
With the notation of Fig. 2.4(b), the de Bruijn diagram is as shown in Fig. 2.12.

The path of least Hamming weight beginning and ending at the state 000 is the path $000 \rightarrow 100 \rightarrow 110 \rightarrow 011 \rightarrow 001 \rightarrow 000$. This has Hamming weight six and is five transitions in length. Hence the code can correct two errors in a sequence comprising five groups of two transmitted digits.

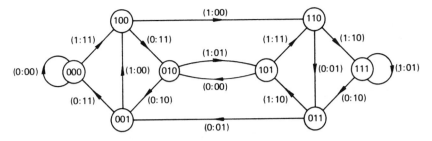

Fig. 2.12 Solution to Exercise 2.24

2.25 Generating function of convolutional code

Construct the de Bruijn diagram for the convolution encoder of Fig. 2.13. Use a generating function to investigate its error-correcting properties. Note that when the data sequence appears unmodified in the transmitted sequence (in this case as alternate digits), the code is said to be 'systematic'.

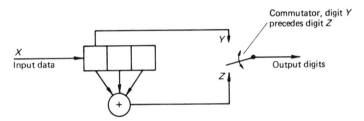

Fig. 2.13 Convolution encoder of Exercise 2.25

Solution outline
The de Bruijn diagram is shown in Fig. 2.14. With the 00 node split as in Section 2.3.7, the node variables are related by:

$$b = D^2a + Dc$$
$$c = Db + d$$
$$d = Db + D^2d$$
$$a' = Dc.$$

Hence

$$\frac{a'}{a} = \frac{D^4(2 - D^2)}{1 - 3D^2 + D^4} = 2D^4 + 5D^6 + \text{higher powers}.$$

The paths of lowest weight are two paths of weight four. The longer of these paths consists

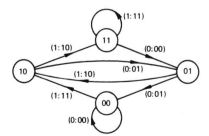

Fig. 2.14 Solution to Exercise 2.25

of four transitions. Hence the code may be expected to correct a single error in a sequence of eight received digits.

2.26 Scrambler

Figure 2.15 shows a simple scrambler and descrambler based on modulo-two adders and three-stage shift registers.

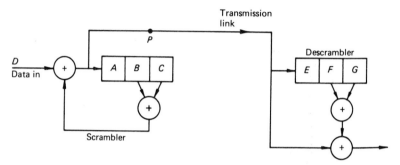

Fig. 2.15 Scrambler and descrambler of Exercise 2.26

(i)   Assuming that A, B and C are initially in the 0 state, construct a table showing the contents of A, B and C and the sequence at P for the first 12 digits of an arbitrary input binary data sequence at D.

(ii)   Verify that if the input data sequence is represented by the coefficients of a data polynomial in descending powers of X, the corresponding polynomial for the sequence at P is obtained by dividing that data polynomial by $1 + X^{-2} + X^{-3}$.

(iii)   Extend the table constructed in (i) to verify that the effect of the scrambler shift register not being initially in the all-zero state is to add the maximal length sequence . . . 1110010 . . . to the sequence at P.

(iv)   State under what conditions a long unbroken sequence of zeros would appear at P.

(v)   Outline how the scrambler should be modified to satisfy the CCITT recommendations that
(a)   the divisor polynomial is $1 + X^{-6} + X^{-7}$,
(b)   repeated patterns of 1, 2, 3, 4, 6, 9 or 12 digits can be detected in the sequence at P.

Solution outline
(i) and (iii)

Number of clock pulse	Assumed data input (D)	State on arrival of data (ABC)	Transmitted message (P) $= D + B + C$	State on arrival of data $A'B'C'$	Transmitted message (P') $= D + B' + C'$	$P + P'$
1	1	000	1	010	0	1
2	1	100	1	001	0	1
3	0	110	1	000	0	1
4	0	111	0	000	0	0
5	0	011	0	000	0	0
6	1	001	0	000	1	1
7	0	000	0	100	0	0
8	1	000	1	010	0	1
9	1	100	1	001	0	1
10	1	110	0	000	1	1
11	0	011	0	100	0	0
12	1	001	0	010	0	0

(ii)   Divide $1 + X^{-1} + X^{-5} + X^{-7} + X^{-8} + X^{-9} + X^{-11}$ by $1 + X^{-2} + X^{-3}$
    to obtain $1 + X^{-1} + X^{-2} + X^{-7} + X^{-8}$.

(iv)   When a long unbroken sequence of data zeros occurs at a time when the scrambler register is in an all-zero state.

(v)   (a)   Extend the register to seven stages. Move the taps to the sixth and seventh stages.

(b)   Extend the scrambler register to 12 stages. A modulo-two adder connected to the input and the ninth stage will give an output continuously zero if there is a repeated pattern of 9, 3 or 1 digits. A modulo-two adder connected to the input and the twelfth stage will give an output continuously zero if there is a repeated pattern of 12, 6, 4, 3, 2, or 1 digits.

For further information see CCITT[24].

3
Transmission of pictorial information

3.1 Facsimile

The purpose of a facsimile system is to transmit a stationary image and produce a hard copy at the receiver.

Traditional facsimile systems are designed for use over telephone channels. The restricted bandwidth of a telephone channel limits the speed at which the image can be scanned to speeds at which mechanical scanning arrangements are practicable. In the usual arrangement[1], the picture to be transmitted is wrapped around a rotating drum and scanned by a photocell moving slowly parallel to the axis of the drum. The transmitted electrical signal varies with the brightness of the picture element immediately below the photocell. The time taken to scan an image is typically several minutes.

The drum in the transmitter is usually driven by a synchronous motor and the drum in the receiver is synchronised to it by means of a transmitted pilot-tone. The starting time for each line is marked by a short break in the transmitted brightness signal.

Transmission of a facsimile signal is restricted to the region of the audio spectrum suffering least delay distortion (typically between 1200 Hz and 2600 Hz). Over audio-frequency circuits amplitude modulation is found to give more acceptable reproduction than frequency modulation[2]. Satisfactory transmission requires a four-wire telephone circuit.

More recently digital transmission links have been used for facsimile transmission. Approaches to the efficient digital encoding of images are outlined in Section 3.4.2.

3.2 Monochrome television

3.2.1 Scanning

The purpose of a television system is to produce at the receiver a transient display which is up-dated sufficiently often for movement to appear smooth.

In conventional television systems an image of the picture to be transmitted is scanned line by line in the camera by a pencil beam of electrons and its luminance (perceived brightness) is represented by a voltage known as the 'video signal'. The picture is reconstructed on the receiver cathode-ray tube by causing a beam of electrons to scan the screen in synchronism with the beam in the camera and with its intensity modulated by the video signal.

There should, obviously, be sufficient scanning lines to give the required quantity of detail in the picture. The number of scanning lines should also be sufficient for them not to be resolvable as separate lines at a convenient viewing distance. The resolving capability of the human eye may be expressed in terms of its visual acuity, which is defined as the reciprocal of the angle in radians subtended at the eye by two small objects which the eye can just resolve. Visual acuity is typically about 3400 reciprocal radians. A convenient viewing distance for television is about eight times the picture height.

Two considerations limit the time which can be taken to scan the picture. The first is the desirability of making any movement within the picture appear smooth, which requires that the receiver display should be up-dated at least about 25 times per second. The second is the flicker caused by the intermittent irradiation of each point of the display due to the scanning process. The number of scans per second necessary for flicker not to be apparent increases with the brightness of the display and is typically between 40 and 70.

The usual method of avoiding flicker without up-dating the picture detail unnecessarily often is to scan all the odd lines and then all the even lines and so on alternately. This technique is known as 'interlacing'. Each of the two sets of lines making up a complete scanning of the picture is called a 'field'. The frequency relevant to the avoidance of flicker is the field frequency, which is twice the frequency at which the complete picture is updated. Problems of mains interference are greatly reduced by making the field frequency nominally equal to the mains frequency.

3.2.2 Bandwidth

An estimate of the bandwidth required for transmission of the video signal may be obtained by arguing that as the scanning spot traverses a line the video signal may change from one extreme value to the other in a distance equal to the distance between the centres of adjacent lines. The period of the highest frequency to be allowed for in the video signal is therefore twice the time taken by the spot to move that distance.

In fact the bandwidth estimated in the above way is usually considered to be unnecessarily large and is in practice modified by multiplication by the Kell factor, which is approximately 0.7. The justification for this reduction is that it is unnecessary to allow for the resolution of patterns of vertical bars spaced more closely than that pattern of horizontal bars which can always be resolved however it lies with respect to the (horizontal) scan lines.

The required bandwidth is usually several MHz (Exercise 3.2). Broadcasting systems usually use vestigial-sideband amplitude modulation to conserve radio-frequency bandwidth.

A comprehensive account of the subjective assessment of transmitted pictures has been given by Allnatt[3].

3.2.3 Synchronisation

The transmitted waveform must contain timing information to enable both horizontal (line) and vertical (field) timebases in the receiver to be synchronised to the transmitter. This is achieved by adding pulses to the video signal during the periods allowed for flyback. The polarity of the pulses is such that the display remains invisible (blanked) throughout the flyback times. The composite video signal comprising video and synchronising information for a typical picture line in British 625 line monochrome television is shown in Fig. 3.1. Field synchronisation is provided by a sequence of 27 μs pulses at the beginning of each field.

Some amplitude-modulation broadcasting systems use what is known as positive modulation, while others use negative modulation. In positive modulation, maximum carrier amplitude corresponds to the white level and minimum (usually zero) carrier amplitude to the sync level. In negative modulation, maximum carrier amplitude corresponds to the sync level and minimum (typically 20%) carrier amplitude to the white level. It is found that spots due to interference are less objectionable when negative modulation is used. Negative modulation has the additional advantage that the occurrence of maximum carrier amplitude during each line is convenient for automatic gain control. On the other hand synchronisation is more affected by interference in negative modulation systems. The blanking level is chosen so that with low received signal

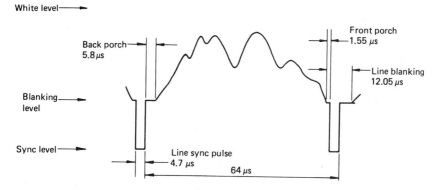

Fig. 3.1 Composite video signal for typical line in British 625 line monochrome transmission

strengths synchronisation is lost at about the signal-to-noise ratio at which the picture disappears.

A lucid introduction to the principles of television systems is given by Smol, Hamer and Hills[4].

3.3 Colour television

3.3.1 Chrominance signals

Colour television is based on the assumption that any perceived colour can be matched by a mixture of three primary colours. A colour television display presents three superposed images, one in each of the primary colours. The primary colours used vary slightly from system to system but are always red, green and blue.

An instrument enabling a given colour to be described in terms of three primary colours is called a colorimeter. In a colorimeter a screen is illuminated by the three primary colours simultaneously and their intensities are adjusted until the screen matches the colour to be described. The instrument has three meters, which display the tri-stimulus values R, G, B of the colour of the screen. The readings R, G, B are proportional to the intensities of the red, green and blue primaries illuminating the screen, the constants of proportionality being set up so that $R = G = B = 1$ matches a specified reference white. When $R = G = B < 1$ the colour described is a white less bright than reference white, i.e. grey. A colour television camera is usually set up to provide three electrical signals E_R, E_G, E_B, which are proportional to R, G, B throughout the scanning of the picture and are equal for reference white.

It is found that some perceivable colours, notably spectrally pure blue-greens, can be matched in a colorimeter only if one of the primaries is added to the colour to be matched and the combination matched by the other two primaries. Clearly such colours cannot be reproduced on a television display based upon the superposition of the three primaries.

Sufficient information for a colour television display could be provided by transmitting E_R, E_G, E_B as separate video signals. It is, however, usual to transmit the information in the form of linear combinations of E_R, E_G, E_B, one of those combinations being a measure of the perceived brightness (luminance) of the picture and therefore appropriate for the control of a monochrome display.

When the three primaries are in the correct proportion for their addition to give white, the electrical signals E_R, E_G, E_B are equal, but their contributions to the total luminance are unequal, being in the proportions $l:m:n$, say, ·where $l + m + n = 1$. It is therefore reasonable to suppose that the luminance of a mixture of the three primaries is proportional to a signal E_Y defined by

$$E_Y = lE_R + mE_G + nE_B.$$

The coefficients l, m, n are called the luminosity coefficients. For most systems $l = 0.30$, $m = 0.59$, $n = 0.11$: thus over half the luminance of a white display is contributed by the green primary. The signals $E_R - E_Y$, $E_G - E_Y$, $E_B - E_Y$ are called the 'chrominance values' of the colour from which they are derived; they are all zero for grey.

Compatibility with monochrome television could be obtained by transmitting E_Y as in monochrome television and supplementing it by two of the chrominance values, transmitted in such a way that a monochrome receiver is unaffected by them. A colour receiver could calculate E_R, E_G, E_B from E_Y and the two chrominance values quite simply.

Unfortunately the luminance of a cathode-ray tube display does not vary linearly with the grid-cathode voltage and to avoid the necessity for non-linear processing in every receiver, video signals in a broadcasting system are pre-distorted before transmission in a process known as γ-correction. For a correct display on a monochrome receiver the transmitted luminance signal should be $E_Y^{1/\gamma}$ (where γ is typically 2.2). If a colour transmission consisted of the three signals $E_Y^{1/\gamma}$, $E_B^{1/\gamma} - E_Y^{1/\gamma}$, $E_R^{1/\gamma} - E_Y^{1/\gamma}$, the required signals $E_B^{1/\gamma}$ and $E_R^{1/\gamma}$ could be derived easily in the receiver, but deriving the other required signal $E_G^{1/\gamma}$ would require non-linear processing and be difficult to implement. Instead it is usual to transmit the following three signals:

$$E'_Y = 0.30\, E_R^{1/\gamma} + 0.59\, E_G^{1/\gamma} + 0.11\, E_B^{1/\gamma}$$
$$E_B^{1/\gamma} - E'_Y$$
$$E_R^{1/\gamma} - E'_Y$$

from which $E_R^{1/\gamma}$, $E_G^{1/\gamma}$, $E_B^{1/\gamma}$ are easily derived. It is considered that the use of E'_Y instead of $E_Y^{1/\gamma}$ does not unacceptably degrade a monochrome display. Note that when $E_R = E_G = E_B$ (as in white or grey) $E'_Y = E_Y^{1/\gamma}$ and the true luminance is displayed on a monochrome receiver.

Thus colour television requires the transmission of three signals, which are usually E'_Y (an approximation to the luminance) and two chrominance signals. The chrominance signals may be either proportional to $E_B^{1/\gamma} - E'_Y$ and $E_R^{1/\gamma} - E'_Y$ or linear combinations of them. Since the eye cannot resolve such fine detail in colour changes as it can in luminance changes, the transmission bandwidth required for the chrominance signals is considerably less than for E'_Y.

3.3.2 Transmission of chrominance signals

In the American NTSC system and in the PAL system derived from it, two chrominance signals are transmitted by quadrature modulation of a colour subcarrier added to the luminance signal. The choice of subcarrier frequency f_{sc} MHz is a compromise between maximising the transmission bandwidth for the chrominance signals and minimising interference between luminance and chrominance. The phase reference necessary for demodulation of the subcarrier is provided by transmitting a short burst of the subcarrier (the colour burst) during the back porch preceding each line.

The signals

$$E'_U = 0.493\,(E_B^{1/\gamma} - E'_Y)$$
$$E'_V = 0.877\,(E_R^{1/\gamma} - E'_Y)$$

feature in both the NTSC and PAL systems. In the NTSC system the chrominance signals modulating the subcarrier are E'_I and E'_Q defined by

$$E'_I = E'_V \cos 33° - E'_U \sin 33°$$
$$E'_Q = E'_V \sin 33° + E'_U \cos 33°.$$

The angle $33°$ has been chosen so that E'_I represents those colour changes for which the eye is most sensitive to detail[5]. The sidebands due to E'_Q are restricted to frequencies between $(f_{sc} - 0.6)$ MHz and $(f_{sc} + 0.6)$ MHz, whereas the lower sideband due to E'_I extends down to $(f_{sc} - 1.3)$ MHz. E'_I represents a shift towards orange or cyan, E'_Q towards purple or green-yellow. In the PAL system, more video bandwidth is available; E'_I and E'_Q are not used and the subcarrier is modulated by E'_U and E'_V, the sidebands of both extending from $(f_{sc} - 1.3)$ MHz to $(f_{sc} + 1.3)$ MHz.

PAL is an abbreviation for 'phase alternation by line' and the principal difference between the NTSC and PAL systems is that in the latter the phase of the subcarrier component carrying E'_V is reversed on alternate lines. In a PAL receiver each chrominance signal is stored for the duration of a line and its average over two successive lines is used to control the chrominance of the display. It may be shown that a phase error ϕ in the received subcarrier reduces the received chrominance signals in a PAL receiver to $E'_U \cos \phi$ and $E'_V \cos \phi$. The effect of this reduction is to add grey to the displayed colour, thus reducing its saturation, but leaving its hue unchanged. The result is less objectionable than the changes of hue which occur in the NTSC system when transmission non-linearities cause the phase shift of the colour subcarrier to be dependent on luminance.

In the SECAM systems used in France and Eastern Europe, each of the chrominance signals

$$D'_R = -1.9(E_R^{1/\gamma} - E'_Y)$$
$$D'_B = 1.5(E_B^{1/\gamma} - E'_Y)$$

is transmitted throughout alternate lines by frequency modulation of one of two colour subcarriers. Each chrominance signal is stored for the duration of one line and used to control the chrominance for two successive lines.

The main characteristics of the various colour television systems have been tabulated by the CCIR[6].

3.4 Digital encoding of pictorial information

3.4.1 Encoding of video signals

The most straightforward method of digital encoding is pulse-code modulation (PCM), which has been described in Section 1.3.1 in connexion with the encoding of speech signals. In the case of video signals, the effect of quantisation is to introduce contours, which can be very distracting to the eye, particularly in areas of low detail. In the case of monochrome television it is found that 128 or 256 quantum levels are necessary to make the contours unobtrusive, far more than would be suggested by other considerations. The necessary number of levels cannot be reduced by instantaneous companding.

Fewer quantum levels need to be used, however, if the spurious contours are blurred by the addition of a random dither waveform to the video signal before encoding. The added dither is less noticeable if it is concentrated at the higher spatial frequencies. Alternatively, the added dither may be pseudorandom, in which case a replica generated in the decoder can be subtracted after decoding so that the dither does not cause any degradation of the picture.

One approach to improving the efficiency of PCM is to encode not the value of each sample itself but the difference between the sample and some predicted value. This method is called differential pulse-code modulation (DPCM). Delta modulation (Section 1.3.2) is a simple form of one digit DPCM and is excellent for encoding luminance signals in areas of gradual change. In its basic form it requires a very high sampling frequency to handle large abrupt changes, but a lower sampling frequency can be used if the delta modulator incorporates double integration or is made adaptive in a manner analogous to the syllabic companding described in Section 1.3.2.

Provided that sufficient digits are used to encode each difference, DPCM can both handle abrupt changes and avoid objectionable contours in low detail areas. The number of digits can be reduced by instantaneous companding, that is, by arranging for the quantum levels to be closer together for small differences than for large differences. With instantaneous companding, 16 quantum levels have been judged sufficient for video-telephone applications.

Some reduction in the required digit rate can be achieved by dual-mode schemes, in which two different encoding techniques are used for different areas of the picture. For example, DPCM might be used for areas containing large variations of luminance and delta modulation for areas of gradual change.

A fundamental disadvantage of DPCM and delta modulation is the persistent effect of transmission errors. Delta-sigma modulation (Section 1.3.2) is, however, an exception.

When the video signal includes a modulated colour subcarrier, a DPCM encoder is in a condition of continuous overload unless each prediction is based on a sample separated by an integral number of subcarrier periods. Undesirable intermodulation effects are reduced[5] by choosing the sampling frequency to be an integral multiple of the subcarrier frequency. Five digit DPCM or eight digit PCM is found to be adequate for most colour television systems. Greater efficiency may be obtained by encoding luminance and chrominance separately, using a lower sampling rate and alternate-line transmission for each chrominance signal.

3.4.2 Efficient encoding of two-tone images

Typical examples of two-tone (black-and-white) images often transmitted are typewritten documents, newspaper pages, weather maps and finger prints. The image may be regarded as a rectangular array of picture elements (pels), each of which may be either black or white. The information content of the image is usually much less than one bit per pel and hence straightforward encoding using one binary digit for each pel is likely to be inefficient.

When there is relatively little detail, run-length encoding may be efficient. In run-length encoding, each line is regarded as a sequence of runs, alternately black and white, and the length of each run is encoded. It may be worthwhile to use an efficient binary code (e.g. the Huffman code of Exercise 3.14) rather than simple binary numbers to encode the run lengths. Run-length coding may be extended to more than two tones if each encoded run length is preceded by a binary number describing its colour.

Run-length coding may be made more efficient by incorporating prediction, transmitting for each line the difference in length of each run compared with the corresponding run in the previous line. 'New-start' and 'merge' symbols are necessary to indicate the beginnings and ends of black areas. Basing predictions on previous lines as well as on previous pels within a line is sometimes called two-dimensional prediction and can be used when encoding television video signals.

When the image consists predominantly of black marks on a white background, other approaches are appropriate, such as white-block skipping and the transmission of boundary points. In white-block skipping, each line of pels is divided into equal blocks, each all-white block is encoded as 0 and each block containing black pels is encoded as 1 followed by the pattern within that block. Alternatively the image can be divided into rectangular blocks of pels or the encoding can be made adaptive by varying the block size according to the amount of detail in the image. A useful extension when transmitting typewritten documents is to encode each all-white line as 0 and each other line as 1 followed by normal white-block skipping for that line.

Information about boundaries can be transmitted as a list of the co-ordinates of every pel on a boundary. Alternatively each boundary can be traced by transmitting to whichever of eight neighbouring pels the boundary moves at each step.

A useful survey of the encoding of two-tone images has been given by Huang[7].

3.5 References

1 Fraser, W., *Telecommunications*, Macdonald, Second Edition, 1967
2 *CCITT Yellow Book*, Vol. III, Recommendation H41, Geneva, 1980
3 Allnatt, J., *Transmitted-picture Assessment*, Wiley-Interscience, 1983
4 Smol, G., Hamer, M. P. R., and Hills, M. T., *Telecommunications: A Systems Approach*, George Allen and Unwin Ltd, 1976
5 Limb, J. O., Rubinstein, C. B., and Thompson, J. E., 'Digital coding of colour video signals—a review', *IEEE Trans*, COM-25, pp 1349–85, November 1977
6 *CCIR*, Vol. XI, Broadcasting Service (Televison) Report 624-2, Geneva, 1982
7 Huang, T. S., 'Coding of two-tone images', *IEEE Trans*, COM-25, pp 1406–24, November 1977

Exercise topics

3.1   Conventional facsimile: speed and bandwidth
3.2   Bandwidth of monochrome television
3.3   Minimum viewing distance for television
3.4   Bandwidth of monochrome television
3.5   Inter-line sound transmission
3.6   Video telephone
3.7   Luminosity coefficients
3.8   Colour triangle
3.9   Relationships between colour-difference signals
3.10 U, V, I and Q signals related to colour triangle
3.11   Interaction of luminance and chrominance (NTSC)
3.12   Effect of differential phase distortion on luminance (PAL)
3.13   Binary encoding of two-tone pictures
3.14   Huffman run-length encoding of two-tone pictures
3.15   Laemmel run-length encoding of two-tone pictures
3.16   White-block-skipping encoding of two-tone pictures
3.17   Optimum block length for white-block-skipping encoding
3.18   Transmission of two-tone pictures as boundary points

Exercises

3.1 Conventional facsimile: speed and bandwidth

In a facsimile system the ratio of the drum diameter to the scanning pitch is 352. (The scanning pitch is the distance moved by the photocell for one revolution of the drum.) Transmission is to be by full double-sideband amplitude modulation and a bandwidth of 2 kHz is available. Calculate the maximum speed of rotation of the drum.

Solution outline
The highest video frequency should not exceed 1 kHz. Hence the surface of the drum should not travel a distance greater than the scanning pitch in 0.5 ms.

Denoting the diameter of the drum by d and the number of revolutions it makes per second by r,

$$\pi dr \times (0.5 \times 10^{-3}) < d/352.$$

Hence $r < 1.8$ rev/s.

3.2 Bandwidth of monochrome television

In a 625 line monochrome television system each field contains $312\frac{1}{2}$ lines and the field frequency is 50 Hz. The width : height ratio of the picture is 4 : 3 and 585 of the 625 lines are visible, each for 52 μs. Calculate the value of the Kell factor which has been assumed in deciding upon a maximum video frequency of 5.5 MHz.

Solution outline
If the Kell factor were unity, the number of half-cycles of the maximum video frequency occurring during 52 μs should be $585 \times \frac{4}{3}$. Hence the maximum video frequency should be

$$\left(585 \times \frac{4}{3} \times \frac{1}{2} \times \frac{1}{52}\right) \text{MHz} = 7.5 \text{ MHz}.$$

Hence the Kell factor used is 5.5/7.5, i.e. 0.73.

3.3 Minimum viewing distance for television

A receiver in the system of Exercise 3.2 has a display with a 600 mm diagonal. How far away should a viewer be if he is not to resolve individual scanning lines?

Solution outline
The height of the picture is ($\frac{3}{5} \times 600$) mm = 360 mm. Adjacent lines must subtend less than 1/3400 rad at the viewer. Hence the minimum viewing distance is

$$3400 \times \frac{360}{585} \text{ mm} = 2.1 \text{ m}.$$

3.4 Bandwidth of monochrome television

An n line monochrome television system sends f pictures per second. The width/height ratio of the pictures is a, line blanking occupies a fraction l of each line time and field blanking occupies a fraction m of each field time. Show that the bandwidth required for the video signal is

$$\frac{1}{2} n^2 \frac{1-m}{1-l} afK \text{ Hz},$$

where K is the Kell factor.

Solution outline
There are $(1-m)n$ visible lines. Hence the number of half cycles in each line of the maximum video frequency should be at least $aK(1-m)n$. The duration of the visible part of each line is $(1-l)/fn$ seconds. Hence the bandwidth required is

$$\frac{1}{2} \times aK(1-m)n \times \frac{fn}{1-l}.$$

3.5   Inter-line sound transmission

In the system of Exercise 3.2 it is proposed to transmit two digitally encoded programme-sound signals as bursts of digits occupying the line-blanking periods. Estimate the minimum rate of transmission during the bursts.

Solution outline
Assuming near-instantaneously companded PCM (Section 1.4.3) in which 10 digit samples are transmitted at 32 kHz together with a 3 digit range indication at 1 kHz, the average bit rate required is 323 000 bit/s for each sound signal. The fraction of the time occupied by line-blanking periods is 12/64. Hence the rate of transmission required during bursts is

$$\left(2 \times 323\,000 \times \frac{64}{12} \right) \text{bit/s} = 3.45 \text{ Mbit/s}.$$

3.6   Video telephone

A proposed 267 line video-telephone system has a square picture format and a field frequency of 50 Hz (with a 2:1 interlace). Each line is blanked for 20% of the line period and 15 lines are blanked in each field. Calculate the bandwidth required for the video signal, taking the Kell factor to be 0.7.

It is proposed to transmit speech as bursts of digits during field-blanking periods. Assuming that the speech is encoded as a continuous 64 kbit/s stream of digits and that during bursts digits are transmitted at a rate in bit/s equal to the bandwidth in Hz of the video signal, calculate how many line periods in each field are occupied by the speech. Estimate the digital storage capacity required in the receiver.

Solution outline
In the formula of Exercise 3.4, $n = 267$, $m = 30/267$, $l = 0.2$, $a = 1$, $f = 25$ and $K = 0.7$. Hence the required video bandwidth is 692.1 kHz.

1280 bits must be transmitted in each field at 692.1 kbit/s. Since each line occupies 149.8 μs, 12.35 line periods are required in each field.

It is necessary to store enough speech for at least 118.5 lines, i.e.

$$\left(1280 \times \frac{118.5}{133.5} \right) \text{bits} = 1136 \text{ bits}.$$

3.7 Luminosity coefficients

A colour television display of a map is to show red lines ($R = 1$, $G = B = 0$), blue lines ($B = 1$, $R = G = 0$), black areas ($R = G = B = 0$) and green areas ($R = B = 0$, $G = G_1$). Calculate G_1 if the luminance of the green areas is to be the average of the luminances of a red line and a blue line.

Solution outline
The luminance of the red lines is 0.30 and the luminance of the blue lines is 0.11. Therefore we require

$$0.59\, G_1 = \frac{1}{2}(0.30 + 0.11)$$

and hence $G_1 = 0.35$.

3.8 Colour triangle

The chromaticity of a given colour is determined by the ratios of its tri-stimulus values R, G, B and may be described by the chromaticity co-ordinates r, g, b defined by:

$$r = \frac{R}{R+G+B}, \quad g = \frac{G}{R+G+B}, \quad b = \frac{B}{R+G+B}.$$

Since $r + g + b = 1$, chromaticity is specified uniquely by a point in a plane with cartesian co-ordinates (r, g).

Denote by W the point in the (r, g)-plane representing the chromaticity of reference white and by S the closed contour in the (r, g)-plane which just encloses all perceivable chromaticities. If the chromaticity of a given colour is represented by the point C, the point C' at which WC (produced) meets S is said to represent the 'hue' of the given colour and the ratio of lengths WC/WC' is said to be its 'saturation'.

(i) Show that when two colours with chromaticities (r_1, g_1) and (r_2, g_2) are added, the chromaticity of their sum lies on the straight line joining (r_1, g_1) and (r_2, g_2).
(ii) Show that for all colours reproducible by addition of the three primaries, the chromaticity (r, g) lies within the triangle with vertices $(0, 0)$, $(1, 0)$, $(0, 1)$.
(iii) Determine (r, g) for reference white.
(iv) Determine (r, g) for each of the complementary colours cyan, magenta and yellow. (A complementary colour is a mixture of two primaries, which can be mixed with the third primary to give reference white.)
(v) Assuming that S passes through $(0, 1)$, calculate the proportions in which reference white and the green primary must be added to produce a colour with saturation 0.5.
(vi) Determine the region of the (r, g)-plane which represents colours reproducible with luminance 80% of that of reference white, given that the intensities of the primaries must not exceed the values reproducing reference white.

Solution outline
(i) For the mixture,

$$r = \frac{R_1 + \lambda R_2}{(R_1 + G_1 + B_1) + \lambda(R_2 + G_2 + B_2)}, \quad g = \frac{G_1 + \lambda G_2}{(R_1 + G_1 + B_1) + \lambda(R_2 + G_2 + B_2)}.$$

Show that $(g - g_1)/(r - r_1)$ is independent of λ, where r_1, g_1 are the values of r, g when $\lambda = 0$.
(ii) $(0, 0)$, $(1, 0)$, $(0, 1)$ represent respectively the chromaticities of the blue, red, green,

primaries. Adding two primaries gives a chromaticity on one side of the triangle, adding the third gives a chromaticity within the triangle.

(iii) $R = G = B = 1$. Hence $r = g = b = \frac{1}{3}$. Hence W is $(\frac{1}{3}, \frac{1}{3})$.

(iv) $(0, 0.5)$, $(0.5, 0)$, $(0.5, 0.5)$.

(v)   For the required mixture, $(r, g) = (\frac{1}{6}, \frac{2}{3})$.
    Adding one part of reference white to λ parts of the green primary gives

$$(r, g) = \left(\frac{1}{3 + \lambda}, \frac{1 + \lambda}{3 + \lambda} \right).$$

Hence $\lambda = 3$.

(vi)   For the required luminance, R, G, B must satisfy

$$0.30R + 0.59G + 0.11B = 0.80.$$

Assume first that $R = 1$. For the correct luminance,

$$0.59G + 0.11B = 0.50.$$

Colours satisfying these conditions have chromaticities lying on the line

$$0.48g - 0.61r = -0.11.$$

Chromaticities below this line can be reproduced only if $R > 1$. Similarly chromaticities above the line

$$0.32g - 0.19r = 0.11$$

can be reproduced only if $G > 1$, and those below the line

$$1.28g + 0.99r = 0.69$$

can be reproduced only if $B > 1$.
    Hence for a chromaticity to be reproducible for $R \leqslant 1, G \leqslant 1$ and $B \leqslant 1$ it must lie inside the triangle formed by these three lines.
    Since negative R, G, B are not permissible, the required region is that which is common to this triangle and the triangle described in (ii), (see Fig. 3.2).

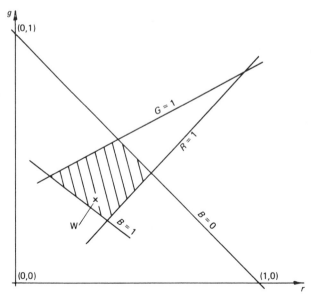

Fig. 3.2   Graph of solution to Exercise 3.8(vi)

3.9 Relationships between colour-difference signals

Show that $E_G^{1/\gamma} - E_Y'$ can be derived from E_U' and E_V' without knowledge of E_Y'.

Solution outline
Re-arranging

$$E_Y' = 0.30 E_R^{1/\gamma} + 0.59 E_G^{1/\gamma} + 0.11 E_B^{1/\gamma},$$

gives

$$0.59 (E_G^{1/\gamma} - E_Y') = -0.30 (E_R^{1/\gamma} - E_Y') - 0.11 (E_B^{1/\gamma} - E_Y').$$

3.10 *U, V, I,* and *Q* signals related to colour triangle

Find equations for the four lines in the (r, g)-plane on which $E_U = 0$, $E_V = 0$, $E_I = 0$ and $E_Q = 0$ and display the lines on a graph. [Assume no γ-correction, i.e. assume that $E_U = 0.493 \ (E_B - E_Y)$, $E_V = 0.877 \ (E_R - E_Y)$, $E_I = E_V \cos 33° - E_U \sin 33°$ and $E_Q = E_V \sin 33° + E_U \cos 33°$.]

Solution outline
$E_U = 0 \Rightarrow E_B = E_Y$. Hence $B = 0.30 R + 0.59 G + 0.11 B$ and, since $r + g + b = 1$, $0.89 (1 - r - g) = 0.30 r + 0.59 g$,

i.e. $1.19 r + 1.48 g = 0.89$.

Similarly, if $E_V = 0$,

$0.81 r - 0.48 g = 0.11$.

If $E_I = 0$, $0.74 (E_R - E_Y) - 0.27 (E_B - E_Y) = 0$, whence

$0.92 r + 0.04 g = 0.32$.

Similarly if $E_Q = 0$,

$0.10 r + 0.84 g = 0.32$.

These four lines are shown in Fig. 3.3.

3.11 Interaction of luminance and chrominance (NTSC)

(i) Explain why the NTSC colour subcarrier in regions of uniform chrominance produces on a monochrome display an array of fine dots which move slowly up the screen. Calculate the time taken for each dot to move a distance equal to the height of the display.
(ii) Show that in the NTSC system the effect on chrominance of a luminance pattern of vertical bars is equal and opposite in successive lines in any field. Show also that the effect is the same in successive lines if the bars are inclined at about 40.6° to the vertical.

Assume that 485 of the 525 lines are visible, each for 83 % of the line period, that the width/height ratio of the displayed picture is 4/3, that the field frequency is 59.94 Hz and that the frequency of the colour subcarrier is 227.5 times the line scan frequency.

Solution outline
(i) The fine pattern of dots is one line higher in each successive field and therefore takes $485/59.94$ s ($= 8.1$ s) to travel the height of the display.
(ii) The phase of the subcarrier is reversed on successive lines. Half a period of the colour subcarrier occupies a length $w/(2 \times 0.83 \times 227.5)$, where w is the width of the display. A luminance pattern of bars inclined at angle θ to the vertical is shifted by a distance

$$\frac{3}{4} \times \frac{2}{485} \times \tan \theta$$

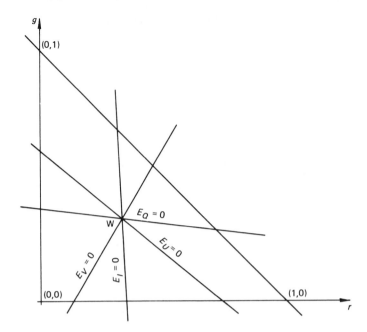

Fig. 3.3   Graph of solution to Exercise 3.10

on successive lines. Equating these expressions shows that the luminance pattern remains in step with the subcarrier if $\theta = 40.57°$.

3.12   Effect of differential phase distortion on luminance (PAL)

Calculate the percentage error in the luminance of the display in a PAL receiver due to a phase error in the colour subcarrier of $+68.6°$ relative to the colour burst when the transmitted signal is green-yellow with $E_R = 0.87$, $E_G = 1.0$ and $E_B = 0$. Take $\gamma = 2.2$.

Solution outline
The luminance of the transmitted colour is $(0.30 \times 0.87) + 0.59 = 0.851$. The transmitted luminance signal is

$$E'_Y = 0.30(0.87)^{1/2.2} + 0.59(1)^{1/2.2} = 0.8716,$$

and the transmitted chrominance signals are

$$E'_U = 0.493\{(0)^{1/2.2} - 0.8716\} = -0.430,$$
and $$E'_V = 0.877\{(0.87)^{1/2.2} - 0.8716\} = 0.0588.$$

The display is derived from $E'_Y = 0.8716$ and averaged values of E'_U and E'_V, which are $-0.430 \cos 68.6°$ and $0.0588 \cos 68.6°$ respectively. Hence the display is derived from

$$E_R^{1/\gamma} = 0.8716 + 0.0215/0.877 = 0.896,$$
$$E_B^{1/\gamma} = 0.8716 - 0.157/0.493 \ \ = 0.553$$
and $$E_G^{1/\gamma} = \{0.8716 - (0.30 \times 0.896) - (0.11 \times 0.553)\}/0.59 = 0.919.$$

Therefore the displayed luminance is

$$0.30(0.896)^{2.2} + 0.59(0.919)^{2.2} + 0.11(0.553)^{2.2} = 0.755,$$

corresponding to a decrease of about 11 %.

3.13 Binary encoding of two-tone pictures

Calculate the time required to transmit a two-tone image occupying an A4 sheet (297.3 mm × 210.2 mm) over a 2400 bit/s digital channel in each of the following cases.
(i) The image is typescript, requiring 6 pels/mm for satisfactory resolution.
(ii) The image is a two-tone simulation of a photograph, in which the impression of shading is given by printing one of 64 different sizes of black dot at each sampling point. There are 3.4 sampling points per millimetre.
  Assume straightforward binary encoding in each case.

Solution outline
(i) The image contains (1784 × 1262) pels, each requiring one binary digit. The transmission time is therefore (1784 × 1262) ÷ 2400 seconds, i.e. about 16 minutes.
(ii) The image contains (1011 × 715) sampling points, each requiring six binary digits. The transmission time is therefore (1011 × 715 × 6) ÷ 2400 seconds, i.e. about 30 minutes.

3.14 Huffman run-length encoding of two-tone pictures

A two-tone approximation to a monochrome photograph is made up of runs of between one and eight pels, the average proportions of black and white pels in a region giving an impression of the darkness of that region. The probability p_k that a run (black or white) is of length k pels is given in Table 3.1. Construct a Huffman code suitable for run-length encoding of the image and calculate the average number of binary digits required to encode one pel.

<table>
<tr><td colspan="2" align="center">**Table 3.1**</td><td colspan="2" align="center">**Table 3.2**</td></tr>
<tr><td>k</td><td>p_k</td><td>Message symbol</td><td>Binary code</td></tr>
<tr><td>1 or 8</td><td>0.05</td><td>A</td><td>00</td></tr>
<tr><td>2 or 7</td><td>0.1</td><td>B</td><td>01</td></tr>
<tr><td>3 or 6</td><td>0.15</td><td>C</td><td>10</td></tr>
<tr><td>4 or 5</td><td>0.2</td><td>D</td><td>110</td></tr>
<tr><td></td><td></td><td>E</td><td>111</td></tr>
</table>

[As an example of the construction of a Huffman code assume that message symbols A, B, C, D, E occur with probabilities 0.35, 0.25, 0.2, 0.15, 0.05. Construct the diagram in Fig. 3.4 (in which probabilities are shown in parentheses), from which the code is seen to be as given in Table 3.2.]

Solution outline
A suitable Huffman code is given in Table 3.3.
  There are other solutions. For this code, the average number of binary digits required to encode a run is

$$2 \times \{(0.05 \times 4) + (0.1 \times 4) + (0.15 \times 3) + (0.2 \times 2)\} = 2.9.$$

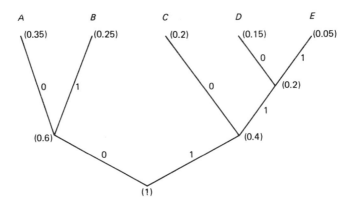

Fig. 3.4   Construction of Huffman code of Exercise 3.14

Table 3.3

k	1	2	3	4	5	6	7	8
	0110	0000	001	10	11	010	0001	0111

The average number of pels in a run is 4.5. Hence the number of binary digits required per pel is

$$2.9/4.5 = 0.64.$$

3.15   Laemmel run-length encoding of two-tone pictures

In a two-tone map regarded as a 1024×1024 array of pels, the probability p_k that a black or white run has length k is given by

$$p_k = (0.1)(0.9)^{k-1}.$$

Determine the average number of binary digits required to encode one scan line of the map by run-length coding with each run represented by a 10 digit binary number.

Repeat the calculation assuming that the run-lengths are represented by the Laemmel code of block-size 2.

[To determine the Laemmel codeword of block-size N representing the positive integer k, express k in the form

$$k = q(2^N - 1) + r,$$

where q is a positive integer or zero and

$$1 \leqslant r \leqslant 2^N - 1.$$

The required codeword is $(q+1)N$ binary digits in length and consists of qN 0s followed by the N digit binary representation of r.]

Solution outline
The average length of a run is

$$0.1 + 2(0.1)(0.9) + 3(0.1)(0.9)^2 + \ldots + 1024(0.1)(0.9)^{1023} \simeq 10.$$

[Note that $1 + 2x + 3x^2 + 4x^3 + \ldots = d/dx \ (1 + x + x^2 + x^3 + x^4 + \ldots) = (1 - x)^{-2}$]
Hence on average there are 102.4 runs in a scan line, requiring 1024 binary digits when binary numbers are used.

Tabulating for the Laemmel code of block-size 2 gives Table 3.4.

Table 3.4

k	q	r	
1	0	1	01
2	0	2	10
3	0	3	11
4	1	1	001
5	1	2	010
6	1	3	011
7	2	1	0001 etc.

The probability of a run-length being encoded with just two binary digits is

$$(0.1) + (0.1)(0.9) + (0.1)(0.9)^2 = 0.271.$$

The probability of a run-length being encoded with just four binary digits is

$$(0.1)(0.9)^3 + (0.1)(0.9)^4 + (0.1)(0.9)^5 = (0.9)^3(0.271), \text{ etc.}$$

Hence the average number of binary digits required to encode a run is

$$(0.271) \times [2 + \{4 \times (0.9)^3\} + \{6 \times (0.9)^6\} + \ldots + 682(0.9)^{1020}]$$

$$\simeq \frac{0.271 \times 2}{\{1 - (0.9)^3\}^2} = 7.38.$$

Thus the average number of binary digits required to encode a scan line when the Laemmel code is used is $7.38 \times 102.4 = 756$.

3.16 White-block-skipping encoding of two-tone pictures

A two-tone diagram has on average one black pel for every 19 white pels. Assuming the black pels to be randomly distributed, determine the optimum block-size for white-block-skipping encoding of the diagram.

Assuming the optimum block size, calculate the efficiency of the transmission in bits of information per transmitted binary digit.

Solution outline
If the block-size is N pels, the probability of one digit being required to transmit the block is $(0.95)^N$ and the probability of $N + 1$ digits being required is $1 - (0.95)^N$. Hence the average number of digits required to transmit one pel is

$$\frac{1}{N}[(0.95)^N + (N + 1)\{1 - (0.95)^N\}].$$

Evaluating this expression for positive integral values of N shows that it has a minimum value of 0.426 when $N = 5$.

The information content per pel is (Appendix 9)

$$-0.05 \log_2(0.05) - 0.95 \log_2(0.95) = 0.286 \text{ bits.}$$

Hence the required efficiency is

$$\frac{0.286}{0.426} = 0.67 \text{ bits per digit.}$$

3.17   Optimum block length for white-block-skipping encoding

In a two-tone diagram, the probability that a block of length N is all white is p_N. Show that when the diagram is encoded by a white-block-skipping code with block length N the average number of binary digits required to encode one pel is b_N given by

$$b_N = 1 + \frac{1}{N} - p_N.$$

Given that the probability that a pel is black is 0.05 if the previous pel is white and 0.7 if the previous pel is black, express b_N as a function of N and hence determine the optimum block length.

Calculate also the average length of run in this diagram.

Solution outline
The average number of binary digits required to encode a block is

$$p_N + (1 - p_N)(N + 1).$$

Hence

$$b_N = \frac{1}{N}\{p_N + (1 - p_N)(N + 1)\} = 1 + \frac{1}{N} - p_N.$$

The overall proportions p_b and p_w of black and white pels respectively are related by

$$p_b + p_w = 1$$

and $0.7p_b + 0.05p_w = p_b,$

whence $p_b = 1/7$ and $p_w = 6/7$.
    Now

$$p_N = \frac{6}{7}(0.95)^{N-1}.$$

Hence

$$b_N = 1 + \frac{1}{N} - \frac{6}{7}(0.95)^{N-1}$$

and the integral value of N minimising b_N may be found to be $N = 5$.
    The probability that a white run is of length k is $(0.05)(0.95)^{k-1}$ and hence the average length of the white runs is

$$(0.05)[1 + 2(0.95) + 3(0.95)^2 + \ldots] = 20$$

(see the solution to Exercise 3.15). Similarly the average length of the black runs is

$$0.3[1 + 2(0.7) + 3(0.7)^2 + \ldots] = 3.33.$$

Since there are equal numbers of black and white runs, the overall average run length is $\frac{1}{2}(20 + 3.33) = 11.7$ pels.

3.18 Transmission of two-tone pictures as boundary points

The resolution of a two-tone image is improved by a factor 3 (i.e. the number of pels is increased by a factor 9). Estimate the increase in the number of binary digits required to encode the image when the encoding is
(i) one pel at a time (two-level PCM),
(ii) by transmission of the co-ordinates of the boundary points,
(iii) by boundary tracing.
In (ii) and (iii) assume that the number of boundary points is increased by the factor 3.

Solution outline
Assume that before the improvement the image contains $M \times N$ pels and B boundary points.
(i) MN binary digits are required. The increase is a factor 9.
(ii) $B[\log_2 M + \log_2 N]$ binary digits are required. This is increased to $3B[\log_2 3M + \log_2 3N] = 3B[\log_2 9 + \log_2 M + \log_2 N]$. If M and N are large the increase is approximately a factor 3.
(iii) $3B$ binary digits are required. This is increased by a factor 3.

4
Exchanges

4.1 Telephone switching systems

4.1.1 Switching methods

Telephone subscribers in any one locality are usually interconnected by connecting each of them to a switching centre called a local exchange. In large systems local exchanges are usually interconnected by a hierarchy of switching centres supplemented by direct connexions between neighbouring local exchanges. In some military systems[1] there is no hierarchy, but links are provided between neighbouring switching centres to form a network offering a choice of routes for any call.

For many years most telephone exchanges were based on the step-by-step rotary selectors patented by Strowger[2]. In the earliest exchanges, rotation of the selectors was controlled directly by current impulses generated by subscribers' dialling, but later systems have incorporated registers which translate dialled numbers into numbers appropriate to the routing of the call to other exchanges, thus enabling the same dialled number to be used to call a given subscriber wherever the call originates[3].

Strowger's rotary selectors provide both the switch contacts to carry the telephone signals and the means of controlling which contacts should be selected when setting up a call. An alternative approach to the design of exchanges is to separate the controlling function from the switching function, using a digital computer to control the opening and closing of contacts in an array of switches. Electronic devices provide satisfactory switches when the telephone signals are in digital form; but inadequate off resistance and other limitations make them unsuitable for switching conventional analogue telephone signals in electronically controlled exchanges, for which purpose small encapsulated reed switches are widely used[3].

An exchange interconnecting digital highways carrying groups of signals in time-division multiplex (Section 6.4.1) must not only switch each incoming signal 'in space' so that it leaves on the correct highway, but also delay it in time so that it occupies the correct time slot on that highway. Space switching is carried out by an array of electronic switches, which operate as necessary for a short period during each time slot. Buffer memories at the inlets and outlets of the array store the contents of each time slot until they are required[3].

4.1.2 Concentrators

The most straightforward way of enabling any of N inlets to be connected to any of N outlets is to use an $N \times N$ matrix of switching elements. In most telephone systems, however, this method makes very inefficient use of the switching elements, since very few of the inlets are in use simultaneously and many of the possible interconnections are never used at all.

The number of switching elements may be substantially reduced by the use of concentrators. A concentrator enables the traffic from a number of inlets to be carried by a smaller number of outlets. A simple illustrative example of the use of concentrators is

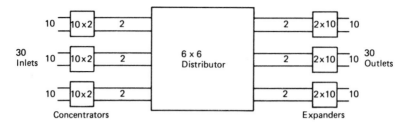

Fig. 4.1   Simple switching structure showing grouping and concentrators

shown in Fig. 4.1. The structure of Fig. 4.1 enables any of 30 inlets to be connected to any of 30 outlets and thus replaces a 30 × 30 matrix of switching elements. The 30 inlets are divided into three groups of 10 and each group is connected to two links to the central 6 × 6 switching matrix by a matrix of switches called a concentrator. In a similar way traffic on six links is distributed among the 30 outlets by three expanders. If the concentrators and expanders are respectively 10 × 2 and 2 × 10 matrices, the whole structure contains 156 switching elements, a considerable saving on the 900 required for a single 30 × 30 matrix. The obvious disadvantage is that since not more than two inlets or outlets in a group can be used simultaneously the introduction of concentrators has introduced the possibility of 'blocking' when the use of more is required. An attempted call is said to be 'blocked' when a suitable interconnexion cannot immediately be made.

The 'availability' of the outlets of a concentrator is the number of outlets which can be connected to an inlet. If a concentrator can always connect any free inlet to any free outlet regardless of connexions already made, 'full availability' is said to exist in the group of inlets served by the concentrator. Full availability within the groups of Fig. 4.1 requires the concentrators to be complete 10 × 2 matrices: an incomplete matrix gives rise to 'limited availability'.

4.1.3   Grading

If the outlets of a concentrator are always searched in the same order when a free outlet is required, as is the case when rotary switches are used, the outlets towards the end of the search are very rarely used and could be shared with other concentrators without significantly increasing the probability of blocking. Such sharing is called 'grading' and is important in making efficient use of expensive Strowger selectors.

A simple illustrative example of grading is shown in Fig. 4.2(a). The inlets are divided into two 'grading groups' and a concentrator is assigned to each. The concentrators consist of rotary switches, one to each inlet, which rotate when required from the homing position through positions 1 to 6 until a free outlet is found. The availability of the outlets is six. A diagrammatic representation of the structure of Fig. 4.2(a) is shown in Fig. 4.2(b). The example of Fig. 4.2 is an example of 'progressive grading', since the number of grading groups sharing an outlet increases towards the later switch positions.

For given numbers of grading groups, outlets and switch positions, the optimum grading arrangement depends on the amount of traffic carried. An empirical rule by O'Dell, given in Bear[8], recommends that under normal conditions there should be an equal number of switch positions associated with each commoning arrangement. For light traffic it is preferable to associate rather more of the outlets with the earlier switch positions.

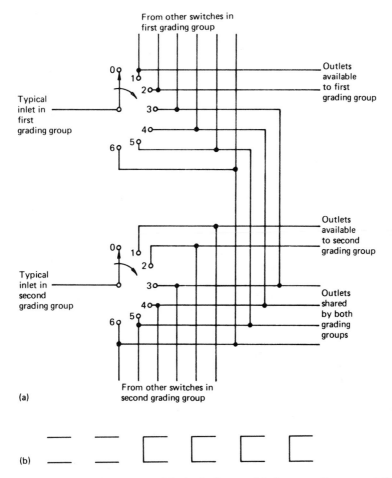

Fig. 4.2 Simple grading structure (a) circuit diagram (b) diagrammatic representation

4.2 Store-and-forward systems

4.2.1 Message switching

In message-switched systems each complete message is transferred to a store in a switching centre, thence to a store in another switching centre, and so on until it reaches its destination. Operator-controlled message switching has existed for many years in the form of torn-tape systems, in which teleprinter messages are stored on punched tape and forwarded as transmission links become available.

Message switching requires only one transmission link to be free at each stage and therefore makes more efficient use of transmission links than is made by those systems (called 'circuit-switched' systems) in which a complete route is established from source to destination before transmission takes place. However, the possibility of delays occurring during transmission makes message switching unsuitable for interactive communication such as telephony.

Each message must obviously contain information enabling switching centres to route it to its destination (or destinations). In some systems the route is predetermined and the message contains instructions to each switching centre along its route. In other systems, greater flexibility is provided by allowing each switching centre a choice of switching centres to which to forward the message, basing each decision on queue lengths and ultimate message destination. Either of these strategies may be used to set up circuits in a circuit-switched network.

4.2.2 Packet switching

Packet switching is a development of message switching in which long digital messages are transmitted as several short messages called 'packets'. A typical packet consists of about 2000 binary message digits preceded by a 'header'[1]. The header contains sufficient information to route the packet to its destination and to locate it correctly within the complete message. Packets travel independently through the network and packets from a single message may travel by different routes.

Queue lengths and therefore delays are shorter when packet switching is used than when message switching is used. Consequently less storage is required at the switching centres and interactive communication becomes a possibility.

It is the usual procedure in packet-switched systems for switching centres to wait for an acknowledgment that each packet has been received by another switching centre before transmitting the next packet. When error protection digits are included in each packet, this procedure can be modified to include the repetition of packets received with parity violations (Section 2.3.1).

Short clear introductions to the concepts of circuit, message and packet switching are given by Marshall[4]. For a fuller discussion of packet switching see Rosner[5]. For further details of an experimental packet-switched system see Neil et al[6, 7].

4.3 Telephone-traffic theory

4.3.1 Traffic statistics

The time for which a telephone call occupies a channel is called the 'holding time' of the call. In some cases, time taken to set up and clear down the channel may cause the holding time to exceed the duration of the conversation. It is often a good approximation to assume that holding times follow a negative-exponential distribution and hence that the future duration of a call is independent of its history (Exercise 4.1).

The traffic in erlangs carried by a transmission link is defined as the average number of calls beginning in unit time multiplied by the average holding time. It is equal to the average number of channels in simultaneous use in the link[8] and is a dimensionless quantity.

The busyness of a telephone system usually follows a well-established diurnal pattern, and to provide a satisfactory service when demand is greatest the design of a system is usually based on estimates of traffic made during the 'busy hour', which is defined as that period of 60 minutes during the day for which traffic is on average the greatest.

The product of the average holding time and the average number of calls presented to a transmission link in unit time is called the 'attempted traffic' or the 'traffic offered' to the link. When sources of calls are connected to a link by way of a concentrator the attempted traffic may exceed the traffic actually carried by the link. The difference between the traffic attempted and traffic carried is referred to as the 'traffic lost.'

In many situations it may be assumed that calls are attempted in a pure-chance manner (i.e. in a Poisson process), the probability of an attempt occurring during a short time interval δt being constant (independent of previous attempts). When that is the case, the probability that x attempts are made in a time interval T is given by the Poisson distribution (Exercise 4.2) and the time intervals between attempts follow a negative-exponential distribution. For a discussion of the interrelation of the Poisson and the negative-exponential distributions see References[1,9]. For pure-chance traffic, the probability of an attempt occurring in a time interval δt is the product of δt and the average number of calls attempted in unit time, provided that δt is sufficiently small for the probability of two or more attempts occurring in the interval to be vanishingly small.

4.3.2 Lost-call probability

Suppose that A erlangs of pure-chance traffic are offered to a concentrator with n outlets and full availability. Callers finding all n outlets busy form a queue and are allocated outlets as they become free. The maximum length of queue is m callers and calls attempted when the queue is at its maximum length are said to be 'lost'. The lost-call probability is the probability that a caller finds the queue at its maximum length. A proof of the important formula for lost-call probability under these conditions is outlined below: fuller explanations can be found in References[3,4,8].

The concentrator may be regarded as having $n+m+1$ possible states and is said to be in the state $[i]$ when the sum of the number of busy outlets and the number of callers in the queue is i. The attempting of a call increases i by one, unless $i = n+m$, in which case that call is lost. If the probability of the concentrator being in the state $[i]$ is denoted by $p(i)$, $p(n+m)$ is the lost-call probability.

The proof is based on the fact that under conditions of statistical equilibrium the probability of a transition to state $[i]$ in a short time interval δt equals the probability of leaving that state in the interval δt. From this it follows that the probability of the occurrence (in δt) of the transition $[i] \rightarrow [i+1]$ must equal that of the transition $[i+1] \rightarrow [i]$ for all integers i satisfying $0 \leqslant i < m+n$.

From Section 4.3.1, the average number of calls attempted in unit time is A/h (where h is the average holding time) and hence the probability of a call being attempted in δt is $A \, \delta t/h$. Thus the probability of the transition $[i] \rightarrow [i+1]$ occurring in δt is $p(i)A\delta t/h$.

Assuming either that holding times are equal or that they have a negative-exponential distribution, it may be shown (Exercise 4.1) that the probability of a given busy output becoming free in δt is $\delta t/h$. The probability of the transition $[i+1] \rightarrow [i]$ occurring in δt is therefore $p(i+1)(i+1)\delta t/h$ if $0 \leqslant i \leqslant n-1$ and $p(i+1)n \, \delta t/h$ if $n \leqslant i < n+m$.

By equating the probability of the transition $[i] \rightarrow [i+1]$ to the probability of the transition $[i+1] \rightarrow [i]$, it may be shown that

$$
p(i) = \begin{cases} \dfrac{A^i}{i!}p(0) & 0 \leqslant i \leqslant n \\[2ex] \dfrac{A^i}{n! \, n^{i-n}}p(0) & n < i \leqslant n+m. \end{cases}
$$

Using the fact that $\displaystyle\sum_{i=0}^{n+m} p(i) = 1$, for $0 \leqslant x \leqslant m$,

$$
p(n+x) = \frac{\dfrac{A^{n+x}}{n! \, n^x}}{1 + A + \dfrac{A^2}{2!} + \dots + \dfrac{A^{n-1}}{(n-1)!} + \dfrac{A^n}{n!}\left(1 + \dfrac{A}{n} + \dfrac{A^2}{n^2} + \dots + \dfrac{A^m}{n^m}\right)}.
$$

Putting $x = m$ gives the required lost-call probability $p(n+m)$.

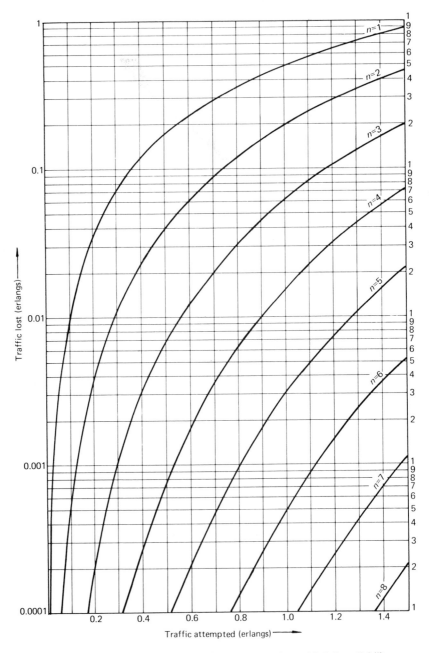

Fig. 4.3 (a) and (b)   Traffic lost in a group sharing *n* outlets with full availability

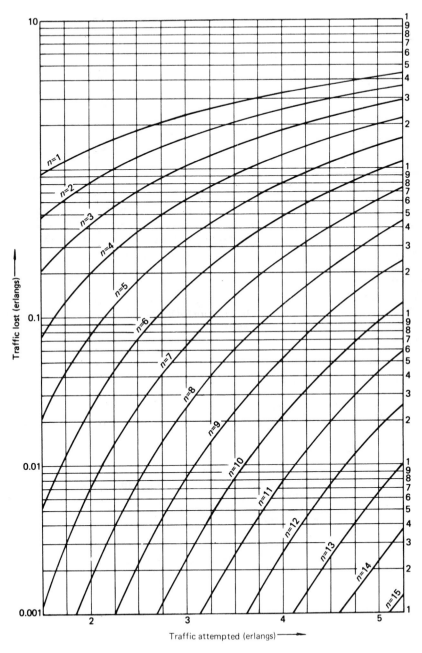

(b)

When there is no storage ($m = 0$) the lost-call probability reduces to

$$\frac{\dfrac{A^n}{n!}}{1 + A + \dfrac{A^2}{2!} + \ldots + \dfrac{A^n}{n!}},$$

which is Erlang's 'lost-calls-cleared' formula. Note that it is assumed that lost calls have zero holding time and that unsuccessful callers do not make repeated attempts. When n is sufficiently large for the value of A involved, the denominator of Erlang's formula is approximately $\exp(A)$.

The graphs of Fig. 4.3 are intended to assist in approximate calculations involving the lost-calls-cleared formula. They display $Ap(n)$, the traffic lost when A erlangs of pure-chance traffic are offered to n outlets.

4.3.3   Distribution of waiting times

The probability $P(t)$ that a caller must wait longer than a time t before being allocated an outlet in the concentrator described in Section 4.3.2 depends on the order in which waiting callers are served (the queue discipline) and we shall assume order-of-arrival service when deriving $P(t)$. We shall also assume that m (the maximum queue length) is infinite and that calls terminate in a pure-chance manner. The latter assumption is justified if the holding times follow a negative-exponential distribution, but may not be justified if they do not.

When there is a queue in existence there are just n calls in progress and therefore if the average holding time is h the average number of terminations occurring in unit time is n/h. Hence the probability that just x terminations occur in a time interval t is (from Exercise 4.2)

$$\left(\frac{nt}{h}\right)^x \frac{1}{x!} \exp\left(-\frac{nt}{h}\right).$$

Hence the probability that a caller finds the concentrator in state $[n]$ and is allocated an outlet before having waited a time t is

$$p(n)\exp\left(-\frac{nt}{h}\right)\left[\frac{nt}{h} + \frac{1}{2!}\left(\frac{nt}{h}\right)^2 + \frac{1}{3!}\left(\frac{nt}{h}\right)^3 + \ldots\right].$$

Similarly the probability that a caller finds the concentrator in state $[n+1]$ and is allocated an outlet before having waited a time t is

$$p(n+1)\exp\left(-\frac{nt}{h}\right)\left[\frac{1}{2!}\left(\frac{nt}{h}\right)^2 + \frac{1}{3!}\left(\frac{nt}{h}\right)^3 + \frac{1}{4!}\left(\frac{nt}{h}\right)^4 + \ldots\right],$$

and similarly for states $[n+2]$, $[n+3]$ etc. Summing these probabilities, noting that

$$p(n+x) = \frac{A^x}{n^x}p(n)$$

(see Section 4.3.2) and regrouping terms leads to the expression

$$p(n)\frac{n}{n-A}\left[1 - \exp\left\{-\frac{t}{h}(n-A)\right\}\right]$$

for the probability that a caller is delayed but allocated an outlet before having waited for a time t. Since the sum of this probability and $P(t)$ must equal the probability of caller being delayed, which is $p(n)n/(n - A)$,

$$P(t) = p(n)\frac{n}{n - A}\exp\left\{ -\frac{t}{h}(n - A) \right\}.$$

By differentiation of $P(t)$, the probability of a caller waiting for a time between t and $t + dt$ before being allocated an outlet is

$$p(n)\frac{n}{h}\exp\left\{ -\frac{t}{h}(n - A) \right\}dt.$$

Hence the average waiting time (averaged over all calls) is

$$\int_0^\infty tp(n)\frac{n}{h}\exp\left\{ -\frac{t}{h}(n - A) \right\}dt = p(n)\left(\frac{n}{n - A}\right)\left(\frac{h}{n - A}\right).$$

This last result may be obtained more directly by noting that a call arriving when the concentrator is in state $[n + x]$ has to wait on average a time xh/n $(x \geqslant 0)$ and writing the average waiting time as

$$p(n)\frac{h}{n} + p(n + 1)\frac{2h}{n} + p(n + 2)\frac{3h}{n} + \ldots.$$

4.4 References

1   Flood, J. E. (Editor), *Telecommunication Networks*, IEE Telecommunications Series 1, Peter Peregrinus Ltd, 1975
2   Atkinson, J., *Telephony*, Vol. 2, Pitman, 1950
3   Smol, G., Hamer, M. P. R., and Hills, M. T., *Telecommunications, a Systems Approach*, George Allen and Unwin Ltd, 1976
4   Marshall, G. J., *Principles of Digital Communications*, McGraw-Hill (UK) Ltd, 1980
5   Rosner, R. D., *Packet Switching*, Lifetime Learning Publications, 1982
6   Neil, W., Spooner, M. J., and Wilson, E. J., 'Experimental packet-switched service: procedures and protocols, Part 1 Packet formats, facilities and switching', *Post Office Electrical Engineers Journal*, **67**, pp. 232–9, January 1975
7   Neil, W., Spooner, M. J., and Wilson, E. J., 'Experimental packet-switched service: procedures and protocols, Part 2 Transmission procedures', *Post Office Electrical Engineers Journal*, **68**, pp 22–8, April 1975
8   Bear, D., *Principles of Telecommunication Traffic Engineering*, IEE Telecommunications Series 2, Peter Peregrinus Ltd, 1976
9   Syski, R., *Introduction to Congestion Theory in Telephone Systems*, Oliver and Boyd, 1960

Exercise topics

4.1   Negative-exponential distribution
4.2   Poisson distribution
4.3   Traffic carried by one outlet
4.4   Exponential approximation to Erlang's formula
4.5   Erlang's formula (use of graphs)
4.6   Grading
4.7   Grading
4.8   Calling rate of unconnected subscriber
4.9   Erlang's formula for few subscribers
4.10   Blocking probability under lost-calls-held conditions
4.11   Digital TASI (time-assigned speech interpolation)
4.12   Lost-call probability with storage
4.13   Probability of delay and average length of queue
4.14   Store-and-forward telegraphy
4.15   Telephony with queue
4.16   TASI with storage
4.17   Random access channel
4.18   Efficiency of aloha and slotted aloha

Exercises

4.1 Negative-exponential distribution

Show that when the probability of the duration of a call exceeding t is $\exp(-kt)$, where k is a constant,
  (i) the mean and variance of the durations are respectively $1/k$ and $1/k^2$,
  (ii) the probability that a call observed to be in progress at $t = 0$ continues to be in progress at time t is $\exp(-kt)$ irrespective of when the call actually started,
  (iii) the probability that a call observed to be in progress at the beginning of a very short time interval δt terminates within that interval is $\delta t/\tau$, where τ is the average call duration.
  Show too that the result (iii) is also true when all calls have equal duration τ.

Solution outline
  (i) By differentiation, the probability density function (p.d.f.) of call duration is $k \exp(-kt)$. Hence the mean duration is

$$\int_0^\infty tk \exp(-kt)\,dt = 1/k,$$

and the variance is

$$\int_0^\infty t^2 k \exp(-kt)\,dt - 1/k^2 = 1/k^2.$$

  (ii) Consider a call starting at $t = t_0$ ($t_0 < 0$). The p.d.f. for its terminating time is $k \exp\{-k(t - t_0)\}$. The probability that it is in progress at $t = 0$ is $\exp(kt_0)$. Denote the p.d.f. for its terminating time when it is known to be in progress at $t = 0$ by $p(t)$ ($p(t) = 0$ for $t < 0$). Hence

$$\exp(kt_0)\, p(t) = k \exp\{-k(t - t_0)\}$$

and therefore

$$p(t) = k \exp\{-kt\},$$

which is independent of t_0. The result follows by integration of $p(t)$.
  (iii) From (ii), the required probability is $p(0)\,dt$, i.e. $k\delta t$. From (i), $k = 1/\tau$.
  When the calls have equal duration τ, the p.d.f. for the terminating time of a call observed to be in progress at a given instant is rectangular and extends over a time interval τ. Hence its value during that interval is $1/\tau$.

4.2 Poisson distribution

When random events occur in a Poisson process, the probability that x events occur in time interval T is

$$\frac{(KT)^x}{x!} \exp(-KT) \qquad \text{(the Poisson distribution)},$$

where K is a constant. Prove that K is the average number of events occurring in unit time.
  Write down an expression for the probability that no event occurs in a time interval t and use it to derive expressions for the mean and variance of the intervals between successive events.

Solution outline
For a unit time interval, $T = 1$. The average number of events occurring in unit time is

therefore

$$\sum_{x=0}^{\infty} x \, \frac{K^x}{x!} \exp(-K) = K.$$

Putting $x = 0$ and $T = t$ in the Poisson distribution, the required probability is $\exp(-Kt)$. This is also the probability that the time interval between two successive events exceeds t. Hence the mean time between successive events is (cf. Exercise 4.1)

$$\int_0^\infty tK \exp(-Kt) \, dt = 1/K$$

and the variance is

$$\int_0^\infty t^2 K \exp(-Kt) \, dt - 1/K^2 = 1/K^2.$$

4.3   Traffic carried by one outlet

A erlangs of pure-chance traffic are offered to a fully available group of outlets. Determine from first principles an expression for the traffic carried by an outlet which carries every call attempted when it is free. Assume a negative-exponential distribution of holding times.

Solution outline
Let the probabilities that the outlet is free and busy be p_0 and p_1 respectively. Let the average holding time be h.

The probability that the outlet becomes busy during a short time interval δt is $p_0 A \, \delta t / h$ and the probability that the outlet becomes free during the interval δt is $p_1 \, \delta t / h$. Equating these probabilities of transition and noting that $p_0 + p_1 = 1$ gives $p_0 = 1/(1 + A)$. The traffic carried by the outlet is $A p_0$, which is therefore $A/(1 + A)$.

4.4   Exponential approximation to Erlang's formula

Traffic from a large number of subscribers is offered to 200 fully available outlets. Estimate the total traffic offered if the lost-call probability is 0.01. Assume lost-calls-cleared conditions and that $200! \simeq 7.89 \times 10^{374}$. Express your estimate in terms of calls attempted per hour if the average holding time is six minutes.

Explain why it may be inaccurate to apply Erlang's formula if the number of subscribers is not large compared with 200.

Solution outline
Assuming the exponential approximation to Erlang's formula (Section 4.3.2), the traffic offered (A) must satisfy

$$\frac{A^{200}}{200! \exp A} = 0.01.$$

Hence $A \simeq 180$ erlangs. If the average holding time is six minutes, this is equivalent to 1800 calls per hour.

The derivation of Erlang's formula assumes that the probability of a call being attempted is independent of the number of calls in progress. Since a subscriber can make only one call at a time, the probability of a call being attempted decreases as the number of calls in progress increases and is therefore a function of the state of the concentrator. The decrease may be significant if the number of subscribers is not large compared with the number of outlets.

4.5 Erlang's formula (use of graphs)

A group of 12 selectors is fully available to a large group of subscribers and the selectors are searched in the same order for every attempted call. Records show that the last selector carries on average 0.12 calls per hour in the busy hour and that the average duration of a call is six minutes. Estimate the traffic offered to the first selector and the lost-call probability during the busy hour.

Solution outline
Search the graphs of Fig. 4.3 for the ordinate on which traffic lost for $n = 11$ is 0.012 erlangs greater than that lost for $n = 12$. This occurs for a total offered traffic of about 4.5 erlangs, for which the traffic lost for $n = 12$ is about 0.007 erlangs. The lost-call probability is therefore about $0.007 \div 4.5 = 0.0015$.

4.6 Grading

Figure 4.4 shows two possible gradings for the connexion of four groups of subscribers to 12 outlets with availability six. Use Erlang's formula to estimate the traffic lost in each case when 2 erlangs of traffic are attempted by each of the four groups.

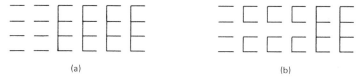

(a)   (b)

Fig. 4.4   Grading diagrams of Exercise 4.6

Solution outline
(a)   Erlang's formula with $n = 2$, $A = 2$ gives 3.2 erlangs for the total traffic offered at the third stage of search (to the first of the shared outlets).

Traffic reaches the first shared outlet only when one of the groups has two calls in progress and is therefore not pure-chance traffic. Hence it is inaccurate to apply Erlang's formula with $n = 4$, $A = 3.2$ to the last four stages. Instead, first use Erlang's formula to determine how much pure-chance traffic offered at the first stage by a single group with full availability would cause 3.2 erlangs to be lost after two stages of search. From Fig. 4.3, this hypothetical offered traffic is about 4.8 erlangs (using the $n = 2$ curve). Finally the 4.8 erlang ordinate crosses the $n = 6$ curve for a lost traffic of about 0.85 erlangs.
(b)   Apply the procedure described in (a). 2.67 erlangs are offered to each outlet at the second stage and this is regarded as due to 3.44 erlangs attempted by each pair of groups. Erlang's formula with $n = 4$, $A = 3.44$ gives 1.74 erlangs offered at the fifth stage of search. This in turn is regarded as due to all the subscribers attempting 4.7 erlangs as a single group. Finally Erlang's formula with $n = 6$, $A = 4.7$ shows that about 0.78 erlangs are lost.

4.7 Grading

300 subscribers, each attempting on average one call every five hours, share 12 outlets with availability four. The average holding time is three minutes. Use Erlang's formula to estimate traffic lost,
 (i)   when the subscribers are grouped into three groups of 100, with four outlets dedicated to each group and full availability within each group, and
(ii)   when the subscribers are grouped into six groups of 50, interconnected as shown in Fig. 4.5.

Fig. 4.5   Grading diagram of Exercise 4.7(ii)

Solution outline

(i)   Each group attempts $100 \times \frac{1}{5} \times \frac{3}{60} = 1$ erlang. Erlang's formula with $n = 4$ and $A = 1$ shows the traffic lost by each group to be $1/65$. Hence total traffic lost is $3/65 = 0.046$ erlangs.

(ii)   These answers are given to greater accuracy than can be readily obtained from Fig. 4.3.

Each group attempts 0.5 erlangs. Therefore 0.333 erlangs are offered to each outlet at the second stage. This is regarded as the traffic lost when 0.768 erlangs of pure-chance traffic have been offered to one outlet. Hence 0.1097 erlangs are lost by each outlet at the second stage.

Thus 0.1645 erlangs are offered to each outlet at the third stage. This is regarded as the traffic lost when 0.916 erlangs have been offered to two outlets in turn. Hence 0.0476 erlangs are lost by each outlet at the third stage.

Thus 0.0952 erlangs are offered to the outlet at the fourth stage. This is regarded as the traffic lost when 1.150 erlangs have been offered to three outlets in turn. The traffic lost by the one outlet at the fourth stage is therefore 0.027 erlangs.

4.8   Calling rate of unconnected subscriber

The calling rate of a subscriber is defined as the probability that he attempts a call in a time equal to the average holding time. Its overall value A is equal to the traffic in erlangs which he attempts. Show that the calling rate of an unconnected subscriber, which is the conditional probability a that he will attempt a call in a time equal to the average holding time if he is free, is given by

$$a = \frac{A}{1 - A(1 - B)},$$

where B is the lost-call probability of the switching centre to which he is connected. Ignore any possible participation in received calls.

Solution outline

The traffic carried from this subscriber is $A(1 - B)$ and hence the probability that he is free is $1 - A(1 - B)$. Hence

$$A = a\{1 - A(1 - B)\}.$$

4.9   Erlang's formula for few subscribers

The derivation of Erlang's formula in Section 4.3.2 assumes that the traffic attempted (A) is independent of the number of calls in progress and is therefore strictly valid only when there is a very large number of subscribers. Modify that derivation to show that when there are M subscribers $(M > n)$ the lost-call probability (in the $m = 0$ case) is

$$\frac{(M-n)\,a^{n+1}\binom{M}{n}}{A\sum\limits_{i=0}^{n}a^{i}\binom{M}{i}},$$

where a is the calling rate of an unconnected subscriber (Exercise 4.8).

Solution outline
Equating transition probabilities,

$$p(i+1)(i+1)\,\delta t/h = p(i)(M-i)\,a\delta t/h.$$

Hence

$$p(i) = \frac{a^{i}}{i!}\,M(M-1)\ldots(M-i+1)\,p(0)$$

and since $\sum\limits_{i=1}^{n} p(i) = 1$,

$$p(n) = \frac{\dfrac{a^{n}}{n!}\,M(M-1)\ldots(M-n+1)}{\sum\limits_{i=0}^{n}\dfrac{a^{i}}{i!}\,M(M-1)\ldots(M-i+1)}.$$

In state $[n]$ the calling rate is $(M-n)a$ and therefore the lost-call probability is

$$\frac{(M-n)a}{A}\,p(n),$$

where A is the overall calling rate. The required result follows.

4.10 Blocking probability under lost-calls-held conditions

Modify the derivation of Erlang's formula (Section 4.3.2) so as to arrive at the 'lost-calls-held' formula for the blocking probability in the case $m = 0$. Under lost-calls-held conditions a call which is blocked is connected for the remainder of its holding time if an outlet becomes available before the expiry of that time.

Use the Poisson distribution (Exercise 4.2) to verify your result for the case of equal holding times.

Solution outline
In this case there is an infinite number of states. It follows that

$$p(i) = \frac{A^{i}}{i}\,p(0) \qquad \text{all } i > 0$$

and $\quad p(0) = \exp(-A)$.

The blocking probability is $\sum\limits_{i=n}^{\infty} p(i)$, which is therefore

$$\exp(-A)\sum\limits_{i=n}^{\infty}\frac{A^{i}}{i!}.$$

When all calls have holding time h, the blocking probability is the probability that n or more calls are attempted in a time interval h. From the Poisson distribution formula it is equal to

$$\sum_{i=n}^{\infty} \frac{(Kh)^i}{i!} \exp(-Kh),$$

where K is the average number of calls attempted in unit time. From the definition of attempted traffic, $Kh = A$.

4.11 Digital TASI (time-assigned speech interpolation)

Two hundred digitally encoded speech waveforms are to be transmitted over eight 51.2 Mbit/s transmission channels in the following way.

The digits representing each speech waveform are compressed and packaged into 10 μs bursts of 512 binary digits, each burst representing on average 1 ms of speech waveform. The bursts arrive at the input of the eight channels in a pure-chance manner. If there is a free channel when a burst arrives, it is transmitted intact. Since there is no storage at the channel inputs, a burst arriving when there is no free channel is lost, unless a channel becomes available before the end of such a burst, in which case it is allocated to that burst for the remainder of its duration.

Calculate the probability that a burst is not transmitted intact.

Solution outline
The system operates under the lost-calls-held conditions assumed in Exercise 4.10. Regarding each packet as a 'call', there are in effect 200 'subscribers', each attempting 10^3 calls per second, and the holding time is 10^{-5} s. Hence the attempted traffic is

$$200 \times 10^3 \times 10^{-5} = 2 \text{ erlangs.}$$

Therefore the blocking probability is

$$\exp(-2) \sum_{i=8}^{\infty} \frac{2^i}{i!} = 0.0011.$$

4.12 Lost-call probability with storage

Packets of data, each containing 1024 bits, arrive at random at an average rate of five per second for transmission over four 2400 bit/s trunks. Calculate the lost-call probability if there is storage for two packets.

Solution outline
The duration of each transmitted packet is $1024/2400 = 0.4267$ s. Hence the traffic attempted is $5 \times 0.4267 = 2.133$ erlangs.

In the formula for $p(n + x)$ in Section 4.3.2, put $x = m = 2$, $n = 4$ and $A = 2.133$, giving a lost-call probability of about 0.029.

4.13 Probability of delay and average length of queue

Obtain an expression for the probability P_D that a call is delayed but not lost when A erlangs of pure-chance traffic are offered to a concentrator with n outlets and facilities for a queue of m callers ($n > A$).

Show that for very large m the average length of queue is

$$P_D \frac{A}{n-A}.$$

Solution outline

With the notation of Section 4.3.2,

$$P_D = \sum_{i=n}^{n+m-1} p(i) = \frac{\dfrac{A^n}{n!} \sum\limits_{i=0}^{m-1} \left(\dfrac{A}{n}\right)^i}{\sum\limits_{i=0}^{n-1} \dfrac{A^i}{i!} + \dfrac{A^n}{n!} \sum\limits_{i=0}^{m} \left(\dfrac{A}{n}\right)^i}.$$

For very large m, the average length of queue is

$$\sum_{i=1}^{\infty} ip(n+i) = p(n)\frac{A}{n}\left\{1 + 2\frac{A}{n} + 3\left(\frac{A}{n}\right)^2 + \ldots\right\}$$

$$= p(n)\frac{(A/n)}{\{1-(A/n)\}^2}$$

$$= P_D \frac{A}{n-A} \qquad \left(\text{since } P_D = \frac{p(n)}{1-(A/n)}\right).$$

4.14 Store-and-forward telegraphy

Telegraph messages with lengths following a negative-exponential distribution are stored on punched tape and handed to an operator in a pure-chance manner at an average rate of one every two minutes for transmission one at a time over a single channel operating at 10 characters per second. The average length of the messages is 1000 characters and the operator transmits them in the order in which they are handed to him. Calculate
(i) the probability that a message is delayed,
(ii) the probability that the operator has 35 or more messages awaiting transmission,
(iii) the probability that a message is delayed by more than one hour,
(iv) the average number of messages awaiting transmission.

Solution outline
(i) Since no traffic is lost, the traffic carried is equal to the attempted traffic, which is $5/6$ erlangs. Hence the probability that the link is occupied and therefore that a message is delayed is $5/6$. This can be confirmed by putting $A = 5/6$ and $n = 1$ and letting $m \to \infty$ in the expression for P_D derived in Exercise 4.13.
(ii) The probability $p(1 + x)$ that the operator has just x messages waiting may be shown by putting $n = 1$ and letting $m \to \infty$ in the analysis of Section 4.3.2 to be given by

$$p(1 + x) = (1 - A)A^{x+1}.$$

The required probability is $\sum\limits_{x=35}^{\infty} p(1 + x)$, which is A^{36} and hence approximately 0.0014 when $A = 5/6$.

(iii) The required probability is $P(t)$ of Section 4.3.3 evaluated for $t/h = 36$. In this case $n = 1$, $A = 5/6$ and $p(n) = A(1 - A)$. Hence the required value is 0.0021.
(iv) Using (i) and the expression derived in Exercise 4.13, the average length of queue is $25/6$ messages.

4.15   Telephony with queue

A large group of telephone subscribers offers 5 erlangs to 10 outlets with full availability. Callers finding all outlets busy form a queue and are allocated outlets on a first-come-first-served basis as calls terminate. Holding times have a negative-exponential distribution with average value three minutes. Calculate
(i)   the probability that a caller has to wait,
(ii) the average waiting time.

Solution outline
(i)   With the notation of Section 4.3.2, the required probability is

$$\sum_{x=0}^{\infty} p(n+x) = p(n)\frac{n}{n-A}$$

evaluated for $n = 10$ and $A = 5$ and letting $m \to \infty$. The result is approximately 0.036.
(ii) Evaluating the formula in Section 4.3.3 for $n = 10$, $A = 5$ and $h = 180$ gives for the average waiting time

$$p(n)\left(\frac{n}{n-A}\right)\left(\frac{h}{n-A}\right) = 0.036 \times \frac{180}{5} = 1.3 \text{ s.}$$

4.16   TASI with storage

Speech signals travelling in the same direction in 32 simultaneous telephone conversations share 16 channels, each burst of speech seizing a free channel if one is available. Bursts of speech occurring when all 16 channels are occupied are stored and transmitted when a channel becomes free on a first-come-first-served basis. Calculate the probability that a burst is delayed by more than 400 ms, assuming that bursts occur in a pure-chance manner at an average rate of one every 5 s for each speech signal and that the lengths of the bursts have a negative-exponential distribution with mean 2 s.

Solution outline
Evaluate $P(t)$ from the formula in Section 4.3.3 for the values $t = 0.4$, $n = 16$, $h = 2$, $A = 32 \times \frac{1}{5} \times 2 = 12.8$. $p(n)$ is approximately 0.061 (Section 4.3.2) and hence $P(0.4)$ is approximately 0.16.

4.17   Random access channel

In a maritime satellite communication system, a ship wishing to transmit a message first requests a channel by transmitting over a common random-access calling channel a request signal of 20 ms duration. Assuming that there are 100 ships, each making on average one request every three minutes, calculate the probability that the calling channel is free when the request signal begins and remains free for the duration of that request signal.

Solution outline
Assume that requests occur in a Poisson process. The probability that no request signal begins in a period of duration Ts is $\exp(-KT)$ (Exercise 4.2), which in this case $K = 100/180$.
   The required probability is the probability that no other request signal begins in a 40 ms period centred on the beginning of a request signal and is therefore

$$\exp\{-(100/180)(0.04)\} = 0.98.$$

4.18 Efficiency of aloha and slotted aloha

(a) **Aloha random access**[5] Packets of data of duration τ are transmitted at random times by n terminals over a shared single-channel radio net. Packets involved in collisions with other packets are re-transmitted after a random delay. Show that if the terminals transmit a total traffic of G erlangs, the traffic S carried by the net without suffering collision is given by

$$S = G \exp(-2G).$$

Hence show that the efficiency of the net cannot exceed 18.4%.

(b) **Slotted aloha**[5] In this case the aloha system in (a) is modified so that the shared channel is divided into time slots of duration τ and the terminals are synchronised so that all transmitted packets just fit into time slots. Show that if each user transmits G/n erlangs,

$$S = G\left(1 - \frac{G}{n}\right)^{n-1}.$$

Hence show that when n is large the efficiency of the net cannot exceed 36.8%.

Solution outline

Note that the proportion of transmitted packets not involved in a collision is S/G and that on average G/τ packets are transmitted per second.

(a) Arguing as in Exercise 4.17,

$$\frac{S}{G} = \exp\left\{-\left(\frac{G}{\tau}\right)(2\tau)\right\}.$$

Differentiation shows that S has a maximum value of $1/2e$ when $G = 0.5$.

(b) In a given time slot the probability that a given terminal is transmitting is G/n. Hence the probability that a packet is not involved in a collision is $\{1 - (G/n)\}^{n-1}$.

As $n \to \infty$, $S \to G \exp(-G)$ and differentiation shows that S has a maximum value of $1/e$ when $G = 1$.

5
Carrier telephony

5.1 The carrier-telephony signal

5.1.1 Single-sideband frequency-division multiplex (SSB/FDM)

Carrier telephony is the simultaneous transmission of a number of speech signals over a common transmission link as modulation of carriers at different frequencies. It is a widely used method of making efficient use of long-distance cables. The number of speech signals sharing one pair of conductors has increased from three in open-wire systems prior to 1920[1] to 10800 in present-day coaxial-cable systems[2]. Since the simultaneous transmission is achieved by division of the frequency band available in the common link, carrier telephony is an example of frequency-division multiplex (FDM).

The method of modulation generally used is single-sideband (SSB) suppressed-carrier amplitude modulation. It makes efficient use of available bandwidth and transmitter power and for telephony it may be satisfactorily demodulated by product demodulation without full synchronisation of the locally generated carrier.

It is recommended[3] that in a carrier-telephony system the voice frequencies transmitted by a telephone channel should extend from 300 Hz to 3400 Hz and therefore the bandwidth of each SSB speech signal need not exceed 3.1 KHz. However, due to practical limitations to the filters used in combining and separating the SSB signals, it is usual to allocate 4 kHz to each speech channel and to locate the carrier frequencies at multiples of 4 kHz. In view of the high cost of submarine cables, an alternative carrier spacing of 3 kHz has been agreed for use in submarine cable systems.

Incomplete sideband suppression gives rise to unintelligible but annoying adjacent-channel crosstalk between channels occupying adjacent frequency bands (Section 1.1.5).

5.1.2 Signal levels and test-tone

For some purposes it is more helpful to know the power level of a transmitted signal relative to the power level that a channel was designed to handle than to know the absolute power level of that signal. Hence it is convenient to denote the levels of signals in decibels relative to a standard test-tone assumed to pass through the channel.

The standard test-tone is defined as follows. A point in the channel is chosen to be the point of zero relative level and the standard test-tone is defined to be a 1 kHz tone at an absolute level of 0 dBm (1 mW) when measured at the point of zero relative level.

In principle any point in a channel could be chosen to be the point of zero relative level, but it is usual to choose that point so that the levels of the test-tone and the speech signals have the approximate relationship described in Section 1.1.4. In fact it is conventional[4] to choose the point of zero relative level to enable the system to be set up so that when averaged over the busy hour the speech power in each channel rarely exceeds a level 15 dB below the standard test-tone (Section 5.2.3). In this way the standard test-tone has some physical significance as approximately that tone which a speech channel must handle satisfactorily if it is to handle the speech waveform. If there is no point in the channel at which the appropriate test-tone has an absolute level of 0 dBm, a virtual point of zero relative level may be established.

Many abbreviations are used to denote levels in carrier telephony. Three of the most common are defined below.

(i) **dBm0** This denotes a signal level in dBm measured at the point of zero relative level. It may be used for signals not actually present at that point by allowing for the gain or loss of the channel between that point and some point at which they do exist. Provided that a signal is subject to the same gains and losses as the test-tone as it progresses through the channel, the level of that signal relative to the test-tone is equal to its level in dBm0. Using this notation, one may say that it is conventional to set up a system so that the average power in a channel in the busy hour is $-15\,\mathrm{dBm0}$.

(ii) **dBm0p** This denotes a noise power measured on a psophometer at the point of zero relative level.

(iii) **dBr** This is used to denote the transmission level at specified points in a channel rather than the level of a specific signal. It denotes the level of a signal at the specified point relative to the level of that signal at the point of zero relative level. Thus for example at a $-20\,\mathrm{dBr}$ point the absolute level of the test-tone is $-20\,\mathrm{dBm}$ and the average busy-hour speech power is $-35\,\mathrm{dBm}$.

5.1.3 Signal statistics

The composite SSB/FDM signal is an information-bearing signal. Therefore it is essentially unpredictable and can be described only in statistical terms.

In describing how an SSB/FDM signal voltage varies with time, it is helpful to begin by distinguishing between fast and slow variations. The fast variations are due to rapid changes of voltage in the component SSB signals and occur at frequencies comparable with the carrier frequencies used. The slow variations are due to changes in (i) the number of channels which are active and (ii) speaker volumes. An analogous distinction between fast and slow changes is made when describing fading in radio systems (Section 7.3.1). We shall first discuss the statistics of the fast variations, assuming that the number of active channels and the speaker volumes remain constant. Then we shall discuss the slow variations, regarding them as gradual changes in the short-term statistical parameters describing the fast variations. Our approach is essentially that of Holbrook and Dixon[5].

It is generally accepted that the voltage obtained by adding n constant-volume SSB speech signal voltages may be regarded as gaussian-distributed with zero mean provided that n is not less than about 60. For $n < 60$, the probability density function is more 'peaky' than the gaussian distribution having the same variance[5]. When considering fast variations we shall assume that $n > 60$ and use the gaussian distribution to calculate the probability that the instantaneous SSB/FDM signal voltage exceeds any value during an interval for which n and the speaker volumes may be regarded as constant.

A one-way telephone channel is said to be 'active' when it is actually carrying speech. It is considered to remain active during the short pauses between the words and syllables of ordinary connected speech, but not during longer pauses such as periods of listening. A channel is, however, regarded as 'busy' and unavailable for use by other callers for the whole duration of a call, including setting-up and clearing times. Hence a channel is not continuously active throughout the time for which it is busy and it is generally accepted that on average each one-way channel is active for 25 % of the time during the busy hour. Thus if the contributions of dialling and other signalling are disregarded, the speech power averaged over the busy hour (Section 5.1.2) is approximately 6 dB below the mean continuous speech power (Section 1.1.4) obtained by averaging over active periods only. The probability $p(r)$ that exactly r channels of an n-channel system are simultaneously active in the busy hour is therefore given by

$$p(r) = \frac{n!}{r!(n-r)!}(0.25)^r(0.75)^{n-r} \tag{5.1}$$

(the binomial distribution). Calculations based on Eq. 5.1 show that increasing n decreases the probability that the proportion of the channels which are simultaneously active has deviated from 0.25 by a given amount.

If it is assumed that speaker volumes are distributed log-normally throughout the population (Section 1.1.4), the average power in r active channels is distributed as the sum of r log-normally distributed variables. An approximation to the behaviour of that sum may be obtained by assuming that it too is log-normally distributed (Exercise 5.2).

5.2   Signal degradations

5.2.1   Overload requirements

Whenever a peak of the SSB/FDM signal is clipped by the overloading of an amplifier, a short burst of interference is introduced into a large number of channels. The maximum acceptable frequency for the occurrence of such bursts is ultimately a subjective matter, but a quantitative approach has been described by Holbrook and Dixon[5]. They consider first the fast variations of signal voltage and draw the empirical conclusion that the interference is acceptable if the probability of overload due to fast variations is less than 0.1 %. They define 'multi-channel peak factor' as the ratio of the instantaneous signal voltage at the threshold of overload to the root-mean-square signal voltage and show that that ratio is close to 13 dB if the number of active channels exceeds about 300. They then consider the slow variations of signal voltage and suggest that the probability of overload due to fast variations should not exceed 0.1 % for more than 1 % of the busy hour. This leads to a suggested value for the root-mean-square voltage of a sinusoidal test signal which the amplifier should handle without overloading. With small modifications the conclusions of Holbrook and Dixon have been incorporated in CCITT recommendations[4] and recommended values for the power of that sinusoidal test signal are listed as 'equivalent peak power' in Table 5.1.

In Table 5.1 the mean power in the SSB/FDM signal is assumed to be the conventional loading $(-15 + 10 \log n)$ dBm0 (Section 5.1.2) and overloading is deemed to have occurred when an increase of 1 dB in the sinusoidal test-tone at the amplifier input causes an increase of 20 dB or more in the absolute level of the third harmonic at the output[4]. Some allowance has been made for the possibility that clips may occur at different times in other amplifiers in the system, but not for pre-emphasis (Section 5.3.3). It may be seen from Table 5.1 that the equivalent peak power is approximately $(10 \log n - 2.5)$ dBm0 for large n.

Table 5.1   Equivalent peak power [based on CCITT]

Number of channels in system (n)	12	24	36	48	60	120	300	600	960	1800	2700
Equivalent peak power (dBm0)	19	19.5	20	20.5	20.8	21.2	23	25	27	30	32

5.2.2   Specification of noise

As in all communication channels, fluctuation noise is added to signals passing through a carrier-telephony link. We shall follow the usual practice in carrier-telephony literature and refer to fluctuation noise of all kinds as 'thermal noise'. Contributions to thermal noise from different parts of the link are usually uncorrelated and can be combined at the output of the link by power addition.

Departures from linearity in a carrier-telephony link result in interaction between individual speech signals to produce intermodulation interference within the frequency bands of other speech signals. When there are many active channels, the intermodulation interference in one channel is the sum of many contributions and therefore has a gaussian distribution and other noise-like properties. In such a case its subjective effect is similar to that of fluctuation noise and it is usually called, loosely, 'intermodulation noise'.

Recommended design objectives for the thermal and intermodulation noise introduced by a carrier-telephony system are described by the CCITT[6] in terms of the total noise in a speech channel at the output of an imaginary channel called a 'hypothetical reference circuit'. The hypothetical reference circuit is 2500 km in length and contains a specified number of frequency translations of specified kinds. It is typical of an international circuit in an area such as Europe. The recommendation is broadly that the total noise in any speech channel should not exceed 10 000 pWp (Section 1.1.3) when measured at a point of zero relative level in a hypothetical reference circuit. It is also recommended that in coaxial-cable systems 7500 pWp (3 pWp/km) should be allocated to line equipment (repeaters) and 2500 pWp to modulation and translating equipment.

The choice of transmitted signal power in a carrier-telephony link is a compromise between using high power for little degradation by thermal noise and low power for little degradation by intermodulation noise. When the dominant intermodulation noise is third-order (i.e. due to cubic non-linearities), the signal-to-noise ratio is maximised by choosing the signal power so that the thermal noise power is twice the intermodulation noise power (Exercise 5.4). Hence it is common practice to allow 1 pWp/km for intermodulation noise and 2 pWp/km for thermal noise when designing line equipment.

5.2.3   White-noise testing

Verification that a carrier-telephony system meets the noise objectives outlined in Section 5.2.2 requires measurements to be made under busy-hour conditions. An accepted method of simulating busy-hour conditions when a system is not actually in service is to substitute for the SSB/FDM signal the output of a wide-band noise generator, using appropriate filters to confine its spectrum to the frequency band of the system under test.

When there is a large number of channels in the system, the power required for white-noise loading is $(10 \log n - 15)$ dBm0, where n is the number of channels (Section 5.1.2). This is the CCITT conventional loading level, recommended[4] as appropriate for white-noise loading and based on experience with many systems. It approximates to the level exceeded by the variance of the rapid variations for less than one per cent of the busy hour (Section 4.3.1). However, when there are not many channels in the system ($n < 240$), two considerations make it necessary for white-noise loading to exceed the average traffic power: firstly fluctuations in activity about the average value of 25 % become significant (Section 5.1.3); and secondly the probability distribution of the SSB/FDM signal departs significantly from the gaussian. Taking these two considerations into account, the recommended[4] white-noise loading when $12 \leqslant n < 240$ is $(4 \log n - 1)$ dBm0. If a significant proportion of the channels are carrying data, a noise power greater than the conventional loading may be necessary[7].

A typical white-noise test set is shown in simplified form in Fig. 5.1. The band-elimination filter may be switched in to remove the noise loading from a narrow frequency band. The pass-band of the band-pass filter in the receiver lies within the stop-band of the band-elimination filter, so that when the latter is in circuit the receiver measures only noise generated in the system under test. That noise consists of intermodulation interference,

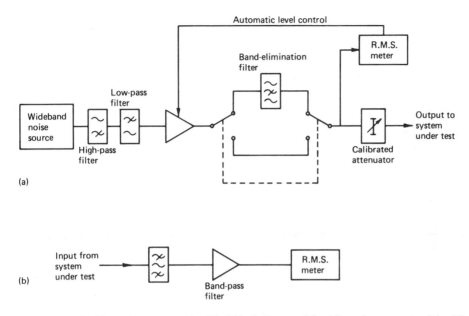

Fig. 5.1 Typical white-noise test set: simplified block diagram (a) white-noise generator (b) white noise receiver

generated by the noise loading, and of thermal noise. If the noise generator is switched off, the receiver measures thermal noise alone. The quality of the system under test is sometimes described by its 'noise-power ratio', which is the ratio of the receiver reading with the band-elimination filter switched out to that with it switched in.

Although Fig. 5.1 shows only one band-elimination filter and one band-pass filter, a white-noise test set usually contains ranges of each so that noise-power ratio can be measured at several frequencies. The narrower the stop-band of each band-elimination filter, the less is the loading disturbed by the insertion of that filter and the more accurate therefore are the results. To ensure inter-operability between transmitters and receivers in different countries, the principal features of filters suitable for most standard systems are specified by the CCITT or Intelsat.

5.3 Line transmission

5.3.1 Cables

Although early carrier-telephony systems operated over open-wire lines or underground twisted pairs, most modern systems operate over coaxial pairs. The outer conductor of the coaxial pair is often in the form of a continuous copper screen to keep crosstalk low; the dielectric is often air with polyethylene discs at intervals to locate the inner conductor, and sometimes the outer has a polyethylene lining to prevent short-circuits if the cable is crushed.

Coaxial pairs of two sizes are commonly used on land. The larger has inner-conductor diameter 2.6 mm, outer-conductor internal diameter 9.5 mm and attenuation 2.32 dB/km at 1 MHz and 290 K. Corresponding figures for the smaller are 1.2 mm, 4.4 mm and 5.3 dB/km. The attenuation in dB/km is approximately proportional to the square root of frequency.

Coaxial cables for use on land usually consist of several coaxial pairs. Submarine cables usually consist of a single coaxial pair in which the inner contains steel wire for tensile strength.

In order to prevent the signal from becoming so far attenuated that it is unacceptably degraded by thermal noise, a long carrier-telephony link is divided into sections. Each section consists of a length of cable and a repeater amplifier designed to compensate as accurately as possible for the attenuation of that length of cable. When a transmission link is required to cross water or difficult terrain it may be advantageous to replace some or all of its sections by radio links (Section 5.4).

5.3.2   Intermodulation interference

Assume that the instantaneous output voltage v_2 of a repeater amplifier may be expressed in terms of the instantaneous input voltage v_1 by the equation

$$v_2 = a_1 v_1 + a_2 v_1^2 + a_3 v_1^3, \tag{5.2}$$

where a_1, a_2, a_3 are constants. When $v_1 = A \cos \alpha t + B \cos \beta t$ the contribution of the v_1^2 term to v_2 is

$$a_2 [\tfrac{1}{2} A^2 (1 + \cos 2\alpha t) + \tfrac{1}{2} B^2 (1 + \cos 2\beta t) + AB \cos (\alpha - \beta) t + AB \cos (\alpha + \beta) t].$$

Hence the v_1^2 term contributes intermodulation products at the second harmonics of the input frequencies and at frequencies which are the sum or the difference of two input frequencies. These contributions constitute second-order intermodulation interference. Note that if $A = B$ and variations with frequency are neglected, intermodulation products at sum or difference frequencies are 6 dB higher in level than those at second-harmonic frequencies. If the n speech signals carried by the repeater amplifier are assumed to be equivalent to n sinewaves at frequencies af_0, $(a + 1)f_0$, $(a + 2)f_0$, ... , bf_0 (where a, b are integers and $b - a = n - 1$), the number of second-order intermodulation products falling in the channel at rf_0 is shown approximately by Fig. 5.2. Clearly when n is large the second harmonic contribution is negligible compared with the $(\alpha + \beta)$ and $(\alpha - \beta)$ contributions.

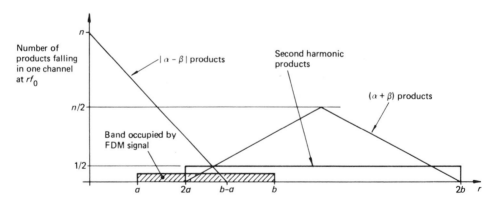

Fig. 5.2   Distribution of second-order intermodulation products

Some indication of the way in which second-order intermodulation interference accumulates from section to section may be obtained from the following argument. Assume that each section of a link introduces a phase-shift $\phi_0 + k\omega$ at angular frequency ω. [Each section thus has uniform envelope delay: see Appendix 2.] Suppose that at the output of one section tones $\cos \alpha t$ and $\cos \beta t$ are accompanied by the intermodulation product $\cos (\alpha - \beta)t$ generated within that section. At the output of the next section these three tones have become $\cos(\alpha t + \phi_0 + \alpha k)$, $\cos(\beta t + \phi_0 + \beta k)$ and $\cos\{(\alpha - \beta)t + \phi_0 + (\alpha - \beta)k\}$ and are accompanied by the $(\alpha - \beta)$ intermodulation product $\cos\{(\alpha - \beta)t + (\alpha - \beta)k\}$ generated within that section. If ϕ_0 is zero or a multiple of 2π, contributions from successive repeaters to the $(\alpha - \beta)$ intermodulation product will be in phase and their voltages should be added. For other values of ϕ_0 the $(\alpha - \beta)$ products will accumulate in an oscillatory manner, their resultant always remaining less than when there is voltage addition. A similar argument may be used for the accumulation of $(\alpha + \beta)$ products.

Consider now the third-order intermodulation products contributed by the v_1^3 term of Eq. 5.2 when $v_1 = A \cos \alpha t + B \cos \beta t + C \cos \gamma t$. Expansion of the term as the sum of sinewaves (Exercise 5.7) shows that there are intermodulation products at angular frequencies such as 3α, $|2\alpha \pm \beta|$ and $|\alpha \pm \beta \pm \gamma|$. This distribution of third-order products with frequency may be derived from the distribution of second-order products (Exercise 5.6). It may also be shown (Exercise 5.5) that products of the form $|\alpha + \beta - \gamma|$ may be expected to accumulate by voltage addition in successive sections, so that at the end of a long-link, third-order intermodulation interference is dominated by products of that form.

5.3.3 Repeater design

The basic requirement of a repeater in a coaxial-cable carrier-telephony system is to compensate for the attenuation of the preceding length of cable. The compensation must be very accurate, since otherwise errors in signal level could accumulate sufficiently to give rise to excessive noise in a long link.

A block diagram of a typical repeater for an underground coaxial-cable system is shown in Fig. 5.3. The negative feedback necessary to ensure sufficient linearity is applied by way of Equaliser A, the attenuation characteristic of which is similar to that of the length of cable preceding the repeater. Equaliser B ensures accurate compensation under specified operating conditions. Equaliser B effectively adds attenuation to the cable and to avoid unnecessary degradation of the system its attenuation should be as low as possible.

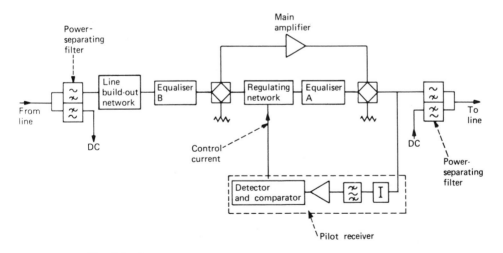

Fig. 5.3 Block diagram of typical cable repeater for carrier telephony

The pilot receiver and regulating network provide automatic adjustment of repeater gain to compensate for changes in cable attenuation. Typically each 1 dB of attenuation is increased by 0.0021 dB for each 1°C increase in cable temperature. A pilot-tone, usually at a frequency just above the highest channel frequency, accompanies the SSB/FDM signal and the gain of the repeater is regulated to keep the level of that pilot-tone constant at the repeater output. The regulating network is designed to give the correct relative changes of gain at other frequencies.

The hybrid transformers isolate the regulating network and Equaliser A from the line. The hybrid transformer at the input is asymmetric to avoid unnecessary degradation of noise factor, and that at the output is asymmetric for the efficient use of amplifier output power.

The line build-out network enables some adjustment of the length of section so that difficult repeater locations can be avoided.

It is common practice for the direct-current supply to active networks in repeaters to be carried by the inner of the coaxial cable. Power-separating filters enable the direct current to be routed as required independently of the signal.

Design considerations for submarine repeaters differ considerably from those for underground repeaters. The relatively small temperature variations ($\pm 0.5°C$) on the ocean floor make regulation easier, but the inaccessibility of repeaters necessitates high reliability, high supply voltages and allowances for increases in section length due to repairs.

Most submarine cable systems use equivalent four-wire operation, in which traffic is carried in both directions simultaneously in one coaxial pair by using separate frequency bands. Two suitable repeater configurations are shown in simplified form in Fig. 5.4. The Cantat II transatlantic system[8] provides 1840 two-way channels and uses two-amplifier

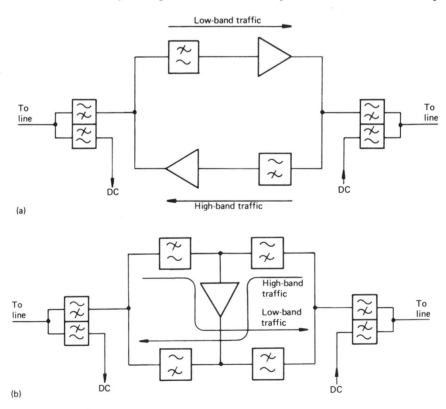

Fig. 5.4 Repeater configurations for equivalent four-wire systems (a) two-amplifier configuration (b) single-amplifier configuration

configuration (a); the Bell SG system[9] provides about 4000 two-way channels and uses single-amplifier configuration (b). Configuration (b) relies on fewer amplifiers and is therefore claimed to be the more reliable.

It does not usually make most efficient use of the output power available from a repeater to distribute that power equally among the channels. If it were distributed equally, the much lower cable attenuation at low frequencies would result in low-frequency channels being much less affected than high-frequency channels by thermal noise. The allocation of more power to high-frequency channels and less to low-frequency channels can ease the problem of meeting thermal-noise requirements at high frequencies without adding significant thermal noise at low frequencies. Hence systems are usually set up to have pre-emphasis, that is, the transmission level at repeater outputs increases with increasing frequency. To assist inter-operability the pre-emphasis characteristic is agreed internationally. In view of the contribution from high-frequency channels to intermodulation interference in low-frequency channels, much less pre-emphasis can be used than would be expected from a consideration of thermal noise alone.

5.3.4 FDM hierarchies

SSB/FDM signals are assembled in the transmitting terminal by forming blocks of several channels, combining several blocks by frequency translation into larger blocks and so on. The numbers of channels and the frequencies used in the basic building blocks have been agreed internationally[10] and are shown in Fig. 5.5.

Three methods of assembling the basic group of 12 channels are shown in Fig. 5.6. The single-modulation method (a) is the most straightforward, but requires the most expensive filters. Suppression of the unwanted sideband is particularly important in reducing adjacent-channel crosstalk (Sections 1.1.5 and 5.1.1). The presence of residual carrier tones in the basic group may give rise to intelligible crosstalk and a maximum level is specified[11].

Submarine cable systems use a basic group in which 16 channels occupy a 48 kHz bandwidth[12]. The closer spacing leads to increased distortion due to filters at the terminals, but the distortion is usually acceptable in routes including only one submarine cable link.

Once assembled, a basic group is likely to pass through several links and undergo several frequency translations before its component channels are separated. In order that its level may be monitored, and in some cases regulated, as it passes through the system, each basic group is accompanied by a tone referred to as a 'group pilot'. In most systems 84.080 kHz has been used, but when a basic group is used for data transmission (Section 5.5.3) 104.080 kHz is used instead. The recommended[13] frequency accuracy for a group pilot is ± 1 Hz and its transmission level is -20 dBm0. There are also supergroup pilots (411.920 kHz or 547.920 kHz), mastergroup pilots (1552 kHz) and supermastergroup pilots (11 096 kHz).

Any inaccuracy in the frequency of a carrier used in frequency translation will result in an audio signal acquiring a frequency off-set on passing through the system. It is recommended[14] that in a hypothetical reference circuit the frequency off-set should not exceed 2 Hz and with that object in view the following accuracies are recommended for the carriers used in a 60 MHz system:

channel translation (formation of group)	$\pm 10^{-6}$
group translation (formation of supergroup)	$\pm 10^{-7}$
supergroup translation (e.g. formation of mastergroup)	$\pm 10^{-7}$
mastergroup and supermastergroup translation	$\pm 10^{-8}$

These recommendations are based on the possibility of parts of the system carrying telegraphy or music and are unnecessarily stringent for telephony.

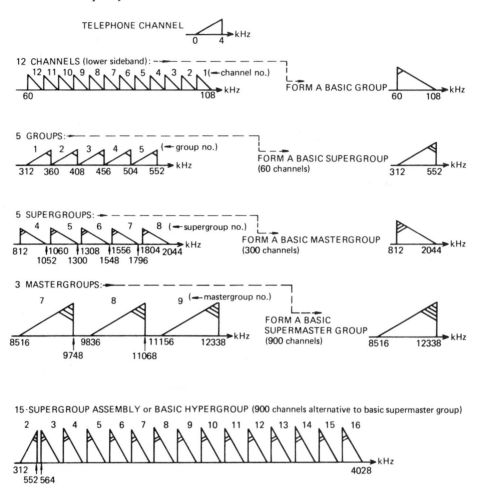

Fig. 5.5   Basic FDM assemblies (in all diagrams the sloping line rises in the direction of increasing voice frequency)

5.4   Radio transmission

5.4.1   Use of frequency modulation (FM)

Microwave frequencies are usually used for the transmission of carrier-telephony signals by radio, partly because of the large bandwidths associated with large numbers of channels, and partly because the possibility of narrow antenna beams enables efficient use of the spectrum.

In most cases, non-linearities associated with microwave devices would introduce

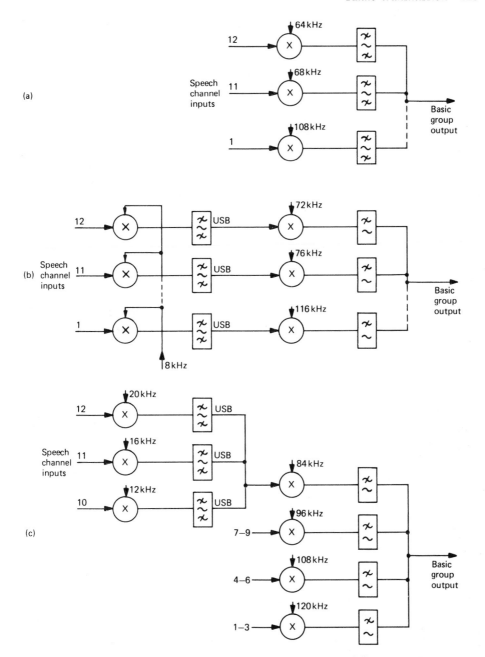

Fig. 5.6 Typical methods of assembling a basic FDM group (channel-translating equipment) (a) single modulation (b) double modulation (c) pre-groups

unacceptable intermodulation interference in an amplitude-modulation system. Frequency modulation, however, is unaffected by transmission non-linearities (Exercise 5.11) and is therefore almost universally used for the transmission of carrier telephony by radio. A wide-band frequency-modulation transmission has the additional advantage of being less affected by noise than is an amplitude-modulation transmission of the same power (Appendix 7).

On the other hand, a frequency-modulation transmission can be severely impaired by transmission deviations, that is, variation with frequency of the gain or the delay of the transmission path. Sources of transmission deviations are filters, microwave amplifiers, echoes in mismatched feeders, multipath propagation and the effect of amplitude variations on the output of FM demodulations. The effect of transmission deviations is to introduce intermodulation interference (Section 5.4.5).

A system in which an SSB/FDM carrier-telephony signal is transmitted by frequency modulation is described as an 'SSB/FDM/FM' system and the modulating SSB/FDM signal is referred to as the 'baseband' signal. The approximately gaussian nature of the baseband signal enables several important properties of the SSB/FDM/FM signal to be derived and some of those properties are discussed below.

5.4.2   SSB/FDM/FM signal spectrum

Consider the angle-modulated wave:

$$v(t) = \cos\{\omega_c t + \phi(t)\}.$$

Assume it to be frequency-modulated by a gaussian baseband signal with a uniform power spectrum extending from f_1 Hz to f_2 Hz. If the root-mean-square frequency deviation is Δf Hz, the 'power' spectrum of the instantaneous frequency deviation is $(\Delta f)^2/(f_2-f_1)$ (Hz)^2/Hz. Hence the power spectrum S_ϕ of the instantaneous phase deviation $\phi(t)$ is given by

$$S_\phi = \frac{(\Delta f)^2}{f^2(f_2-f_1)} \text{ (rad)}^2/\text{Hz},$$

since integration in the time domain is equivalent to division by $2\pi f$ in the frequency domain. Integration of S_ϕ from f_1 to f_2 shows that the mean-square phase deviation is $(\Delta f)^2/f_1 f_2$ (rad)^2.

The power spectrum $G_v(f)$ of $v(t)$ can be found by finding the Fourier transform of $R_v(\tau)$, the autocorrelation of $v(t)$. In this case $\phi(t)$ has a gaussian distribution and hence it can be shown[15, 16] that

$$R_v(\tau) = \tfrac{1}{2}\exp\{-R_\phi(0)+R_\phi(\tau)\}\cos\omega_c\tau,$$

where $R_\phi(\tau)$ is the autocorrelation of $\phi(\tau)$. Hence

$$G_v(f) = \tfrac{1}{2}[G_L(f-f_c)+G_L(f+f_c)],$$

where $G_L(f)$ is the equivalent low-pass power spectrum of $v(t)$ and is the Fourier transform of $\tfrac{1}{2}\exp\{-R_\phi(0)+R_\phi(\tau)\}$. Expanding $\exp\{-R_\phi(\tau)\}$ as a power series in $R_\phi(\tau)$[15] shows that $G_L(f)$ has a spectral line at $f=0$ and hence that $G_v(f)$ has spectral lines at $f=\pm f_c$, each containing power $\tfrac{1}{4}\exp\{-R_\phi(0)\}$. In the narrow-band case $(R_\phi(\tau)\ll 1)$ most of the transmitted power lies in this carrier component. For the results of numerical evaluations of $G_L(f)$, the reader is referred to more specialised texts[16].

Alternatively, provided that the bandwidths of $\cos\phi(t)$ and $\sin\phi(t)$ are small compared with f_c, the power spectrum $G'_v(f)$ of $v(t)$ about the carrier frequency may be expressed[18] as:

$$G'_v(f) = \tfrac{1}{2}[C_\phi(f)+S_\phi(f)], \tag{5.3}$$

where $C_\phi(f)$ and $S_\phi(f)$ are the power spectra of $\cos\phi(t)$ and $\sin\phi(t)$ respectively. Note that $G'_v(f)$ denotes the sum of the power spectral density of $v(t)$ at f_c+f and that at $-f_c-f$.

Since the spectrum of an SSB/FDM/FM signal extends to infinite frequencies, no practical radio link transmits the entire spectrum and a simple rule for estimating an adequate transmission bandwidth is obviously useful. Such a rule originated in the Bell

Laboratories in 1939. It is known as Carson's rule and states that an adequate bandwidth for an angle-modulation transmission is $2(f_m + f_d)$, where f_m is the highest modulation frequency and f_d the peak frequency deviation. When the modulating signal is a gaussian process it is sometimes accepted[16,17] that an appropriate value for f_d is about $\sqrt{10}\,\Delta f$ where Δf is the root-mean-square frequency deviation.

When the angle-modulation is such that the rate at which the frequency deviation varies is rather less than the frequency deviation itself, it is a reasonable approximation to assume that the power at a given frequency in the spectrum of an angle-modulated wave is proportional to the probability that the baseband signal has caused the carrier to deviate to that frequency. Thus in SSB/FDM/FM systems in which $\Delta f > f_m$ it is reasonable to regard the power spectrum of the transmitted signal as gaussian in shape.

5.4.3 Phase-thermal noise

Provided that the signal-to-noise ratio is not small, the effect of adding gaussian noise with one-sided power spectral density N_0 W/Hz to an angle-modulated sinusoid of power C W is to add random phase modulation with one-sided power spectral density N_0/C $(\mathrm{rad})^2/\mathrm{Hz}$. This is equivalent to random frequency modulation with one-sided power spectral density $N_0 f^2/C$ $(\mathrm{Hz})^2/\mathrm{Hz}$ at f Hz. It follows that the noise power at a point of zero relative level in an SSB channel of bandwidth b Hz at a baseband frequency f_m Hz is $N_0 b f_m^2/C(\delta f)^2$ mW0, where δf is the root-mean-square frequency deviation due to a test-tone in the channel under consideration. Noise introduced in this way is often called 'phase-thermal' noise.

It is usually an advantage for there to be approximately the same phase-thermal noise power in each speech channel and this suggests pre-emphasis of the baseband signal before modulation. Theoretically pre-emphasis with power gain proportional to f^2 would be ideal for this purpose; but in practice noise introduced elsewhere in the system makes it preferable for the pre-emphasis characteristic to level out at the lower baseband frequencies.

5.4.4 Interference from FM transmissions

Consider the sum $v(t)$ of the wanted signal $\cos\{\omega_1 t + \phi(t)]$ and the interfering signal $r\cos\{(\omega_1 + q)t + \theta(t)\}$. If $r \ll 1$, $v(t)$ may be written

$$v(t) = m(t)\cos\{\omega_1 t + \phi(t) + \mu(t)\},$$

where $\mu(t)$ is the additional phase modulation due to the interference and is given by

$$\mu(t) \simeq r\sin\{qt + \theta(t) - \phi(t)\}. \tag{5.4}$$

Derivation of the power spectrum of $\mu(t)$ is based on the following expressions for the power spectrum $S_{x+y}(f)$ of $\sin\{x(t) + y(t)\}$ and the power spectrum $C_{x+y}(f)$ of $\cos\{x(t) + y(t)\}$[18]:

$$S_{x+y}(f) = \{S_x(f) * C_y(f)\} + \{C_x(f) * S_y(f)\} \tag{5.5}$$

and $\quad C_{x+y}(f) = \{C_x(f) * C_y(f)\} + \{S_x(f) * S_y(f)\}, \tag{5.6}$

where $S_x(f), S_y(f), C_x(f), C_y(f)$ are the power spectra of $\sin x(t)$, $\sin y(t)$, $\cos x(t)$, $\cos y(t)$ and the symbol $*$ denotes convolution. $x(t)$ and $y(t)$ are assumed to be independent random processes.

When $x(t) = \omega_c t$, $C_x(f) = S_x(f) = \frac{1}{4}\{\delta(f - f_c) + \delta(f + f_c)\}$ and hence putting $x(t) = \omega_c t$ in Eqs (5.5) and (5.6) gives for the power spectra of $\sin\{\omega_c t + y(t)\}$ and $\cos\{\omega_c t + y(t)\}$:

$$\tfrac{1}{4}\{C_y(f - f_c) + S_y(f - f_c) + C_y(f + f_c) + S_y(f + f_c)\}. \tag{5.7}$$

It may be noted in passing that if the bandwidths of $\sin y(t)$ and $\cos y(t)$ are small compared with f_c, Eq. (5.7) reduces to Eq. (5.3).

The power spectrum $G_\mu(f)$ of $\mu(t)$ is given by Eq. (5.7) if f_c is replaced by q' $(= q/2\pi)$ and $y(t)$ by $\theta(t) - \phi(t)$. If Eqs (5.5) and (5.6) are then used to expand S_y and C_y it follows[18] that

$$G_\mu(f) = \tfrac{1}{4}r^2\{G_{u+w}(f-q') + G_{u+w}(f+q')\}, \tag{5.8}$$

where

$$G_{u+w}(f) = G_u(f) * G_w(f),$$

$G_w(f)$ being the power spectrum of the wanted signal about f_1 and $G_u(f)$ that of the interfering signal about $f_1 + q'$. If q' is large compared with the spectral width of $G_{u+w}(f)$, the power spectrum of $\mu(t)$ about q' is simply $\tfrac{1}{2}r^2 G_{u+w}(f)$.

5.4.5   Transmission deviations

Consider a carrier at frequency f_c modulated in frequency by a single tone and suppose that one of its spectral components is altered in magnitude or delay relative to the other spectral components during transmission. If that component is at frequency $f_c + f_n$, its alteration is equivalent to the addition of an interfering sinewave at frequency $f_c + f_n$ and therefore produces at the demodulator output an interfering tone at frequency f_n. Hence variations with frequency of the gain or the delay experienced by a SSB/FDM/FM signal introduce interference.

Analysis of the interference may be based on the assumption that the transfer function $H(\omega)$ of the transmission path is given by

$$H(\omega) = \{1 + g_1(\omega - \omega_c) + g_2(\omega - \omega_c)^2 + g_3(\omega - \omega_c)^3\}$$
$$\exp[j\{b_2(\omega - \omega_c)^2 + b_3(\omega - \omega_c)^3\}].$$

No linear term is included in the argument of the exponential function since such a term would represent a constant group delay and therefore would not contribute interference. It may be shown[19] that the effect of $H(\omega)$ on the angle-modulated signal $\cos\{\omega_c t + \phi(t)\}$ is to introduce additional angle modulation $\theta(t)$ given approximately by

$$\theta(t) \simeq g_2\ddot{\phi}(t) + b_2\dot{\phi}^2(t) + (b_3 + g_1 b_2)\{\dot{\phi}^3(t) - \dddot{\phi}(t)\} - 3g_3\dot{\phi}(t)\ddot{\phi}(t).$$

The linear terms $-g_2\ddot{\phi}(t)$ and $-(b_3 + g_1 b_2)\dddot{\phi}(t)$ represent 'equalisable' distortion, removable by equalisation after demodulation. The interference represented by the remaining terms cannot be removed after demodulation.

In many SSB/FDM/FM systems it is interference due to transmission deviations which limits the narrowness of the transmission bandwidth and has given rise to Carson's rule. Sources of transmission deviation include filters, microwave amplifiers, echoes in mismatched feeders, multipath propagation and the sensitivity of angle-demodulators to amplitude variations.

5.5   Use of carrier-telephony systems for signals other than speech

5.5.1   Telegraphy

Section 2.4.2 has outlined the possibility of transmitting an FDM assembly of VFT signals over a telephone channel and carrier-telephony systems are usually designed with this possibility in mind.

VFT signals are much less tolerant of frequency offset than are speech signals and the accuracies recommended in Section 5.3.4 for carrier frequencies are greater than would be necessary for the transmission of speech alone.

The conventional loading figure of $(10 \log n - 15)$ dBm0 (Section 5.1.2) allowed for the replacement of speech by a $- 10$ dBm0 VFT signal is not more than five per cent of the telephony channels.

5.5.2   Facsimile

Frequency modulation is recommended[20] for the transmission of facsimile over carrier circuits, since it has advantages over amplitude modulation in that it does not overload channels and is less influenced by sudden variations of level or by noise.

Facsimile transmissions can be severely degraded by delay distortion, particularly when frequency modulation is used, and it is therefore recommended[20] that the extreme channels (1 and 12, Fig. 5.5) of a basic group should not be used.

The group delay of a telephone circuit used for facsimile transmission should not[20] vary by more than $1/2f_m$ over the frequency band of the transmitted signal, where f_m is the maximum modulating frequency. Hence the maximum scanning speed for given picture definition is limited by the delay distortion of the telephone circuit used, decreasing as the number of frequency translations is increased.

An average transmitted power of $- 13$ dBm0 is appropriate for a frequency modulation facsimile signal occupying one telephone channel.

5.5.3   Data transmission

Modems (Section 2.4.3) for the transmission of digital data over an individual telephone circuit are usually designed to be tolerant of the additional distortion (particularly frequency-offset and delay distortion) contributed by carrier-telephony links in a typical circuit. They must also be designed so that they do not degrade the service provided for other users. In particular, the presence of sustained tones in the transmitted signal may give rise to intermodulation of an annoying kind and the presence of strong digital signals in the SSB/FDM signal may invalidate the statistical assumptions on which overload requirements are calculated. It is therefore recommended[21] that the level of FSK or PSK data transmission in a telephone circuit should not exceed $- 13$ dBm0 ($- 20$ dBm0 in the absence of data). These levels may have to be reduced if there is a significant increase in the proportion of circuits used to carry data.

The rate at which data can be reliably transmitted over a telephone circuit depends on the quality of that circuit, but even in a very good circuit it is unlikely to exceed 9.6 kbit/s. Higher rates of transmission can be achieved by using transmission links intended for FDM assemblies of speech signals. For example, a basic 12 channel group may be replaced by a 48 kbit/s data signal occupying the frequency band from 60 kHz to 108 kHz. For this purpose vestigial-sideband modulation of a 100 kHz carrier is generally used and the group pilot is moved from 84.08 kHz to 104.08 kHz so as to be outside the band of the digital signal. Channel 1 (Fig. 5.5) is available for accompanying speech and therefore the transmitted level is recommended[22] to be equivalent to the conventional loading of 11 channels ($- 5$ dBm0). Distortion-free demodulation of a vestigial-sideband transmission requires the extraction of a jitter-free carrier. Hence it is common practice to remove low-frequency components of the data signal before modulation and to restore them after demodulation, although the restoration process introduces some propagation of errors[23, 22].

5.5.4   Programme sound

Sometimes a link intended for a basic 12 channel group is used for the transmission of two 15 kHz programme-sound signals. In the recommended[24] system each signal is accompanied by a 16.8 kHz pilot-tone at −12.0 dBm0 and transmitted as single-sideband amplitude modulation. The frequency bands occupied by the transmitted signal are chosen so as to avoid the 84.08 kHz pilot and the extremes of the group frequency band. The pilots are used for frequency and phase correction in the demodulator (particularly important when the two signals form one stereophonic transmission) and for compensation for transmission-loss deviations.

The variations with frequency of the power spectrum of programme sound and of the sensitivity of the human ear make it desirable to increase the higher frequencies of programme-sound signals before modulation for transmission over a group link. The recommended[25] pre-emphasis is to attenuate the signal by

$$10 \log \left\{ \frac{75 + \left(\dfrac{\omega}{3000}\right)^2}{1 + \left(\dfrac{\omega}{3000}\right)^2} \right\} \text{ dB}$$

at ω rad/s. After modulation the signal is compressed for transmission[24]. In some systems the expander is controlled by a separate FM control channel.

In order that replacement of a group by two programme-sound channels should not alter the loading of the system, each programme channel is set up so that its overload level is +9 dBm0.

5.5.5 Television

Large capacity carrier-telephony systems are often designed so that large blocks of telephone channels may be replaced by television channels. As in television broadcasting, the presence of very low frequencies in the video signal and the need to conserve bandwidth lead to the use of vestigial-sideband amplitude modulation. In this case, however, it is not important to use simple envelope detection and therefore it is usual to improve signal-to-noise ratio by using a modulation index greater than unity, the amplitude of the transmitted signal at the white level being equal to that at the blanking level (Fig. 3.1)[26].

Echoes can give severe degradation to a television picture, but their effect is usually insignificant if there is a return loss of at least 20 dB between all cables and repeaters.

The use of a carrier-telephony link for television requires more careful control of some aspects of its performance than is necessary for telephony. Standard test signals such as the 'pulse and bar' are used to ensure that distortion is not unacceptable.

5.6 References

1   Starr, A. T., *Telecommunications*, Pitman, 1954
2   *The Post Office Electrical Engineers' Journal*, **66**, Part 3, October 1973
3   *CCITT Yellow Book*, Vol. III, Recommendation G151, Geneva, 1980
4   *CCITT Yellow Book*, Vol. III, Recommendation G223, Geneva 1980
5   Holbrook, B. D., and Dixon, J. T., 'Load rating theory for multichannel amplifiers', *Bell System Technical Journal*, **18**, pp. 624–44, October 1939
6   *CCITT Yellow Book*, Vol. III, Recommendation G222, Geneva, 1980
7   Tant, M. J., *The White Noise Book*, Marconi Instruments Ltd, 1974
8   Davies, A. P., and Vincent, A. W. H., 'The Cantat 2 Cable system: evolution and design', *The Post Office Electrical Engineers' Journal*, **67**, Part 3, pp. 142–7, October 1974
9   *The Bell System Technical Journal*, **57**, No 7, Part 1, September 1978 (Special Issue on the SG Undersea Cable System)
10 *CCITT Yellow Book*, Vol. III, Recommendation G211, Geneva, 1980

11 *CCITT Yellow Book*, Vol. III, Recommendation G232, Geneva, 1980
12 *CCITT Yellow Book*, Vol. III, Recommendation G235, Geneva, 1980
13 *CCITT Yellow Book*, Vol. III, Recommendation G241, Geneva, 1980
14 *CCITT Yellow Book*, Vol. III, Recommendation G225, Geneva, 1980
15 Schwartz, M., Bennett, W. R., and Stein, S., *Communication Systems and Techniques*, McGraw-Hill, 1966
16 Roberts, J. H., *Angle-modulation*, Peter Peregrinus Ltd, IEE Telecommunications Series 5, 1977
17 Halliwell, B. J., (Editor) STC Monograph: *Advanced Communication Systems*, Chapter 2, Newnes-Butterworth, 1974
18 Johns, P. B., and Rowbotham, T. R., *Communication Systems Analysis*, Butterworth, 1972
19 *Transmission Systems for Communications*, by the Staff of the Bell Telephone Laboratories, Bell Laboratories Inc., Third Edition, 1964
20 *CCITT Yellow Book*, Vol. III, Recommendation H41, Geneva, 1980
21 *CCITT Yellow Book*, Vol. VIII, Recommendation V2, Geneva, 1980
22 *CCITT Yellow Book*, Vol. VIII, Recommendation V35, Geneva, 1980
23 Lucky, R. W., Salz J., and Weldon, E. J., *Prinicples of Data Communication*, McGraw-Hill, 1968
24 *CCITT Yellow Book*, Vol. III, Recommendation J31, Geneva, 1980
25 *CCITT Yellow Book*, Vol. III, Recommendation J17, Geneva, 1980
26 *CCITT Orange Book*, Vol. III, Recommendation J72, Geneva, 1976

Exercise topics

5.1　Distribution of mean speaker powers
5.2　Mean power of SSB/FDM signal
5.3　Number of active channels
5.4　Intermodulation and thermal noise allocation
5.5　Voltage addition of $\alpha + \beta - \gamma$ intermodulation products
5.6　Distribution of third-order intermodulation products
5.7　Relative magnitudes of third-order intermodulation products
5.8　Accumulated third-order intermodulation
5.9　Negative-feedback repeater
5.10　Design of cable link
5.11　Non-linearities in FM transmission
5.12　Channel noise in SSB/FDM/FM transmission
5.13　Pre-emphasis in SSB/FDM/FM transmission
5.14　Interference between two FM signals
5.15　White-noise testing
5.16　Transmission of SSB/FDM signal by PCM
5.17　Data replacing basic group by quadrature VSB modulation

Exercises

5.1 Distribution of mean speaker powers

Measurements of the mean power in an active telephone channel are made for a large number of different talkers. When expressed in dBm the results are found to follow a normal distribution with standard deviation 5.8 dB. Show that
   (i)   the mean power exceeded by 1 % of the talkers is approximately 13.5 dB above the median mean power,
   (ii)   the average measured mean power is approximately 3.9 dB above the median mean power,
   (iii)   the ratio of the standard deviation of the measured mean power (in mW) to the average measured mean power (in mW) is approximately 3.5 dB.

Solution outline
Denote the mean power due to a typical talker by x dBm and its probability density function by $p(x)$. The mean power may then be written as $\exp(ax)$ mW (where $a = 0.1 \ln 10 \simeq 0.230$) and $p(x)$ is given by

$$p(x) = \frac{1}{\sigma\sqrt{2\pi}}\exp\left\{\frac{-(x-x_0)^2}{2\sigma^2}\right\},$$

where $\sigma = 5.8$ and x_0 dBm is the median mean power.
   (i)   Two per cent of the area under $p(x)$ lies outside values of x given by

$$x - x_0 = \pm 1.64\sigma\sqrt{2} = 13.5 \text{ dB}$$

(see Appendix 4).
   (ii)   The mean power averaged over many talkers is x_m mW, where

$$x_m = \int_{-\infty}^{\infty} \exp(ax)p(x)\mathrm{d}x.$$

Evaluation of the integral gives

$$x_m = \exp(ax_0)\exp(a^2\sigma^2/2).$$

The first factor is the median mean power. The second factor is about 3.9 dB.
   (iii)   The variance of the mean speech power is

$$\int_{-\infty}^{\infty} \exp(2ax)p(x)\mathrm{d}x - \{\exp(ax_0)\exp(a^2\sigma^2/2)\}^2,$$

which may be evaluated to give

$$x_m^2\{\exp(a^2\sigma^2) - 1\}.$$

The required ratio is therefore the square root of the second factor, which is about 3.5 dB.

5.2 Mean power of SSB/FDM signal

It is sometimes assumed that the sum of log-normally distributed variables is also log-normally distributed and has mean and variance equal respectively to the sum of the means of the variables summed and to the sum of their variances. Use that assumption to show that for an SSB/FDM signal in which n contributing channels are active with mean powers log-normally distributed as in Exercise 5.1,
   (i)   the standard deviation σ_T of the normal distribution describing the slowly varying

mean power expressed in dBm is given approximately by

$$\sigma_T = 4.3 \sqrt{(\ln{(1 + 4.95/n)})} \, \text{dB}$$

(ii) when $n = 240$, the value which the slowly varying mean power exceeds for one per cent of the time is approximately 1.4 dB above the average value.

Solution outline
 (i) From the stated assumption and with the notation used in the Solution of Exercise 5.1, the average power of the SSB/FDM signal is nx_m. Hence, using the expression derived in Part (iii) of that Solution, the variance of the slowly varying mean power of the SSB/FDM signal is $(nx_m)^2 \{\exp{(a^2 \sigma_T^2)} - 1\}$.

It also follows from the stated assumptions and the same expression that that variance may be written $nx_m^2 \{\exp{(a^2 \sigma^2)} - 1\}$.

Equating these two expressions for variance, putting $a = 0.230$ and $\sigma = 5.8$ and solving for σ_T gives the required result.
(ii) When $n = 240$, $\sigma_T = 0.62$ dB. Hence the value exceeded by the slowly varying mean power for one per cent of the time is $(1.64 \times 0.62 \times \sqrt{2})$ dB (i.e. 1.44 dB) above its median value. (See Appendix 4.)

From the expression derived in Part (ii) of the Solution of Exercise 5.1 the average value of the slowly varying mean exceeds the median value by a factor $\exp{(a^2 \sigma_T^2/2)}$, which is approximately 0.04 dB.

The required result is therefore $(1.44 - 0.04)$ dB.

5.3   Number of active channels

Verify that in a 12 channel system in which each channel has a 25 % probability of being active, the probability of more than seven channels being simultaneously active in less than 1 %.

Solution outline

Evaluate $\sum_{r=8}^{12} p(r)$ from Eq. 5.1 with $n = 12$. It is approximately 0.0028.

5.4   Intermodulation and thermal noise allocation

Assuming that the intermodulation interference is due entirely to third-order products, show that when the transmission level in a carrier-telephony link is adjusted to maximise the output signal-to-noise ratio the thermal noise power at the output is twice the intermodulation noise power.

Solution outline
Denote the thermal noise power at the output by b and the signal power by x. Since intermodulation is third-order, the intermodulation noise power at the output may be written hx^3, where h is independent of x.

The signal-to-noise ratio at the output is therefore $x(b + hx^3)^{-1}$, which has a maximum value when $x = (b/2h)^{1/3}$. Substituting that value for x in hx^3 shows the intermodulation noise power to be $b/2$.

5.5   Voltage addition of $\alpha + \beta - \gamma$ intermodulation products

Each section of a carrier-telephony cable link is equalised to have a phase-shift $\phi_0 + k\omega$ at angular frequency ω, where ϕ_0 and k are constants. Show that $\alpha + \beta - \gamma$ intermodulation products contributed by individual repeaters reach the end of the link in phase, whereas $\alpha + \beta + \gamma$ and $\alpha - \beta - \gamma$ products in general do not.

Solution outline
Suppose that at the first repeater three tones $\cos \alpha t$, $\cos \beta t$ and $\cos \gamma t$ produce the intermodulation contribution $\cos (\alpha + \beta - \gamma)t$. The tones reach the second repeater as $\cos (\alpha t + \phi_0 + k\alpha)$, $\cos (\beta t + \phi_0 + k\beta)$ and $\cos (\gamma t + \phi_0 + k\gamma)$ and produce there the intermodulation contribution $\cos\{(\alpha t + \phi_0 + k\alpha) + (\beta t + \phi_0 + k\beta) - (\gamma t + \phi_0 + k\gamma)\}$. This may be written $\cos \{(\alpha + \beta - \gamma)t + \phi_0 + (\alpha + \beta - \gamma)k\}$ and is therefore in phase with the contribution arriving from the first repeater. The result does not hold for $\alpha + \beta + \gamma$ or $\alpha - \beta - \gamma$ products except for certain values of ϕ_0.

5.6 Distribution of third-order intermodulation products

A carrier-telephony link has n channels occupying the frequency range from 10 MHz to 60 MHz. Assuming that the signal in each channel is a sinewave, sketch on one graph the distributions with frequency of the $\alpha + \beta + \gamma$, $\alpha + \beta - \gamma$ and $\alpha - \beta - \gamma$ intermodulation products falling within the transmission band.

Solution outline
The approximate solution shown in Fig. 5.7(a) may be obtained from Fig. 5.2. For example, the $\alpha + \beta + \gamma$ distribution may be obtained by summing n $\alpha + \beta$ distributions as indicated in Fig. 5.7(b). The distribution is parabolic in shape, increasing from zero at 30 MHz to approximately $\frac{1}{3}$ $(\frac{1}{2} + 1 + 1\frac{1}{2} + 2 + \dots + \frac{1}{2}n) \simeq n^2/12$ at 80 MHz. Figure 5.7(a) is consistent with published charts[19]. Note the asymmetry of the $\alpha + \beta - \gamma$ curve, which may be regarded as due to spectral folding about zero frequency.

5.7 Relative magnitudes of third-order intermodulation products

Show that when three tones with equal amplitudes are applied to an amplifier with its input and output voltages related by Eq. 5.2, $\alpha \pm \beta \pm \gamma$ products are 15.6 dB above the third harmonics and $2\alpha \pm \beta$ products are 9.5 dB above the third harmonics.

Solution outline
Obtain the identity:

$$(\cos \alpha + \cos \beta + \cos \gamma)^3$$
$$\equiv \tfrac{15}{4}(\cos \alpha + \cos \beta + \cos \gamma)$$
$$+ \tfrac{3}{4}\{\cos (2\alpha - \beta) + \cos (2\alpha + \beta) + \text{similar terms}\}$$
$$+ \tfrac{3}{2}\{\cos (\alpha + \beta + \gamma) + \cos (\alpha + \beta - \gamma) + \cos (\alpha - \beta + \gamma) + \cos (\alpha - \beta - \gamma)\}$$
$$+ \tfrac{1}{4}\{\cos 3\alpha + \cos 3\beta + \cos 3\gamma\}.$$

Hence amplitudes of $\alpha \pm \beta \pm \gamma$ products and amplitudes of third harmonics are in the ratio $\frac{3}{2} : \frac{1}{4}$ i.e. 6:1 or 15.6 dB. For $2\alpha \pm \beta$ products the ratio is $\frac{3}{4} : \frac{1}{4}$, i.e. 3:1 or 9.5 dB.

5.8 Accumulated third-order intermodulation

A coaxial-cable carrier-telephony link of 100 sections has 12 500 channels occupying the frequency range from 10 MHz to 60 MHz. Measurements on individual repeaters show that in each repeater a -10 dBm tone generates a third harmonic at -130 dBm (the levels of both tone and third harmonic being measured at the repeater output). Using the results

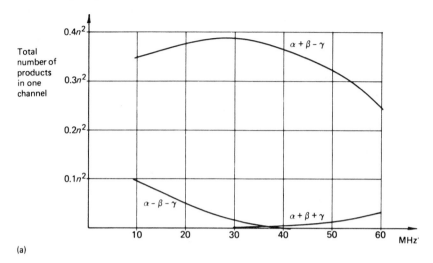

(a)

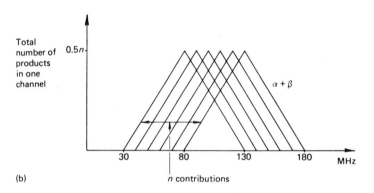

(b)

Fig. 5.7   Solution to Exercise 5.6

of previous Exercises, calculate the accumulated intermodulation noise contributed by $\alpha + \beta - \gamma$ products at a point of zero relative level at the output of the link in the channel in which the contribution of $\alpha + \beta - \gamma$ products is greatest. Assume conventional loading, repeater output power -2.5 dBm and no pre-emphasis.

Solution outline
Regard the power in each channel at a repeater output as a tone of power $(-2.5-10 \log 12\,500)$ dBm, i.e. -43.5 dBm, 33.5 dB below the level used for measurement.

The level of each $\alpha + \beta - \gamma$ product generated in one repeater by three such tones is therefore $\{-130-(3 \times 33.5)+15.6\}$ dBm, i.e. -214.9 dBm.

The contribution of $\alpha + \beta - \gamma$ products is greatest in channels near to 35 MHz, in each of which there are approximately $\frac{3}{8}(12\,500)^2$ contributing products. Hence the $\alpha + \beta - \gamma$ products contribute $[-214.9 + 10 \log\{\frac{3}{8}(12\,500)^2\}]$ dBm, i.e. -137.2 dBm at each repeater.

Assuming voltage addition, intermodulation noise due to $\alpha + \beta - \gamma$ products accumulates to -97.2 dBm at the output of the 100th repeater.

Since loading is conventional, -2.5 dBm corresponds to $(-15 + 10 \log 12\,500)$ dBm0. Hence -97.2 dBm corresponds to -68.7 dBm0, i.e. 134 pW at a point of zero relative level.

5.9 Negative-feedback repeater

In the repeater configuration of Fig. 5.8, A is a three-stage common-emitter amplifier with a power gain of 60 dB when operating between (matching) 1 kΩ source and load resistances. Determine R_1 and R_2 if the loop gain is to be independent of the impedances of the lines. Assuming those values of R_1 and R_2, determine the input and output impedances of the repeater, and the source and load impedances 'seen' by the network N.

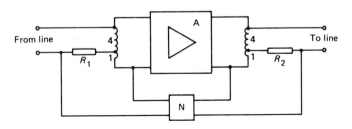

Fig. 5.8 Repeater configuration of Exercise 5.9

Assuming that N matches its source and load and has 20 dB loss, and assuming that the repeater is matched to the impedances of the lines, determine the power gain of the repeater.

Solution outline
Redraw the input end as in Fig. 5.9.

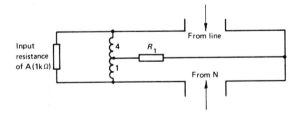

Fig. 5.9 Diagram of solution to Exercise 5.9

For loop transmission to be independent of line impedance, a source in N must produce no current in the line. This requires R_1 to be 160 Ω. Hence N 'sees' 200 Ω and the line 'sees' 800 Ω. A similar argument may be applied to the output end.

Denote the input and output voltages by V_{IN} and V_{OUT} and the voltage at the input-end port of N by V_F. Consideration of the input end shows that the input to the amplifier is $V_{IN} - V_F$. Hence, by consideration of the output end,

$$V_{OUT} = \tfrac{4}{5} \times 10^3 \times (V_{IN} - V_F).$$

By considering the output end and N,

$$V_F = \tfrac{1}{4} \times \tfrac{1}{10} \times V_{OUT}.$$

Elimination of V_F from those two equations gives

$$\frac{V_{OUT}}{V_{IN}} = \frac{800}{21} \quad \text{i.e. 31.6 dB.}$$

5.10   Design of cable link

A 2700 channel carrier-telephony link is to be established between two cities 1000 km apart to the following specification.

   (i)   The bandwidth of each channel is to be 3.1 kHz and the channels should occupy the frequency band from 312 kHz to 12 336 kHz.

   (ii)   Standard 2.6 mm/9.5 mm coaxial cable is to be used, with attenuation

$$[0.013 + 2.3\sqrt{f} + 0.003f^{1/3}]\ \text{dB/km}$$

at 290 K and f MHz.

   (iii)   At 290 K each repeater should compensate accurately for the attenuation of the section of cable preceding it.

   (iv)   To avoid difficult repeater locations, each section is to be built out to an electrical length 200 m greater than the average physical length by means of a network at the input of each repeater.

   (v)   CCITT conventional loading should be used and the total thermal and intermodulation noise at a point of zero relative level at the output of the link should not exceed 3000 pW (psophometrically weighted) in any channel.

   (vi)   The gain of no repeater should exceed 36 dB at the frequency of any channel.

   (vii)   The DC supply to the repeaters is to be 49 mA from a constant-current source and carried by the inner conductor of the coaxial cable. At no point on the link should the mean voltage on the inner conductor differ by more than 300 V from the voltage on the (grounded) outer conductor.

   (viii)   There is to be no emphasis. There should be equal test-tone levels at all repeater outputs and for all channels.

   The following assumptions should be made.

(a)   The noise factor of each repeater is 3 dB when there is no attenuation in the line-build-out network.

(b)   The efficiency of each repeater is two per cent. In other words, at its overload point a repeater delivers a sinewave with power two per cent of the power drawn from the DC supply.

(c)   Intermodulation interference is dominated by $(\alpha + \beta - \gamma)$ products accumulating by voltage addition. Allocate 2000 pW to thermal noise and 1000 pW to intermodulation interference in each channel and do not attempt to take advantage of the fact that the channel with the most intermodulation noise is not that with the most thermal noise.

   Use the above data to calculate

A   the minimum number of repeaters needed,

B   the minimum overload point for a repeater,

C   the psophometrically weighted thermal noise power in the 312–316 kHz channel at a point of zero relative level at the output of the link (assuming it to be 2000 pWp in the 12 332–12 336 kHz channel),

D   the minimum number of intermediate repeaters at which DC power is fed to the link,

E   an approximate specification for the third harmonic generated when a single tone is applied to one repeater.

Solution outline

A   At 12.336 MHz the attenuation of the cable is 8.10 dB/km and therefore the electrical length of each section should not exceed 4.445 km. Thus the average physical length of the sections should not exceed 4.245 km, so that 236 repeaters are required (including one at the receiving terminal).

B   Since there are 236 sections, the average physical length of the sections is 4.237 km and therefore the gain of each repeater (excluding the line-build-out network) should be

35.93 dB at 12.336 MHz. The thermal noise power in a 3.1 kHz bandwidth available from a source at 290 K is -139.06 dBm (Appendix 6). Therefore the thermal noise in the noisiest channel at the output of the link is

$$(-139.06 + 3 + 35.93 + 10 \log 236) \,\mathrm{dBm} = -76.40 \,\mathrm{dBm}.$$

At a point of zero relative level at the output of the link the overload point must be at least $(-2.5 + 10 \log 2700) \,\mathrm{dBm0}$ and the thermal noise at most $(-56.99 + 2.5) \,\mathrm{dBm0}$ (allowing 2.5 dB for psophometric weighting, see Section 1.1.3). Therefore the overload point must be at least 86.30 dB above the thermal noise and at a repeater output must be at least 9.90 dBm.

C At 312 kHz the gain of a repeater is 5.77 dB, 30.16 dB less than at 12.336 MHz. Hence at the output of the link the thermal noise in the 312–316 kHz channel is

$$(-56.99 - 30.16) \,\mathrm{dBm0p} = -87.15 \,\mathrm{dBm0p}.$$

D The overload point is 9.772 mW. Hence each repeater requires (50×9.772) mW of DC power and therefore a DC voltage $(50 \times 9.772 \div 49)\mathrm{V} = 9.97$ V. Hence 300 V can supply 30 repeaters in series and so three intermediate power-feeding points are required in addition to the terminals.

E Consider $\alpha + \beta - \gamma$ products only. In this system the effect of spectral folding will be more severe than in the system of Exercise 5.6 and the maximum number of products in one channel is about $0.5 n^2$.

Assume that the traffic fully occupies the frequency band and represent it by 3006 tones, each at a level of -37.4 dBm.

The accumulated intermodulation noise must not exceed -79.4 dBm (half the thermal noise) at the output of the link and this requires the contribution from each repeater to be less than

$$(-79.4 - 20 \log 236) \,\mathrm{dBm} = -126.9 \,\mathrm{dBm}$$

(assuming voltage addition). This contribution is made up of about $0.5 \times (3006)^2$ products and therefore the level of each product should be less than about -193.4 dBm.

Third harmonics are about 15.6 dB below $\alpha + \beta - \gamma$ products and hence the third harmonic generated in a one repeater by a single tone at an output level of -37.4 dBm should not exceed -209.0 dBm. In terms of more measurable levels, the requirement is that the third harmonic generated in one repeater by a tone at 0 dBm should not exceed about -97 dBm.

5.11 Non-linearities in FM transmission

Derive an expression for the output voltage of an amplifier described by Eq. 5.2 when the input voltage is the angle-modulated wave

$$A \cos \{\omega t + \phi(t)\}.$$

Hence justify the statement that frequency modulation is unaffected by transmission non-linearities.

Solution outline
The output voltage is

$$a_1 A\cos\{\omega t + \phi(t)\} + a_2 A^2\cos^2\{\omega t + \phi(t)\} + a_3 A^3 \cos^3\{\omega t + \phi(t)\},$$

which may be written

$$\tfrac{1}{2}a_2 A^2 + (a_1 A + \tfrac{3}{4}a_3 A^3)\cos\{\omega t + \phi(t)\} + \tfrac{1}{2}a_2 A^2\cos\{2\omega t + 2\phi(t)\}$$
$$+ \tfrac{1}{4}a_3 A^3\cos\{3\omega t + 3\phi(t)\}.$$

The second term represents the original carrier frequency with the original angle modulation. The other terms represent either DC or angle-modulated waves at harmonics of the original carrier frequency which would usually lie well outside the range of frequencies to which a receiver is sensitive.

5.12   Channel noise in SSB/FDM/FM transmission

A basic supergroup of 60 telephone channels occupying the frequency band from 60 kHz to 300 kHz is transmitted by frequency modulation of a 2 GHz carrier. There is no pre-emphasis and the r.m.s. frequency deviation is 1 MHz. Calculate the r.m.s. phase deviation.

Determine the Carson bandwidth of the transmission and estimate the percentage of the sideband power lying outside the Carson bandwidth.

At the input to the demodulator in the receiver the signal power is 10 mW and the signal is accompanied by 0.1 μW of additive gaussian noise. Assuming that the r.m.s. frequency deviation due to a test-tone in one channel is 250 kHz, determine the ratio of test-tone to noise in the highest-frequency channel after demodulation.

Solution outline
From a formula derived in Section 5.4.2, r.m.s. phase deviation is

$$\frac{10^6}{\sqrt{60\,000 \times 300\,000}} = 7.45 \text{ rad.}$$

Assuming 3.162 to be an appropriate peak-to-r.m.s. factor in this case, the Carson bandwidth is

$$2(0.3 + 3.162 \times 1) \text{ MHz} = 6.9 \text{ MHz.}$$

Since the r.m.s. frequency deviation is several times greater than the maximum baseband frequency, the spectrum of the frequency-modulated wave is approximately gaussian about a mean of 2 GHz. Hence the power spectral density f MHz from 2 GHz is proportional to

$$\frac{1}{\sigma\sqrt{2\pi}}\exp\left(-\frac{f^2}{2\sigma^2}\right),$$

where σ is the r.m.s. value of the frequency deviation, given as 1 MHz. The proportion of the power outside $f = \pm 6.9/2$ MHz (i.e. ± 3.45 MHz) is hence

$$\text{erfc}\left(\frac{3.45}{1 \times \sqrt{2}}\right),$$

which is approximately 0.000 56 or 0.056% (Appendix 4).

Substituting $N_0 = 10^{-7} \div (6.9 \times 10^6)$, $b = 3100$, $f_m = 3 \times 10^5$, $C = 10^{-2}$ and $\delta f = 2.5 \times 10^5$ into the formula established in Section 5.4.3 shows that the required ratio is 81.9 dBm0.

5.13   Pre-emphasis in SSB/FDM/FM transmission

The system of Exercise 5.12 is modified by the inclusion of pre-emphasis with power gain $(f/0.3)^2$, where f is frequency in MHz. Determine the modified r.m.s. values of phase deviation and frequency deviation.

Solution outline
In the notation of Section 5.4.2, the power spectrum of the instantaneous frequency deviation $\frac{1}{2\pi} \, \mathrm{d}\phi/\mathrm{d}t$ is modified to $(\Delta f)^2 f^2/(f_2-f_1)f_2^2$ $(\mathrm{Hz})^2/\mathrm{Hz}$. Hence the power spectrum of $\phi(t)$ becomes $(\Delta f)^2/(f_2-f_1)f_2^2$ $(\mathrm{rad})^2/\mathrm{Hz}$, which is flat.

Hence the r.m.s. phase deviation is $\Delta f/f_2$ rad ($= 3.3$ rad). The mean-square frequency deviation is

$$\int_{f_1}^{f_2} \frac{(\Delta f)^2 f^2}{(f_2-f_1)f_2^2} \mathrm{d}f.$$

Evaluation of this integral and taking its square root gives the r.m.s. frequency deviation to be 643 kHz.

5.14 Interference between two FM signals

Two 132 channel SSB/FDM/FM telephony transmissions have carriers separated by 10 MHz and are received with equal powers. In each transmission, the baseband extends from 12 kHz to 552 kHz, there is no emphasis and the Carson bandwidth is 10 MHz. For the channel at the highest baseband frequency in one of the transmissions estimate the ratio of the test-tone to the interference from the other transmission. Assume that the spectrum of the interfering transmission is a δ-function.

Solution outline
By Carson's rule the r.m.s. frequency deviation is 1.407 MHz. Assuming that for n channels the loading is $(-1 + 4 \log n)$ dBm0 (Section 5.2.3), the r.m.s. frequency deviation due to a test-tone in one channel is 0.594 MHz.

The power spectrum about 10 MHz of the additional phase deviation due to interference will be the convolution of the power spectra about their carriers of the wanted and the interfering transmissions. Since the spectrum of the interfering transmission is a δ-function, the convolution is a gaussian curve with mean value 10 MHz and standard deviation 1.407 MHz. Since the powers of the wanted and the interfering transmissions are equal, the mean-square value of the additional phase deviation is 0.5 $(\mathrm{rad})^2$ (Eq. 5.4) and therefore the area under the gaussian curve is 0.5. The interference in the baseband channel at 0.552 MHz is therefore

$$0.5 \times 0.0031 \times \frac{1}{1.407\sqrt{2\pi}} \exp\left\{-\left(\frac{9.448}{1.407\sqrt{2}}\right)^2\right\} (\mathrm{rad})^2 = 7.1 \times 10^{-14} \, (\mathrm{rad})^2.$$

The contribution of this interference to the power spectrum of the frequency deviation is

$$7.1 \times 10^{-14} \times (552\,000)^2 \, (\mathrm{Hz})^2 = 0.0216 \, (\mathrm{Hz})^2.$$

Hence relative to the test-tone the interference is

$$10 \log \frac{0.0216}{(0.594 \times 10^6)^2} \, \mathrm{dB} = -132.1 \, \mathrm{dB}.$$

5.15 White-noise testing

White-noise testing is to be carried out on a 500 km coaxial-cable link with bandwidth 240 kHz to assess its suitability for a 60 channel carrier-telephony system. The test-tone level is to be -10 dBm in each channel at the input and output of the link. Calculate
(i) the noise power required to simulate busy-hour loading,
(ii) the minimum noise-power ratio which ensures that the link would meet the recommended noise objective of 3 pW0p/km.

Solution outline
(i)   The appropriate noise power loading is $(-1 + 4 \log 60)$ dB = 6.1 dB relative to the test-tone level. Hence the required noise power is

$$(-10 + 6.1) \, \text{dBm} = -3.9 \, \text{dBm}.$$

(ii)   The thermal noise and intermodulation noise in one channel should not add to more than -58.2 dBm0p (1500 pW0p) at the output of the link. This corresponds to a spectral density

$$(-58.2 + 2.5 - 10 \log 3100) \, \text{dBm0/Hz}.$$

The spectral density of the noise loading is

$$(6.1 - 10 \log 240\,000) \, \text{dBm0/Hz}.$$

The noise power ratio must not be less than the difference between these spectral densities, which is 42.9 dB.

5.16   Transmission of SSB/FDM signal by PCM

A basic supergroup SSB/FDM signal is to be directly encoded by PCM. Justify the choice of 576 kHz for the sampling frequency and determine the number of bits per sample if the quantisation distortion in one channel should not exceed -60 dBm0p. Assume that the quantisation is uniform with step δ and that the quantisation distortion power $(\delta^2/12)$ is uniformly distributed over the frequency band from 288 kHz to 576 kHz.

Solution outline
The spectrum of the sampled signal consists of the baseband contribution (312–552 kHz) and sidebands about the sampling frequency and its harmonics. The choice of 576 kHz gives equal 48 kHz gaps between the baseband and neighbouring sidebands.
    Let there be m steps covering the voltage range from -1 to $+1$. The ratio of the quantisation noise power in one channel to the power in the sinewave which just overloads the quantiser is therefore

$$\frac{1}{12}\left(\frac{2}{m}\right)^2 \frac{3.1}{288} \div \frac{1}{2}.$$

Since the equivalent peak power for a 60-channel system is 20.8 dBm0 (Table 5.1) and allowing 2.5 dB for psophometric weighting, this ratio should be less than

$$(-60 + 2.5 - 20.8) \, \text{dB} = -78.3 \, \text{dB}.$$

Hence $m > 697$ and 10 digit PCM is necessary.

5.17   Data replacing basic group by quadrature VSB modulation

Two data signals A and B are to be transmitted simultaneously over a 12 channel group as vestigial-sideband (VSB) modulation of two 84 kHz carriers in quadrature. The transmitted spectrum of signal A extends from 64 kHz to 92 kHz and that of signal B from 76 kHz to 104 kHz. Show that filtering by replicas of the transmitter VSB filters followed by coherent product demodulation will give complete separation of signals A and B in the receiver. Neglect any phase shift in the VSB filters.
    Suggest frequency-response characteristics for the VSB filters which will give rise to no attenuation distortion of the data signals.

Solution outline
Vestigial-sideband filters are usually assumed to have the symmetry shown in Fig. 5.10.

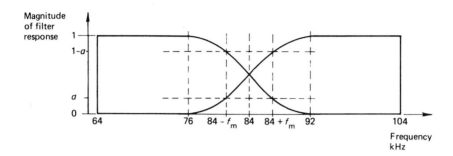

Fig. 5.10 Characteristics of VSB filters of Exercise 5.17

Modulation of $\cos \omega_c t$ by a component $\cos \omega_m t$ of signal A contributes after filtering

$$v(t) = \tfrac{1}{2}(1-a)\cos(\omega_c - \omega_m)t + \tfrac{1}{2}a\cos(\omega_c + \omega_m)t$$

to the transmitted signal. After further filtering in the receiver and multiplication by $\cos \omega_c t$, $v(t)$ becomes

$$[\tfrac{1}{2}(1-a)^2 \cos(\omega_c - \omega_m)t + \tfrac{1}{2}a^2\cos(\omega_c + \omega_m)t]\cos \omega_c t,$$

i.e.

$$\tfrac{1}{4}\{(1-a)^2 + a^2\}\cos \omega_m t + \text{terms at higher frequencies.}$$

After filtering in the receiver and multiplication by $\sin \omega_c t$, $v(t)$ becomes

$$[\tfrac{1}{2}(1-a)a\cos(\omega_c - \omega_m)t + \tfrac{1}{2}a(1-a)\cos(\omega_c + \omega_m)t]\sin \omega_c t,$$

which has no component at ω_m. A similar argument may be applied to a component of signal B modulating $\sin \omega_c t$. Hence it follows that signals A and B are separated by the receiver.

Since $\{(1-a)^2 + a^2\}$ varies with ω_m, the double filtering introduces attenuation distortion. Attenuation distortion will not occur if the filters are modified so that the *squares* of their responses have the symmetry required for VSB filters. A cosine shape for each filter is appropriate[23].

6
Topics in digital line transmission

6.1 Intersymbol interference

6.1.1 Nyquist rate

In synchronous digital transmission systems decisions are usually based on samples of the received waveform taken at regular intervals, one sample for each signalling element (Section 2.1.4). If the received waveform representing any signalling element is of sufficient duration to contribute to more than one sample, 'intersymbol interference' (ISI) is said to occur.

The received waveform representing a signalling element is always bounded in frequency and therefore unbounded in time, but ISI can in principle be avoided by designing that waveform to have zero-crossings at all sampling instants but one. The waveform $V \operatorname{sinc}(t/T)$ has zero-crossings appropriate for there to be no ISI at a signalling speed of $1/T$ baud and has its spectrum restricted to frequencies below $1/2T$ Hz. By the sampling theorem[1,2,3], there is no other waveform which has its spectrum restricted to frequencies below $1/2T$ Hz and has all the zero-crossings appropriate to a signalling speed of $1/T$ baud. Hence if transmission is restricted to frequencies below W Hz, the maximum possible signalling speed for there to be no ISI is $2W$ baud, the 'Nyquist transmission rate'[1,4] for the bandwidth W.

The existence of abrupt transitions in their spectra makes $V\operatorname{sinc}(t/T)$ pulses unrealisable, and, even if they were realisable, the slow decay ($\propto t^{-1}$) of their time side-lobes would make the avoidance of ISI critically dependent on accurate sampling times. Hence it is usual to use waveforms with more smoothly shaped spectra and more rapidly decaying side-lobes, even though the signalling speed is less than the Nyquist rate for the bandwidth they occupy.

An approach to the smoothing of a rectangular spectrum to a shape more nearly attainable in practice without destroying zero-crossings of the $V\operatorname{sinc}(t/T)$ waveform is suggested by Nyquist's theorem on vestigial symmetry (Exercise 6.2)[1,4]. An important example of a spectrum arrived at in this way is the raised-cosine spectrum of Exercise 6.3. The waveform with a real raised-cosine spectrum has rapidly decaying time side-lobes ($\propto t^{-3}$) and itself resembles a raised cosine sufficiently closely to be regarded as a raised-cosine pulse for many practical purposes.

The signalling speed for a given bandwidth may be increased by using 'partial-response' signalling[2] in which a controlled amount of ISI is accepted and allowed for when interpreting the received waveform. Examples of partial-response signalling are the duobinary[3,4,5] schemes described in Exercises 6.6 and 6.7.

6.1.2 Eye patterns

Intersymbol interference is usually displayed in the laboratory as an 'eye pattern'[6]. An eye pattern is obtained by displaying the received waveform on an oscilloscope synchronised to the sampling frequency. When there is intersymbol interference, the display of a random sequence of digits will appear to be made up of broadened lines as shown symbolically for binary transmission in Fig. 6.1. Error-free decisions are possible for all possible sequences of digits only if the 'eye' of the pattern is open. Sampling should be timed to take place at those instants at which the eye opening is greatest.

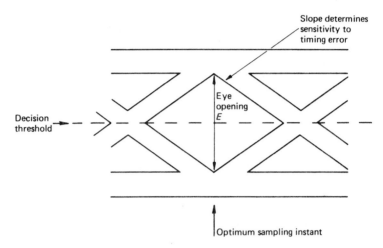

Fig. 6.1　Binary eye pattern

The value E assigned to the eye opening in Fig. 6.1 assumes that the distance between the received levels at the instant of sampling with no intersymbol interference has been normalised to unity. Thus the factor $1/E$ represents the amount by which the presence of intersymbol interference makes it necessary to increase the signal-to-noise voltage ratio if a given digit error rate is to be maintained with the most unfavourable sequence of digits.

In multilevel systems, there must be a 'stack' of open eyes, one associated with each threshold.

6.1.3　Adaptive equalisation

The effect of ISI in a synchronous system may be reduced by a transversal filter[1,2] (Fig. 6.2). A transversal filter may be thought of as adding and subtracting weighted time-shifted replicas of the received pulse, the weightings chosen so that ISI is largely cancelled out. The filter can be made adaptive by making provision for the weighting factors to be

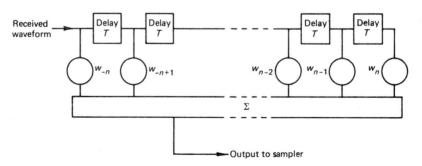

Fig. 6.2　Transversal filter (sampling rate $1/T$)

adjustable to suit different channel characteristics. The adjustment can be made automatic by incorporating a correlation process based on the delay elements of the filter[2]. Provided that there are no errors, ISI may be completely removed by decision-feedback equalisation[7], in which corrections to the received waveform are derived from 'clean' pulses at the output of the decision circuit.

It may be shown[5] by extending the theory of matched filters to cover overlapping pulses, that the optimum receiver consists of a matched filter (Section 2.2.1) followed by a filter having a frequency response periodic in frequency with a period equal to the sampling frequency I/T. The latter filter may be conveniently realised by the structure of Fig. 6.2. Note that when the transversal filter follows a matched filter its input pulses are symmetrical in time and hence one may choose $w_i = w_{-i}$ with no degradation in performance.

6.2 Spectra of digital signals

6.2.1 Derivation of power spectrum from autocorrelation

The power spectrum of a random process may be obtained by finding the Fourier transform of the autocorrelation (Appendix 1.6). The technique is illustrated here by applying it to a random binary sequence transmitted as a train of spaces and RZ (return-to-zero), half-width, rectangular pulses.

A typical length of the signal is shown in Fig. 6.3(a). The autocorrelation $R(\tau)$ is obviously made up of linear segments of duration $T/2$, $R(0)$ is the average signal power and, since the binary message sequence is random and equiprobable, $R(\tau)$ at other peaks is $R(0)/2$. $R(\tau)$ is shown in Fig. 6.3(b).

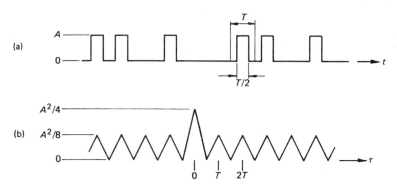

Fig. 6.3 (a) RZ binary waveform (b) autocorrelation of (a)

$R(\tau)$ may be regarded as the sum of a single triangle centred on $\tau = 0$ and a periodic triangular function. The Fourier transform of the single triangle is (Appendix 1.1)

$$\frac{A^2 T}{16} \text{sinc}^2 \left(\frac{fT}{2} \right).$$

The periodic component of $R(\tau)$ may be written

$$\frac{A^2}{16} + \frac{A^2}{2\pi^2} \left\{ \cos \left(\frac{2\pi\tau}{T} \right) + \frac{1}{9} \cos \left(\frac{6\pi\tau}{T} \right) + \frac{1}{25} \cos \left(\frac{10\pi\tau}{T} \right) + \ldots \right\}$$

and hence the power in the fundamental component of its line spectrum is $A^2/2\pi^2$.

An alternative approach is to regard the waveform itself as the sum of a random part with no periodic components and a periodic part (the ensemble average, Appendix 1.6) as shown in Fig. 6.4. The autocorrelation of the random component is, as before, a single triangle of height $A^2/8$. The periodic component may be written

$$\frac{A}{4} + \frac{A}{\pi}\left\{\cos\left(\frac{2\pi t}{T}\right) - \frac{1}{3}\cos\left(\frac{6\pi t}{T}\right) + \frac{1}{5}\cos\left(\frac{10\pi t}{T}\right) - \ldots\right\}.$$

Again, the power in the fundamental component of the line spectrum is $A^2/2\pi^2$.

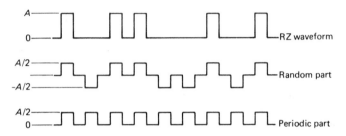

Fig. 6.4   Random and periodic parts of RZ waveform

6.2.2   Spectrum shaping

Figure 6.5 represents a digital transmission system in which all transmitted pulses have the same shape but may have different amplitudes and polarity. Transmitted pulses are assumed to be generated by applying to a transmit filter a train of impulses representing the data. By 'spectrum shaping' is meant the choice of a suitable transmitted power spectrum at the channel input when the channel transfer function $C(f)$ is given.

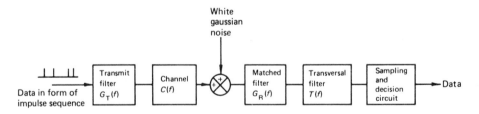

Fig. 6.5   Digital transmission system

Considerations which may make it desirable to control the shape of the transmitted power spectrum include the following.

(i)   If the channel contains power-separating filters (Section 6.3.1) or transformers, problems associated with the loss of low-frequency components are to a large extent avoided if the power spectrum approaches zero as $f \to 0$.

(ii)   Efficient use of signal power requires that it be restricted to frequencies at which $|C(f)|$ is not small.

(iii)   Crosstalk in nearby equipment is reduced if the transmitted power spectrum is reduced at high frequencies.

The transmitted power spectrum at the channel input is the product of $|G_T(f)|^2$ and the power spectrum of the input sequence of impulses representing the data. Hence there are two approaches to the shaping of the transmitted power spectrum:

(a)   modification of $|G_T(f)|^2$, which changes the shape of the individual transmitted pulses;

(b)   introducing redundancy (i.e. interrelations between the impulses) into the input sequence and hence modifying the autocorrelation of that sequence.

An example of (a) is the modification of the NRZ polar pulse train of Exercise 6.10 to the Manchester code of Exercise 6.12 to remove low frequencies from the spectrum. An example of (b) is the AMI encoding described in Exercise 6.13, which removes low frequencies and lines from the spectrum of the RZ pulse train analysed in Section 6.2.1. The existence of interrelations between the impulses of the input sequence implies that that sequence carries less information than it otherwise could. AMI encoding and the encoding described in Exercise 6.14 are examples of pseudo-ternary encoding, so called because although three transmitted symbols are used the information carried is no more than that of the binary sequence from which they are derived. A further example of (b) is the introduction of spectral nulls by interleaving (Exercise 6.15).

Matched-filter theory shows that to maximise the signal-to-noise ratio at the sampling instants in the system of Fig. 6.5 $G_R(f)$ should be chosen to be equal to $G_T^*(f)C^*(f)$, in which case the Fourier transform of a single pulse at the output of the matched filter is $|G_T(f)|^2|C(f)|^2$. Hence an additional consideration in the choice of $G_T(f)$ is to ensure that $|G_T(f)|^2|C(f)|^2$ is the Fourier transform of a pulse which does not give rise to excessive ISI.

6.3   Digital routes with repeaters

6.3.1   The regenerative repeater

So that signals do not become too attenuated and too distorted to be received with an acceptably low error rate, long digital routes are divided into 'sections', with a regenerative repeater inserted at the end of each. Not only does a regenerative repeater provide gain to compensate for the attenuation of the section of line preceding it, but it also re-shapes and re-times the transmitted pulses, removing noise and distortion introduced in the section. If the pulses were perfectly reconstructed, there would be no limit to the number of sections in a route and therefore no limit to its length. By contrast, noise and distortion acquired by an analogue signal cannot be removed and their accumulation limits the length of a repeatered route.

In practice the reconstruction of the transmitted pulse train is never quite perfect. There is always a finite probability that a regenerative repeater will make an erroneous decision and there is always some error (jitter) in the timing of the reconstituted pulse train. Since a small decrease in section length gives rise to a proportionately large decrease in error probability, it may be possible, by using more but slightly shorter sections, to lengthen a given route without increasing the error rate at its output. The number of sections is, however, often limited by jitter, which will be discussed further in Section 6.3.3.

A block diagram of a typical regenerative repeater for digital line transmission is shown in Fig. 6.6. As in carrier telephony, the power-separating filters are necessitated by the practice of carrying the direct-current supply to the repeaters on the conductors carrying the signal. The automatic gain control modifies the frequency response of the amplifier in an appropriate way to compensate for variations in cable temperature.

Accurate timing pulses are required in a regenerative repeater both for synchronous detection and for the generation of accurately timed output pulses. As shown in Fig. 6.6, timing is usually derived from the received signal. If the received signal contains a spectral

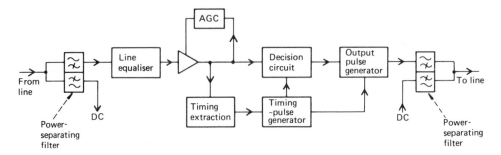

Fig. 6.6   Block diagram of typical regenerative repeater for line digital transmission

line at a suitable frequency, a sinewave can be extracted by a tuned circuit or a phase-locked loop and used to generate the timing pulses. If the received signal does not contain a spectral line, it is necessary to generate one from it by non-linear processing. A simple example of such non-linear processing is the conversion of an AMI RZ bipolar waveform (Exercise 6.13) to an RZ unipolar waveform (Section 6.2.1) by full-wave rectification.

6.3.2   Limitations on section length

If it is assumed that ISI can always be reduced to an acceptable level by equalisation (Section 6.1.3), the degradations likely to limit section length are crosstalk (Section 1.1.5) and thermal noise. As in Section 5.2.2, the latter term is assumed to include fluctuation noise of all kinds.

Far-end crosstalk (FEXT) from signals travelling in the same direction on adjacent cables may be expected from basic considerations to increase with the square of frequency. Hence the ratio of FEXT to signal usually increases with frequency at 6 dB per octave. In addition, the ratio of FEXT to signal is usually proportional to section length (Exercise 6.17(i)). In routes in which FEXT is the dominant impairment (for example multi-pair telephone cable shared by several PCM transmissions travelling in one direction), binary transmission has an advantage over multi-level transmission (Exercise 6.17(ii)).

At the end of a long section of cable, the ratio of near-end crosstalk (NEXT) to signal increases with frequency at 4.5 dB per octave (Exercise 6.17(iii)). In general NEXT would be a more severe limitation than FEXT if transmissions in both directions were allowed to share a multi-pair telephone cable (Exercise 6.17(iv)).

In routes in which the dominant impairment is thermal noise (for example most coaxial-cable routes), ternary transmission is often preferred to binary transmission (Exercise 6.19). This is because the increase in noise margin due to the smaller bandwidth required exceeds the decrease in noise margin due to the greater number of thresholds. Although even larger noise margins could be obtained by transmitting more than three levels, the necessary equalisation would make it uneconomical to do so.

6.3.3   Jitter

Jitter may be defined[8] as short-term variations of the significant instants of a digital signal from their ideal positions in time. It arises in the output of a regenerative repeater when there is phase modulation of the sinewave from which the timing pulses are derived. Accumulation of jitter limits the number of regenerations to which a digital signal may be subjected.

A sinewave derived from the received signal always has residual modulation. Some of it is due to noise accompanying the signal and gives rise to 'non-systematic' jitter, which adds incoherently at successive repeaters. Some of the modulation is due to signal power at frequencies very close to the frequency of the extracted sinewave and gives rise to 'systematic' or 'pattern-induced' jitter, which is derived from the information pattern being transmitted and adds coherently in successive repeaters. It is found that after N repeaters the root-mean-square phase angle of jitter is proportional to $\sqrt{N}$ in the systematic case and to $\sqrt[4]{N}$ in the non-systematic case[9] (see Exercises 6.22 and 6.21). Hence after many regenerations systematic jitter will predominate and is more likely than non-systematic jitter to limit the length of a route. We shall therefore consider only systematic jitter here.

It may be shown (Exercise 6.23) that at frequencies close to the digit rate the continuous part of the spectrum of the digital signal contributes only amplitude modulation to a sinewave extracted for timing purposes. Amplitude modulation of a sinewave does not disturb its zero-crossings, and therefore, if the timing-extraction circuit is accurately set up, no systematic jitter should be introduced. Inaccurate setting-up may, however, cause the amplitude modulation to be converted in part to phase modulation, thus introducing jitter. In particular, two mechanisms of amplitude-to-phase conversion have been studied: mis-tuning of the timing-extraction circuit and misalignment of the threshold voltage at which it operates.

W. R. Bennett[10] showed that when a random binary sequence of impulses at digit rate $\omega_c/2\pi$ is applied to a resonant circuit tuned to ω_0 rad/s, the zero-crossings of the resulting timing wave have mean-square phase error $x_0^2 \pi Q$, where $x_0 [= (\omega_c - \omega_0)/\omega_0 \ll 1]$ is the fractional mis-tuning and Q is the Q-factor of the resonant circuit. Bennett's derivation was based on the summation of impulses. Exercise 6.25 is based on an alternative frequency-domain approach used by Manley[11]. It not only leads to Bennett's formula, but also shows that the spectrum of jitter due to mis-tuning approaches zero at low jitter frequencies.

Because of the low-pass-filter effect of many successive regenerations (Exercise 6.22), jitter due to mis-tuning does not accumulate without limit in a long route (Exercise 6.26) and in practice does not usually limit route length. On the other hand, the spectrum of jitter due to threshold misalignment approaches a uniform level at low jitter frequencies (Exercise 6.27) and accumulates without limit with the number of regenerations (Exercise 6.28), eventually dominating in a very long route.

Since increasing the Q-factor of resonant circuits used for the timing extraction reduces the bandwidth of the accumulated threshold-misalignment jitter, and hence the accumulated jitter power (Exercise 6.28), a high Q-factor should be used. However, the maximum usable Q-factor is limited by the stability of the resonant-circuit components and narrower bandwidths may be obtained by replacing each resonant circuit by a phase-locked loop.

6.3.4 Line codes

It is frequently advantageous not to transmit binary messages directly as baseband pulses or as direct modulation of a carrier, but to carry out some encoding before transmission[6]. Examples of such encoding have already been mentioned: binary encoding to protect against errors (Section 2.3) and pseudo-ternary encoding to remove low frequencies from the transmitted spectrum (Section 6.2.2). Another requirement to be considered in the choice of encoding is the provision of adequate timing content (for digits and words). It is usually necessary for the line code to be 'transparent' (i.e. it should transmit all possible message sequences). An example of lack of transparency is loss of synchronisation when an all-zero message sequence is transmitted in AMI form.

Many of the desirable features of a line code can be obtained by binary encoding. For example, a block code can combine the functions of error control and removal of low frequencies if only those blocks are used which contain balanced numbers of marks and spaces. However, when binary encoding is used for such purposes the rate at which digits are transmitted becomes significantly greater than the rate at which information is transmitted and in many systems it is advantageous to use ternary encoding.

Early PCM transmission over audio cables used AMI pseudo-ternary encoding. This removed low frequencies from the spectrum and enabled the digit error rate to be monitored, since errors introduced bipolar violations (i.e. they destroyed the alternation of positive and negative marks). However, it did not maintain synchronisation during long all-zero sequences. Hence it has been superseded by high-density bipolar[12] (HDB) encoding (Exercise 6.30) or by bi-polar six-zero substitution (B6ZS) encoding (Exercise 6.31). In both of these, marks are inserted into long unbroken sequences of zeros in such a way that they can be recognised by the bipolar violations they create.

Unfortunately encoding a binary message in pseudo-ternary form does not reduce the bandwidth required and therefore decreases the noise margin in systems limited by thermal noise. Hence ternary encoding suitable for coaxial-cable systems is less redundant than pseudo-ternary encoding, although some redundancy may be retained to enable low frequencies to be removed. Suitable redundant-ternary encoding is provided by 4B3T codes, in which groups of four binary message digits are transmitted as groups of three ternary digits. In a 4B3T code there are several alternative ternary representations for some of the binary groups. The alternative chosen is that which minimises the cumulative digital algebraic sum of positive and negative marks. Thus low frequencies are removed and the digit error rate can be monitored by calculating the cumulative digital sum at the receiver. The 000 ternary word is not used: this enables timing to be maintained and helps to separate the received ternary sequence into groups (since any received 000 pattern must straddle the boundary between two groups). A disadvantage of the 4B3T code is that it exhibits error extension; a single error in the ternary transmission may cause as many as four binary digits to be in error after decoding.

6.4 Digital multiplexing

6.4.1 Time-division multiplex

In time-division multiplex (TDM) a digital transmission link is shared between a number of independent signals by being assigned for short periods of time to each signal in turn.

The simplest sharing arrangement to implement is probably that of 'digit interleaving', in which the shared link is switched from one signal to another at the digit rate of the link, each signal contributing just one digit in each time slot allocated to it. However, when the digits of each signal fall naturally into groups (such as the digits representing one telegraph character or the digits representing one sample in PCM telephony), 'character inter-leaving' (or 'word interleaving' or 'byte interleaving') is usually preferred, in which each signal contributes a complete group of digits in each time slot allocated to it.

A length of the transmitted waveform covering one contribution from each of the signals sharing it is called a 'frame'.

As a typical example of a TDM frame, consider the 2048 kbit/s primary PCM multiplex equipment for telephony. Each frame occupies 125 μs (the reciprocal of the sampling frequency for each speech signal, see Section 1.3.1) and contains 256 digits. It is divided into 32 time slots numbered 0 to 31, each containing eight digits. Time slots 1 to 15 and 17 to 31 are assigned to 30 telephone channels and each time slot carries one eight digit speech sample. Time slots 0 and 16 are used for synchronisation and signalling.

6.4.2 Frame alignment

For individual digits in a TDM transmission to be assigned to their correct signals and correctly routed by a receiver, de-multiplexing circuits in the receiver must be 'aligned' with the incoming signal. Hence a receiver must have some means of detecting whether or not it is aligned and some procedure for achieving alignment if it is not.

Hence a TDM waveform must contain some systematic feature which a receiver can use to acquire alignment. It is usually necessary to ensure a systematic feature by inserting the same pattern in certain digits of some frames. Those digits may be distributed throughout the frame or bunched in a 'frame-alignment word' (FAW). For example the 2048 kbit/s PCM system mentioned in Section 6.4.1 includes the pattern 0011011 in the last seven digits of time slot 0 in alternate frames.

Since there is always some risk that a received FAW is corrupted by errors, a receiver will not usually start a re-alignment procedure until it has failed to detect a correct FAW in the appropriate time slot on perhaps four consecutive occasions. Similarly, since there is always a risk that message digits will produce a correct FAW in a misaligned receiver, a receiver will not usually assume alignment on detecting just one correct FAW. The former risk is reduced by decreasing the length of the FAW and the latter by increasing it: hence the choice of length is a compromise.

An approach to calculating the proportion of time for which a receiver is misaligned is described in Exercise 6.33.

When choosing a suitable FAW it is desirable to ensure that message digits can never produce a false indication of alignment when the receiver is out of alignment by a few digits, and to minimise the probability that such a false indication is produced by errors in transmission. Good FAWs are provided by the Barker sequences[13] 110, 1110010 and 11100010010, their time-inverses or their complements. A list of suitable FAWs of other lengths is given by Bylanski and Ingram[6]. Some of the considerations involved in the choice of a FAW are illustrated in Exercises 6.34, 6.35 and 6.36.

6.4.3 Digital hierarchies

In networks handling signals in digital form, time-division-multiplex signals may themselves be combined by TDM for transmission over high-capacity links. In the case of PCM telephony, hierarchies are evolving which are analogous to the FDM hierarchy described in Section 5.3.4. For example, four 30 channel 2048 kbit/s primary multiplex signals may be interleaved to form one second-order multiplex signal at 8448 kbit/s, four second-order signals may be interleaved to form one third-order signal at 34 368 kbit/s and so on.

When a TDM signal is to be formed at a switching centre from lower-order signals arriving by different tributary channels, the timing of the signals must be compatible. Basically there are two approaches to ensuring compatibility of digit rates in a network: synchronism and plesiosynchronism.

Synchronism may be imposed by a hierarchy of master clocks distributed over the network or by arranging for a clock at each node to take up the average phase of the clocks at the nodes connected to it. There are problems when nodes are moving or when link delays are varying with environmental conditions.

In a plesiosynchronous system, nodes use independent clocks to time their transmissions and those clocks are designed to remain within carefully specified frequency limits. Most digital systems are plesiosynchronous. Small differences in digit rate between tributary signals are allowed for in plesiosynchronous systems by a technique known as 'justification' or 'pulse stuffing', in which the information content of the combined signal can be varied to accommodate changes in the digit rates of the tributaries.

The simplest form of justification is 'positive' justification. In positive justification the digit rate of the combined signal is greater than the sum of the digit rates of the tributaries and, when necessary, specified digits of the combined signal are 'justified', that is they carry no information and are disregarded by the receiver. In general, differences between tributary digit rates are small and one justifiable digit per tributary per frame of the combined signal is sufficient. The combined signal must, of course, contain control digits indicating the presence of justification, and those control digits are usually protected from errors. Frequently the presence of justification is determined from a majority decision based on three digits distributed throughout a frame.

Justification causes small fluctuations in the rate of transmission of message digits and hence contributes a source of jitter additional to those sources discussed in Section 6.3.3. The waveform of that justification jitter approximates to a sawtooth (with a line spectrum) plus a random train of rectangular pulses due to variations in the waiting time between the recognition that justification is necessary and the opportunity to implement it. The spectrum of the random component extends to very low frequencies.

All but the very-low-frequency components of justification jitter can be reduced by a 'dejitteriser', in which message digits are stored as they are received and read from the store at a rate synchronised to the average rate of their arrival.

Some details of the CCITT 8448 kbit/s second-order multiplex equipment[14] (which uses positive justification) are given in Exercise 6.37. For further information on the analysis of justification jitter, see in the first instance Bylanski and Ingram[6].

6.5 References

1   Carlson, A. B., *Communication Systems*, McGraw-Hill, Kogakusha, 2nd Edition, 1975
2   Haykin, S., *Communication Systems*, Wiley, Second Edition, 1983
3   Shanmugan, K. S., *Digital and Analog Communication Systems*, Wiley, 1979
4   Bennett, W. R., and Davey, J. R., *Data Transmission*, McGraw-Hill, 1965
5   Lucky, R. W., Salz, J., and Weldon, E. J., *Principles of Data Communication*, McGraw-Hill, 1968
6   Bylanski, P., and Ingram, D. G. W., *Digital Transmission Systems*, IEE Telecommunications Series 4, Peter Peregrinus Ltd, Second Edition, 1980
7   Spilker, J. J., *Digital Communication by Satellite*, Prentice-Hall, 1977
8   *CCITT Yellow Book*, Vol. III, Recommendation G702, Geneva, 1980
9   Byrne, C. J., Karafin, B. J., and Robinson, D. B., 'Systematic jitter in a chain of digital regenerators', *Bell System Technical Journal*, **42**, pp. 2679–714, November 1963
10 Bennett, W. R., 'Statistics of regenerative digital transmission, *Bell System Technical Journal*, **37**, pp. 1501–42, November 1958
11 Manley, J. M., 'The generation and accumulation of timing noise in PCM systems—an experimental and theoretical study', *Bell System Technical Journal*, **48**, pp. 541–613, March 1969
12 Moore, T. A., 'Digital transmission codes: properties of HDB3 and related ternary codes with reference to broadcast signal distribution', *The Radio and Electronic Engineer*, **44**, No. 8, pp. 421–7, August 1974
13 Barker, R. H., 'Group Synchronizing of Binary Digital Systems', in *Communication Theory*, edited by Willis Jackson, Butterworth 1953, pp 273–87
14 *CCITT Yellow Book*, Vol. III, Recommendation G742, Geneva, 1980

Exercise topics

6.1   Rectangular-spectrum pulse
6.2   Nyquist's theorem on vestigial symmetry
6.3   Raised-cosine-spectrum pulse
6.4   Symmetrical additions to raised-cosine spectrum
6.5   Cosine-spectrum pulse
6.6   Duobinary transmission
6.7   Lender's duobinary scheme
6.8   Eye diagram
6.9   Transversal filter
6.10   Power spectrum of NRZ polar pulse train
6.11   Power spectrum of dipulse train
6.12   Power spectrum of Manchester-code pulse train
6.13   Power spectrum of AMI pulse train
6.14   Pseudo-ternary encoding to reduce bandwidth
6.15   Interleaved AMI
6.16   Spacing of regenerative repeaters
6.17   Crosstalk-limited line transmission
6.18   Far-end crosstalk ratios
6.19   Ternary encoding for noise-limited line transmission
6.20   Information capacity of repeatered route
6.21   Accumulation of non-systematic jitter
6.22   Accumulation of systematic jitter
6.23   AM of timing wave due to information
6.24   Modulation index of pattern-induced AM
6.25   Amplitude-to-phase conversion due to mistuning
6.26   Accumulation of jitter due to mistuning
6.27   Amplitude-to-phase conversion due to threshold misalignment
6.28   Accumulation of jitter due to threshold misalignment
6.29   Accumulated response of timing wave to step in pattern
6.30   HDB3 encoding
6.31   B6ZS encoding
6.32   Efficiency of line codes
6.33   Frame alignment, state-transition diagram
6.34   Frame alignment, probability of false synchronisation
6.35   Choice of 4-digit FAW
6.36   Possible 5-digit FAWs
6.37   Justification in second-orderTDM

Exercises

6.1 Rectangular-spectrum pulse

Obtain an expression for the waveform $g(t)$ which has the Fourier transform $G(f)$ given by

$$G(f) = \begin{cases} G_0 \exp(-j2\pi f t_0) & |f| \leqslant f_1 \\ 0 & |f| > f_1 \end{cases}.$$

Show that the maximum signalling speed possible without intersymbol interference for a synchronous binary signal comprising marks of the form of $g(t)$ and spaces is $2f_1$.

Solution outline
Using the inverse Fourier transform (Appendix 1.1),

$$g(t) = \int_{-f_1}^{f_1} G_0 \exp(-j2\pi f t_0) \exp(j2\pi f t)\, df$$
$$= 2f_1 G_0 \frac{\sin\{2\pi f_1 (t - t_0)\}}{2\pi f_1 (t - t_0)} = 2f_1 G_0 \operatorname{sinc}\{2f_1 (t - t_0)\}.$$

Alternatively, write $G(f) = G_0 \operatorname{rect}(f/2f_1) \exp(-j2\pi f t_0)$ and use Appendix 3 and 1.2 (iii). Zero-crossings of $g(t)$ occur at $t = t_0 + k/2f_1$, where k is an integer other than zero.

6.2 Nyquist's theorem on vestigial symmetry

The waveform $g(t)$ has the Fourier transform $G(f)$ defined by

$$G(f) = \begin{cases} 1 & |f| \leqslant f_1 \\ 0 & |f| > f_1 \end{cases}.$$

Show that the waveform $g_1(t)$ with the real Fourier transform $G_1(f)$ derived from $G(f)$ by the addition and subtraction of the equal shapes shown in Fig. 6.7 retains the zero-crossings of $g(t)$.

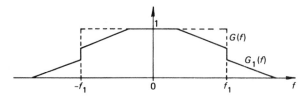

Fig. 6.7 Diagram of Exercise 6.2

Solution outline
$Y(f) = G_1(f) - G(f)$ is an even function of f zero for $|f| \geqslant 2f_1$ and having the (skew-symmetrical) property that

$$-Y(f_1 - x) = Y(f_1 + x) \qquad \text{for } 0 \leqslant x \leqslant f_1.$$

Hence

$$g_1(t) = g(t) + 2\int_0^{2f_1} Y(f) \cos(2\pi f t)\, df$$

$$= g(t) + 2\int_0^{f_1} Y(f) \cos(2\pi f t)\, df + 2\int_{f_1}^{2f_1} Y(f) \cos(2\pi f t)\, df.$$

Putting $f = f_1 - x$ in the first integral and $f = f_1 + x$ in the second,

$$g_1(t) = g(t) + 2 \int_0^{f_1} Y(f_1 - x) \cos \{2\pi (f_1 - x)t\} \, dx$$

$$+ 2 \int_0^{f_1} Y(f_1 + x) \cos \{2\pi (f_1 + x)t\} \, dx$$

$$= g(t) + 2 \int_0^{f_1} Y(f_1 - x) [\cos \{2\pi (f_1 - x)t\} - \cos \{2\pi (f_1 + x)t\}] \, dx$$

$$= g(t) + 4 \sin (2\pi f_1 t) \int_0^{f_1} Y(f_1 - x) \sin (2\pi x t) \, dx.$$

Since the zero-crossings of $\sin 2\pi f_1 t$ include those of $g(t)$ (see Exercise 6.1), zero-crossings of $g(t)$ are also zero-crossings of $g_1(t)$.

6.3 Raised-cosine-spectrum pulse

Obtain an expression for the waveform $g(t)$ which has the 'raised-cosine' spectrum $G(f)$ defined by

$$G(f) = \begin{cases} \frac{1}{2} \{1 + \cos (\pi f/2f_1)\} & 0 \leqslant |f| \leqslant 2f_1 \\ 0 & |f| > 2f_1 \end{cases}.$$

Sketch $g(t)$ for $|t| \leqslant 1/f_1$, showing the values of t for which $g(t) = 0$ and the value of $g(0)$. Show that $2g(1/4f_1) = g(0)$.

Solution outline
Noting that $G(f)$ is even and taking its inverse Fourier transform,

$$g(t) = 2 \int_0^{2f_1} \frac{1}{2} \{1 + \cos (\pi f/2f_1)\} \cos (2\pi ft) \, df = \frac{\sin (4\pi f_1 t)}{2\pi (4f_1)^2 t \left\{ \dfrac{1}{(4f_1)^2} - t^2 \right\}}.$$

To obtain $g(0)$ and $g(1/4f_1)$, differentiate numerator and denominator before taking limits. $g(t)$ is sketched in Fig. 6.8.

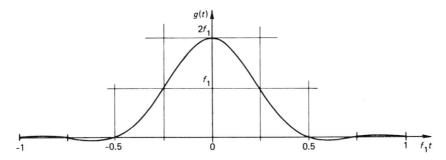

Fig. 6.8 Raised-cosine-spectrum pulse of Exercise 6.3

6.4 Symmetrical additions to raised-cosine spectrum

Show that additions symmetrical about $f = f_1$ and $f = -f_1$ to the raised-cosine spectrum $G(f)$ of Exercise 6.3 do not disturb alternate zero-crossings in the time domain.

Solution outline
Proceed as in Exercise 6.2 except that in this case $Y(f)$ has the property

$$Y(f_1 - x) = Y(f_1 + x) \qquad 0 \leqslant x \leqslant f_1.$$

The inverse Fourier transform of the additions is

$$2 \int_0^{f_1} Y(f_1 - x) \left[\cos \{2\pi (f_1 - x) t\} + \cos \{2\pi (f_1 + x) t\} \right] dx,$$

which equals

$$4 \cos (2\pi f_1 t) \int_0^{f_1} Y(f_1 - x) \cos (xt) \, dx.$$

The values

$$t = \pm 3/4f_1, \ \pm 5/4f_1, \ \ldots \text{ etc.}$$

are zeros both of $g(t)$ of Exercise 6.3 and $\cos 2\pi f_1 t$.

6.5 Cosine-spectrum pulse

Obtain an expression for the waveform $g(t)$ which has the 'cosine' spectrum $G(f)$ defined by

$$G(f) = \begin{cases} \cos (\pi f/2f_1) & 0 \leqslant |f| \leqslant f_1 \\ 0 & |f| > f_1 \end{cases}.$$

Sketch $g(t)$ for $|t| \leqslant 5/4f_1$, showing the values of t for which $g(t) = 0$ and the values of $g(t)$ when $t = 0$, $t = \pm 1/4f_1$ and $t = 1/2f_1$.

Solution outline
Since $G(f)$ is even,

$$g(t) = 2 \int_0^{f_1} \cos \left(\frac{\pi f}{2f_1} \right) \cos (2\pi ft) \, df = \frac{\cos (2\pi f_1 t)}{4\pi f_1 \left\{ \left(\dfrac{1}{4f_1} \right)^2 - t^2 \right\}}.$$

$g(t)$ is sketched in Fig. 6.9. $g(t) = 0$ when $f_1 t = \pm 0.75, \pm 1.25, \pm 1.75$, etc. Note that when $f_1 t = \pm 0.25$, $g(t) = f_1$.

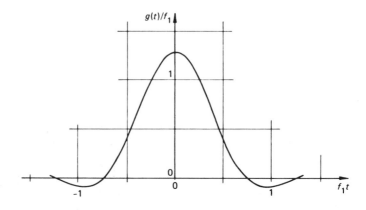

Fig. 6.9 Cosine-spectrum pulse of Exercise 6.5

6.6 Duobinary transmission

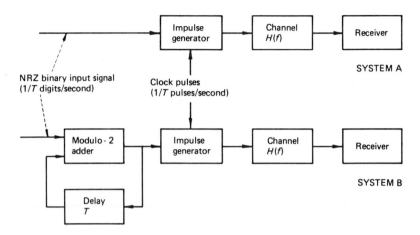

Fig. 6.10 Duobinary transmission systems of Exercise 6.6

Figure 6.10 shows two duobinary transmission systems. In each system,
(i)   each clock pulse causes an impulse to be transmitted if and only if the input signal is a binary 1,
(ii) the frequency response $H(f)$ of the channel is given by

$$H(f) = \begin{cases} \cos(\pi fT) & |f| \leqslant 1/2T \\ 0 & |f| > 1/2T, \end{cases}$$

(iii) the receiver samples its input midway between the times at which the peak in the response to an individual impulse can occur.

For an input binary sequence chosen at random, compile for each system a table showing the values of successive samples of the channel output and state logical rules by which the original binary signal can be derived from those samples. What advantage does System B have over System A?

Solution outline
From Exercise 6.5 (with $f_1 = 1/2T$) each sample is influenced by just two digits and can have one of the three values $0, 1/2T$ and $1/T$. In Table 6.1 these values are denoted by $0, 1$ and 2.

Table 6.1

Input binary sequence	0	0 0 0 1 0 0 1 0 1 1 0 1 1 1 0 0 1
Adder output in B	0*	0 0 0 1 1 1 0 0 1 0 0 1 0 1 1 1 0
Sample values in B		0 0 0 1 2 2 1 0 1 1 0 1 1 1 2 2 1
Sample values in A		0 0 0 1 1 0 1 1 1 2 1 1 2 2 1 0 1

*If a 1 is assumed here, the subsequent sequence is the complement but the following logical rules remain valid.

The correct output is obtained from sample values in B if a 0 or a 2 is interpreted as binary 0 and a 1 as a binary 1. The sample values in A give the correct output if a 0 is interpreted as a binary 0, 2 as a binary 1 and 1 as a change. System A can exhibit error extension, i.e. one error in the channel output can cause more than one error in the derived binary output sequence.

6.7   Lender's duobinary scheme

In Lender's duobinary scheme the transmitted train of marks and spaces is derived from the data sequence by the following rules.

(i)   Transmit a data 1 as a repetition of the previous transmitted symbol.

(ii)   Transmit a data 0 as a change of transmitted symbol (from mark to space or vice versa).

Marks are received as raised-cosine-spectrum pulses with peak value 1 V and the signalling rate is twice the bandwidth of the pulses. The receiver is synchronous and samples its input waveform midway between those times at which peaks of individual marks may occur. Sketch the received waveform when the data sequence is 01100010011110 and hence verify that the correct output may be obtained by interpreting samples between 0.5 V and 1.5 V as 0s and other samples as 1s. Assume that a mark makes a significant contribution to a sample only at the four sampling instants nearest to its peak.

Solution outline

Data sequence   0   1   1   0   0   0   1   0   0   1   1   1   1   0

Transmitted symbol S*   M   M   M   S   M   S   S   M   S   S   S   S   S   M

*assumed previous symbol

The received waveform is sketched in Fig. 6.11.

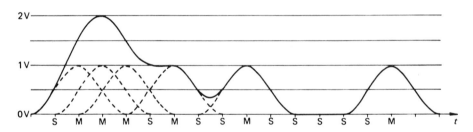

Fig. 6.11   Received waveform of Exercise 6.7

6.8   Eye diagram

A waveform consists of a sequence of marks and spaces. The marks are raised-cosine pulses of the form

$$v(t) = \begin{cases} 5\{1 + \cos(\pi t/T)\} \, \text{V} & |t| \leqslant T \\ 0 \, \text{V} & |t| > T \end{cases}.$$

The digit rate is 25 % greater than the Nyquist rate $(1/T)$. Calculate the eye opening and the reduction in noise margin compared with the noise margin at the Nyquist rate. (The noise margin is the amount in dB by which the noise power may be increased before the error rate becomes unacceptable.)

Solution outline

Intersymbol interference is contributed by at most two neighbouring pulses. With the

10 V peak of the pulse normalised to unity, the eye opening is

$$1 - 2 \times \tfrac{1}{2}\{1 + \cos(0.8\pi)\} = 0.809.$$

Hence the noise margin is decreased by $-20 \log 0.809$ dB, i.e. by 1.84 dB.

6.9 Transversal filter

A transversal filter of the form of Fig. 6.2 has $n = 1$, $w_{-1} = w_1 = a$ and $w_0 = 1$. The input consists of a sequence of marks and spaces, the marks being pulses of the form

$$v(t) = \begin{cases} \tfrac{1}{4}\{1 + \cos(\pi t/2T)\}^2 & |t| \leqslant 2T \\ 0 & |t| > 2T. \end{cases}$$

The digit rate is $1/T$. Denote by R the ratio of the square of the voltage of a sample at the centre of the output waveform to the total squared intersymbol interference at other sampling instants when the filter input is a single mark.
(i) Calculate the value of a for which intersymbol interference between pulses at adjacent sampling instants is completely eliminated and calculate R for that value of a.
(ii) Calculate the value of a which maximises R and calculate the maximum value of R. Write down expressions for the frequency response and the impulse response of the filter.

Solution outline
(i) Since $v(\pm T) = \tfrac{1}{4}$, the required value of a is $-\tfrac{1}{4}$. At the filter output, one pulse contributes the following at five successive sampling instants: $\tfrac{a}{4}$, $a + \tfrac{1}{4}$, $1 + \tfrac{a}{2}$, $a + \tfrac{1}{4}$, $\tfrac{a}{4}$. Hence

$$R = \frac{\left(1 + \dfrac{a}{2}\right)^2}{\left(\dfrac{a}{4}\right)^2 + \left(a + \dfrac{1}{4}\right)^2 + \left(a + \dfrac{1}{4}\right)^2 + \left(\dfrac{a}{4}\right)^2}$$

$$= \frac{2a^2 + 8a + 8}{17a^2 + 8a + 1}.$$

When $a = -\tfrac{1}{4}$, $R = 98$.
(ii) By differentiation, R can be shown to have a maximum value when $a = -7/30$. Putting $a = -7/30$ shows the maximum value of R to be 106. Since a delay T is equivalent to multiplication of the spectrum by $\exp(-j\omega T)$ (Appendix 1.2), the frequency response of the filter is

$$1 + a\{\exp(j\omega T) + \exp(-j\omega T)\} = 1 + 2a \cos \omega T.$$

The impulse response is

$$\delta(t) + a\{\delta(t + T) + \delta(t - T)\}.$$

6.10 Power spectrum of NRZ polar pulse train

Find the power spectrum of an NRZ polar pulse train modulated by an equiprobable random binary sequence. Each pulse is either at $+A$ or $-A$ for the full duration (T) of a digit.

Solution outline
The power in the pulse train is A^2. Since the modulating sequence is equiprobable, the autocorrelation is a single triangle, falling from A^2 at $\tau = 0$ to zero at $\tau = \pm T$. There is no

periodic part to the autocorrelation and therefore there are no lines in the spectrum. The power spectrum is the Fourier transform of the single triangle and may be shown to be $A^2 T \operatorname{sinc}^2 (fT)$ (Appendix 1.1).

6.11   Power spectrum of dipulse train

An equiprobable binary sequence is transmitted by dipulse code, in which 1s are transmitted as dipulses and 0s as spaces, as in the example of Fig. 6.12. Find an expression for the continuous part of the power spectrum and sketch it for $0 \leqslant f \leqslant 4/T$. Determine also the power in the spectral lines lying in that frequency range.

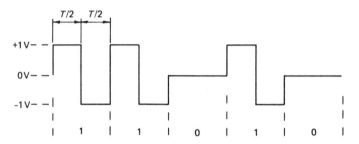

Fig. 6.12   Typical example of dipulse coding for Exercise 6.11

Solution outline
The periodic part of the transmitted signal is a square wave alternating between $+0.5$ V and -0.5 V with frequency $1/T$. With appropriate choice of time origin its Fourier series may be written

$$\frac{2}{\pi} \left\{ \cos \frac{2\pi t}{T} - \frac{1}{3} \cos \frac{6\pi t}{T} + \frac{1}{5} \cos \frac{10\pi t}{T} - \dots \right\}.$$

Hence there are spectral lines at $f = 1/T$ and $f = 3/T$ having (one-sided) powers $2/\pi^2$ and $2/9\pi^2$ respectively.

   The autocorrelation of the random part of the signal is shown in Fig. 6.13 and may be regarded as the sum of three triangles. Hence the continuous part of the power spectrum is (Appendix 1.2):

$$G(f) \{1 - \tfrac{1}{2} \exp(j\pi fT) - \tfrac{1}{2} \exp(-j\pi fT)\} = G(f) \{1 - \cos(\pi fT)\},$$

where (Appendix 1.1 and 3)

$$G(f) = \frac{T}{8} \operatorname{sinc}^2 \left(\frac{fT}{2} \right).$$

6.12   Power spectrum of Manchester-code pulse train

A random equiprobable binary sequence is transmitted in Manchester code, i.e. as a sequence of dipulses as shown for a typical sequence in Fig. 6.14. Obtain an expression for the power spectrum of the transmitted waveform.

Solution outline
The transmitted waveform has no periodic part.
   The autocorrelation of the random part resembles that in Exercise 6.11 except that the

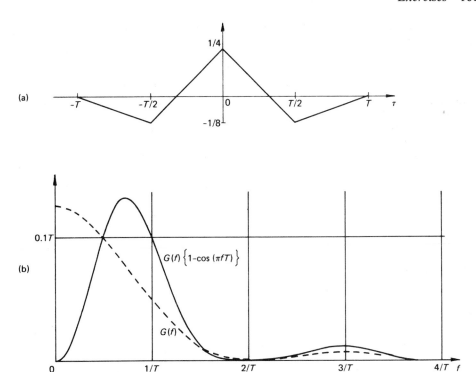

Fig. 6.13 Autocorrelation (a) and power spectrum (b) of dipulse code (solution to Exercise 6.11)

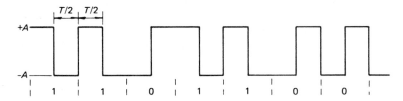

Fig. 6.14 Typical example of Manchester coding of Exercise 6.12

vertical scale is modified so that that autocorrelation is A^2 at $\tau = 0$. (This is the power in the waveform, all of which resides in the random part.) Hence the power spectrum is

$$(A^2T/2)\,\mathrm{sinc}^2\,(fT/2)\{1 - \cos(\pi fT)\}.$$

6.13 Power spectrum of AMI pulse train

In alternate-mark-inversion (AMI) encoding a binary sequence is encoded for transmission as a pseudo-ternary waveform in accordance with the following rules.
(i)   Each binary 0 is encoded as a space.
(ii) Each binary 1 is encoded as a pulse of opposite polarity to the previous pulse.
Assuming that the binary sequence is random and equiprobable, that the digit rate is $1/T$ and that the pulses are RZ rectangular pulses of height $\pm A$ and duration $T/2$, show that

the power spectrum of the pseudo-ternary waveform may be written as the product of two factors thus

$$[(A^2 T^2/4)\, \text{sinc}^2\, (f T/2)] \times [\{1 - \cos (2\pi f T)\}/2T].$$

Verify that the first factor is the energy spectrum of a single transmitted pulse and that the second factor is the power spectrum of an impulse train representing the transmitted ternary sequence.

Solution outline

For $|\tau| \leqslant T/2$ the autocorrelation is identical to that of the unipolar RZ pulse train (Fig. 6.3(a)).

For $T/2 < |\tau| \leqslant 3T/2$ the autocorrelation is made up of contributions from pairs of adjacent pulses. There is a probability of 0.5 that a given pulse is followed by a pulse, and when that occurs the pulses are of opposite polarity. Hence for each of these ranges of τ, the autocorrelation is an inverted triangle with height $A^2/8$ (half the height of the triangle centred on $\tau = 0$).

Since the binary sequence is equiprobable, pulses in the AMI waveform with one digit or more between them are as likely to be of the same polarity as of opposite polarity, and hence for $|\tau| > 3T/2$ the autocorrelation is zero.

The autocorrelation of the RZ AMI waveform is as shown in Fig. 6.15(a) and its Fourier transform gives the required result.

That the first factor is the energy spectrum of a single pulse is easily verified by squaring the modulus of the Fourier transform of the pulse (Appendix 1.1) or by taking the transform of its autocorrelation (Appendix 1.2 and 1.1).

If the pulses in the transmitted waveform were replaced by unit impulses, the autocorrelation of the transmitted waveform would be as shown in Fig. 6.15(b) and hence the power spectrum would be (Appendix 1.6 and 1.3)

$$\frac{1}{2T} - \frac{1}{4T} \exp (j2\pi f T) - \frac{1}{4T} \exp (-j2\pi f T),$$

which is equal to the second factor.

Figure 6.15(a) is of course the convolution of the autocorrelation of Fig. 6.15(b) with the autocorrelation of a single pulse.

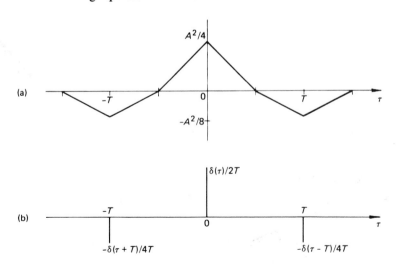

Fig. 6.15 Autocorrelation of (a) RZ AMI waveform, and (b) AMI waveform consisting of unit impulses (solution to Exercise 6.13)

6.14 Pseudo-ternary encoding to reduce bandwidth

A random equiprobable binary sequence is encoded as a pseudo-ternary waveform comprising full-length rectangular pulses of height $\pm A$ and spaces in accordance with the following rules.

(i)   A binary 0 is encoded as a space.

(ii)   A binary 1 is encoded as a pulse of opposite sign to the previous pulse if separated from it by an odd number of spaces.

(iii)   A binary 1 is encoded as a pulse of the same sign as the previous pulse if adjacent to it or separated from it by an even number of spaces.

Obtain an expression for the power spectrum of the pseudo-ternary waveform. Sketch the power spectrum for frequencies below the signalling speed and for comparison sketch on the same diagram the power spectrum of an NRZ unipolar waveform with pulse height A.

Solution outline

The energy spectrum of one rectangular pulse of width T is (Appendix 1.1 and 3)

$$A^2 T^2 \operatorname{sinc}^2 (fT).$$

If the transmitted pulses were unit impulses, the autocorrelation of the pseudo-ternary waveform would be as shown in Fig. 6.16(a) and therefore the power spectrum would be (Exercise 6.13)

$$\{1 + \cos (2\pi fT)\}/2T.$$

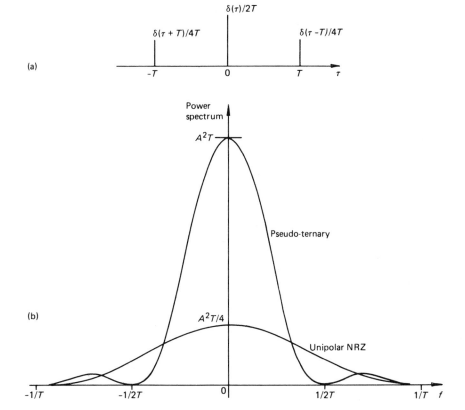

Fig. 6.16 Solution to Exercise 6.14 (a) autocorrelation of pseudoternary impulse train (b) power spectra

The required power spectrum is therefore the product

$$[A^2T^2 \text{sinc}^2(fT)] \times [\{1 + \cos(2\pi fT)\}/2T].$$

The power spectrum of the NRZ waveform is

$$(A^2T/4)\text{sinc}^2(fT)$$

together with a DC term of magnitude $A^2/4$. This may be derived from the Solution to Exercise 6.10.

The two power spectra are sketched in Fig. 6.16(b).

6.15   Interleaved AMI

In interleaved AMI encoding the transmitted pseudo-ternary sequence may be regarded as N interleaved component sequences, each encoded by AMI. Show that the transmitted waveform has spectral nulls at multiples of $1/NT$, where $1/T$ is the rate at which digits are transmitted. Assume the binary message sequence to be random and equiprobable.

Solution outline
The autocorrelation of the impulse train representing the pseudo-ternary encoding of each component consists of an impulse of magnitude $1/2T$ at $\tau = 0$ and impulses of magnitude $-1/4T$ at $\tau = \pm NT$. Hence the power spectrum has as a factor:

$$\frac{1}{2T}\{1 - \cos(2\pi fNT)\},$$

which is zero when f is a multiple of $1/NT$.

6.16   Spacing of regenerative repeaters

Figure 6.17 shows the probability that additive gaussian noise at the input of a certain type of regenerative repeater will introduce an error at the output of that repeater. One hundred of these repeaters are spaced at 2 km intervals to form a 200 km repeatered route. The probability of error at the end of the route is 10^{-4} and the cable attenuation is equalised to 35 dB/km. Errors due to crosstalk and intersymbol interference may be disregarded.

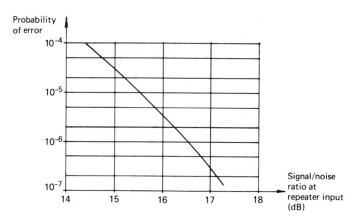

Fig. 6.17   Diagram of Exercise 6.16

(i) After installation of the route, earthworks cause one of the 2 km cable sections to be lengthened by 50 m. Determine the new probability of error at the end of the route.

(ii) A new route of length 1000 km is to be installed using cable with the same characteristics and repeaters of the same type. Calculate the minimum number of repeaters required to obtain a probability of error of 10^{-4} at the end of the route.

Solution outline

(i) The digit error probability for a single repeater is approximately 10^{-6} and hence the signal-to-noise ratio at the input of each repeater is approximately 16.5 dB. Lengthening a section by 50 m increases its equalised attenuation by (35×0.05) dB and hence the signal-to-noise ratio at the input of the following repeater is reduced by 1.75 dB to 14.75 dB. The digit error probability for that repeater is increased to about 5×10^{-5}. Hence the digit error probability for the whole route is increased to about 1.5×10^{-4}.

(ii) Since digit error probability is very sensitive to changes in section length, for a first estimate assume that the digit error probability for each repeater should be $10^{-4} \div 500 = 2 \times 10^{-7}$. This requires a signal-to-noise ratio of 17.1 dB and therefore the repeater spacing should be $2 - (17.1 - 16.5)/35$ km $= 1.98$ km. Hence $1000/1.98 = 505$ repeaters are necessary. Refinement of the answer by assuming that the digit error probability for each repeater should be $10^{-4} \div 505$ does not significantly modify the estimate of 505 repeaters.

6.17 Crosstalk-limited line transmission

Figure 6.18 shows two digital transmission systems operating in the same direction over two pairs of a multi-pair cable. The loss of a length x of a pair is $\exp(ax\sqrt{f})$ at frequency f, where a is a constant. Each amplifier has gain $\exp(al\sqrt{f})$ to compensate for cable loss. Crosstalk between pair elements of length δx is represented by a path of gain $bf^2\delta x$, where b is a constant. The signal power P is uniformly distributed from 0 Hz to B Hz.

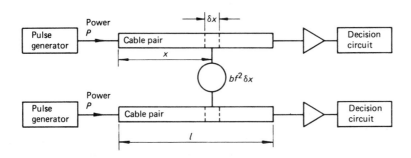

Fig. 6.18 Diagram of Exercise 6.17

(i) Obtain an expression for the signal-to-FEXT ratio at the input to a decision circuit.

(ii) Use the expression derived in (i) to show that the noise margin is decreased if more than two levels are used to transmit information at a given rate when FEXT is the dominant impairment.

(iii) Assuming that the direction of transmission of one of the systems is reversed, obtain an expression for the NEXT power spectral density at the input to a decision circuit.

(iv) Show that at the input to a decision circuit the NEXT power spectral density is not less than the FEXT power spectral density at all frequencies, and determine the ratio of those spectral densities at the frequency at which the loss of one section of a pair is 40 dB.

Solution outline
(i)   At the input to a decision circuit the signal power is P. The FEXT power spectral density (one-sided) is

$$\int_0^{\cdot l} \frac{P}{B} \exp(-al\sqrt{f})bf^2 \exp(al\sqrt{f})\,dx = \frac{Pblf^2}{B}.$$

Hence the FEXT power is

$$\int_0^{\cdot B} \frac{Pblf^2}{B}\,df = \frac{PblB^2}{3}.$$

Hence the required ratio is $3/blB^2$.
(ii)   If L levels are used,
   (a)   the voltage difference between adjacent levels is proportional to $(L-1)^{-1}$,
   (b)   the bandwidth required is proportional to $(\log_2 L)^{-1}$ and hence from (i) the r.m.s. FEXT voltage is proportional to $(\log_2 L)^{-1}$.
Hence the noise margin is proportional to $(\log_2 L)^2(L-1)^{-2}$, which decreases with L for $L \geqslant 2$.
(iii)   The NEXT power spectral density (one-sided) at the input to a decision circuit is

$$\int_0^{\cdot l} \frac{P}{B} \exp(-2ax\sqrt{f})bf^2 \exp(al\sqrt{f})\,dx = \frac{Pb}{aB}f^{3/2}\sinh(al\sqrt{f}).$$

Note that when $al\sqrt{f}$ is large, the sinh term is approximately half the gain of the amplifier.
(iv)   The ratio of NEXT power spectral density to that of FEXT is

$$\frac{Pb}{aB}f^{3/2}\sinh(al\sqrt{f})\bigg/\frac{Pblf^2}{B} = \frac{\sinh(al\sqrt{f})}{al\sqrt{f}}.$$

This ratio is always greater than unity. When the section loss is 40 dB,

$$\exp(al\sqrt{f}) = 10^4.$$

Hence

$$al\sqrt{f} = 9.210.$$

Hence the ratio is 542.9, i.e. 27.3 dB.

6.18   Far-end crosstalk ratios

A multi-core cable consisting of a large number of balanced pairs carries baseband digital signals of equal mean power in the same direction. Far-end crosstalk (FEXT) at 10 kHz accumulates over a 1 km length to a level 90 dB below the signal. The cable is divided into sections of length 2 km separated by regenerative receivers and in each repeater the decision circuits are preceded by equalising amplifiers which compensate for the attenuation of the cable at all frequencies used. The transmitted signal power is uniformly distributed from 0 Hz to 2 MHz and the attenuation of a section of cable at 2 MHz is 75 dB. Assuming that six near neighbours contribute equally to the crosstalk in any pair, calculate
   (i)   the ratio of the power spectral density of the signal to that of FEXT at the output of an amplifier at 2 MHz,
   (ii)   the ratio of signal power to FEXT power at the output of an amplifier,
   (iii)   the ratio of signal power to FEXT power at the input of an amplifier.

Solution outline
With the notation of Exercise 6.17, $blf^2 = 10^{-9}$ when $l = 10^3$ and $f = 10^4$. Hence

$b = 10^{-20}$. Also $\exp(al\sqrt{f}) = 10^{7.5}$ when $l = 2000$ and $f = 2 \times 10^6$. Hence $a = 6.11 \times 10^{-6}$. Also $B = 2 \times 10^6$.

(i)   The power spectral density of the signal at the output of an amplifier is P/B. The power spectral density of FEXT at the output of an amplifier is $6Pbf^2l/B$. Hence the ratio is $1/6bf^2l$, which is 33.2 dB at 2 MHz.

(ii)   The signal power at the output of an amplifier is P. The FEXT power at the output of an amplifier is

$$\int_0^B \frac{6Pbf^2l}{B}\,\mathrm{d}f = 2PbB^2l.$$

Hence the ratio is $1/2bB^2l$, which is 38.0 dB.

(iii)   At the input of an amplifier the signal power is

$$\int_0^B \frac{P}{B}\exp(-al\sqrt{f})\,\mathrm{d}f.$$

The integration may be performed by substituting y for $al\sqrt{f}$ and then integrating by parts. The result is

$$(2P/a^2l^2B)\{1 - (1 + al\sqrt{B})\exp(-al\sqrt{B})\},$$

which is approximately $2P/a^2l^2B$, since $\exp(-al\sqrt{B}) = 10^{-7.5}$. The FEXT power at the input of an amplifier is

$$\int_0^B \frac{6Pbf^2l}{B}\exp(-al\sqrt{f})\,\mathrm{d}f.$$

By putting $y = al\sqrt{f}$ and integrating by parts several times, the FEXT power may be shown to be approximately

$$1440\, Pb/Ba^6l^5.$$

Hence the ratio is 1.55×10^6 (61.9 dB).

6.19   Ternary encoding for noise-limited line transmission

Each section of a digital route consists of a length l of coaxial cable with loss $\exp(al\sqrt{f})$ at f Hz (where a is a constant) followed by an equalising amplifier with gain $\exp(al\sqrt{f})$. The noise temperature of the amplifier referred to its input is T at all frequencies.

(i)   Prove that at the amplifier output the noise power in the frequency band extending from 0 Hz to B Hz is approximately

$$(2kT\sqrt{B}/al)\exp(al\sqrt{B}),$$

provided that $al\sqrt{B} \gg 1$.

(ii)   Assuming that when binary transmission is used the attenuation of a section of cable at the highest frequency transmitted is 75 dB, show that a greater noise margin may be obtained by using ternary transmission to convey information at the same rate. Disregard all crosstalk and intersymbol interference.

Solution outline

(i)   The noise power at the amplifier output is

$$kT\int_0^B \exp(al\sqrt{f})\,\mathrm{d}f.$$

Putting y for $al\sqrt{f}$ and integrating by parts shows this integral to be

$$(2kT/a^2l^2)\{(al\sqrt{B}-1)\exp(al\sqrt{B})+1\},$$

which approximates to the required expression when $al\sqrt{B} \gg 1$.

(ii)   If L levels are used,

   (a)   the voltage difference between adjacent levels is proportional to $(L-1)^{-1}$,
   (b)   the bandwidth required is $B(\log_2 L)^{-1}$ (where B is required in the binary case) and hence from (i) the noise power is equal to

$$\left\{\frac{2kT}{al}\sqrt{\frac{B}{\log_2 L}}\right\}\exp\left\{al\sqrt{\frac{B}{\log_2 L}}\right\}.$$

Hence the noise margin is proportional to

$$(L-1)^{-2}(\log_2 L)^{1/2}\exp\left\{-al\sqrt{\frac{B}{\log_2 L}}\right\}.$$

Since $\exp(al\sqrt{B}) = 10^{7.5}$, $al\sqrt{B} = 17.3$. Hence the above factor is found to be

$$3.2 \times 10^{-8} \text{ when } L = 2$$
$$\text{and } \quad 3.5 \times 10^{-7} \text{ when } L = 3,$$

showing that ternary transmission increases the noise margin by more than 10 dB.

6.20   Information capacity of repeatered route

An existing small-bore, coaxial-cable, carrier-telephony installation is to be modified to carry digital signals. Each repeater will be replaced by a regenerative repeater in which the decision circuit is preceded by an equalising amplifier. Each equalising amplifier compensates accurately for the attenuation of the preceding length of cable at frequencies (in Hz) less than the signalling rate (in baud) and stops all higher frequencies. Estimate the maximum rate at which information may be transmitted by binary transmission assuming the following.

(i)   The cable attenuation is $5.3\sqrt{f}$ dB/km at f MHz.
(ii)   The section length is 2.2 km.
(iii)   The noise temperature of an equalising amplifier and its source is 1500 K at all frequencies.
(iv)   The minimum acceptable signal-to-noise ratio at a decision circuit is 20 dB.
 (v)   The mean transmitted power cannot exceed 100 mW.
(vi)   Crosstalk and intersymbol interference are negligible.

   Estimate also the maximum rate at which information may be transmitted by redundant ternary 4B3T transmission assuming that the same noise margin is required. [Boltzmann's constant $k = 1.38 \times 10^{-23}$ J/K.]

Solution outline
Calculate first the maximum bandwidth for which the signal-to-noise ratio at a decision circuit is acceptable.

   With the notation of Exercise 6.19 and using the expression for noise power derived in its solution, an acceptable signal-to-noise ratio requires

$$\frac{2kT}{a^2l^2}\{(al\sqrt{B}-1)\exp(al\sqrt{B})+1\} < 10^{-3} \text{ W}.$$

From the data given, $T = 1500$, $l = 2200$, $k = 1.38 \times 10^{-23}$ and $a = 1.22 \times 10^{-6}$. The maximum value for $al\sqrt{B}$ may be found to be approximately 22.80. Hence the maximum bandwidth (B) is 72.2 MHz.

Since the bandwidth of the link is equal to the signalling rate, information may be transmitted by binary transmission at 72.2 Mbit/s.

To obtain the same noise margin with ternary transmission, a signal-to-noise ratio of 26 dB is required at the decision circuit and the noise must therefore not exceed 2.5×10^{-4} W. Solving again for B gives the maximum signalling rate to be 6.40×10^7 baud. Since three signalling elements carry a maximum of four bits of information, information may be transmitted by 4B3T transmission at 85.4 Mbit/s.

6.21 Accumulation of non-systematic jitter

The accumulation of non-systematic jitter in successive regenerative repeaters may be estimated from a model based on the following assumptions.

(i)   The jitter injected by a repeater is equivalent to injecting random jitter with uniform power spectral density Φ at the input of that repeater.
(ii)   The jitter injected by a repeater adds linearly on a power basis to jitter passed on from previous repeaters.
(iii)   The effect of a repeater on jitter at its input is equivalent to passing that jitter through a single-pole low-pass filter with its pole at a rad/s (half the bandwidth of the tuned circuit extracting the timing sinewave).

Show that after N repeaters the power spectral density of the accumulated jitter is $\Phi_N(\omega)$ given by:

$$\Phi_N(\omega) = \Phi \sum_{n=1}^{N} \left(1 + \frac{\omega^2}{a^2}\right)^{-n}.$$

Hence show that $\Phi_N(\omega) \to N\Phi$ as $\omega \to 0$ and the half-power bandwidth of the accumulated jitter is approximately $a\sqrt{1.6/N}$ rad/s.

Solution outline
The frequency response of a repeater to jitter may be written $(1 + j\omega/a)^{-1}$. The expression for $\Phi_N(\omega)$ follows. Noting that the expression is the sum of a geometric series,

$$\Phi_N(\omega) = \Phi \frac{a^2}{\omega^2} \left\{1 - \left(1 + \frac{\omega^2}{a^2}\right)^{-N}\right\}.$$

Since $(1 + \omega^2/a^2)^{-N} \sim \exp(-N\omega^2/a^2)$ as $\omega \to 0$, $\Phi_N(\omega) \to \Phi N$ as $\omega \to 0$.
For the half-power bandwidth, find ω which satisfies

$$\frac{a^2}{\omega^2} \left\{1 - \exp\left(-\frac{N\omega^2}{a^2}\right)\right\} = \frac{N}{2}.$$

Trial-and-error solution gives $N\omega^2/a^2 = 1.6$.

Note that the product of $\Phi_N(0)$ and the half-power bandwidth is proportional to $\sqrt{N}$, suggesting that for large N the r.m.s. jitter is proportional to $\sqrt[4]{N}$ as stated in Section 6.3.3.

6.22 Accumulation of systematic jitter

Use the basic model outlined in Exercise 6.21 to show that when identical systematic jitter $\theta(t)$ is injected at each repeater input the power spectral density of the accumulated jitter after N repeaters is $\Phi_N(\omega)$ given by

$$\Phi_N(\omega) = \Phi \frac{a^2}{\omega^2} \left|1 + \left(1 + \frac{j\omega}{a}\right)^{-N}\right|^2,$$

where Φ is the (uniform) power spectral density of $\theta(t)$.

Hence show that $\Phi_N(\omega) \to \Phi N^2$ as $\omega \to 0$ and that the half-power bandwidth of the accumulated jitter is approximately $2.8\,a/N$ rad/s.

Solution outline

If $\Theta(s)$ is the Fourier transform of $\theta(t)$, the transform of the accumulated jitter is

$$\Theta(s) \sum_{n=1}^{N} \left(1 + \frac{s}{a}\right)^{-n},$$

which is the sum of a geometric progression and may be written

$$\Theta(s)\frac{a}{s}\left\{1 - \left(1 + \frac{s}{a}\right)^{-N}\right\}.$$

Since $|\Theta(j\omega)|^2$ is equal to Φ for all ω, the required result follows. For small ω,

$$\Phi_N(\omega) \simeq \Phi\frac{a^2}{\omega^2}\left|1 - \exp\left(-\frac{j\omega N}{a}\right)\right|^2$$

$$= \Phi\frac{4a^2}{\omega^2}\sin^2\left(\frac{N\omega}{2a}\right).$$

The low-frequency value of $\Phi_N(\omega)$ is therefore approximately ΦN^2 and when $\omega = 2.8a/N$, $\Phi_N(\omega)$ is approximately $\Phi N^2/2$.

Note that the product of $\Phi_N(0)$ and the half-power bandwidth is proportional to N, suggesting that for large N the r.m.s. jitter is proportional to $\sqrt{N}$ as stated in Section 6.3.3.

6.23   AM of timing wave due to information

An equiprobable binary sequence is transmitted synchronously as a sequence of marks and spaces. The marks are pulses of identical but unspecified shape. A timing wave is extracted from the transmitted waveform by a high-Q resonant circuit tuned to the digit rate. By regarding the transmitted waveform as the product of a periodic pulse train and a random NRZ waveform, show that the effect of the information pattern on the timing wave approximates to amplitude modulation.

Solution outline

Apart from a line at zero frequency, the power spectrum of the random NRZ waveform is of the form derived in Exercise 6.10 and therefore has nulls at the digit rate and its harmonics. The periodic pulse train has spectral lines at the digit rate and its harmonics. In the spectrum of the product, each spectral line has continuous sidebands, the nulls in which ensure that near to the spectral line at the digit rate there is only a very small contribution from sidebands of other spectral lines. Hence the extracted timing wave consists almost entirely of the product of a sinewave and the random waveform and therefore approximates to an amplitude-modulated sinewave.

6.24   Modulation index of pattern-induced AM

A random equiprobable binary sequence is transmitted with digit rate $1/T$ as a sequence of marks and spaces, the marks being rectangular RZ pulses of duration $T/2$. Show that the power in two small elements δq of the transmitted (one-sided) spectrum situated q rad/s above and below the spectral line at the digit rate is equal to the sideband power associated with sinusoidal amplitude modulation of that spectral line with modulation index $\sqrt{2T\delta q/\pi}$.

Solution outline

From Section 6.2.1, when the pulse height is A, the amplitude of the spectral line is A/π and the one-sided power spectral density in the elements is approximately $A^2 T/2\pi^2$. Hence the total power in the elements is

$$2\left(\frac{\delta q}{2\pi}\right)\left(\frac{A^2 T}{2\pi^2}\right).$$

If the waveform of the sinusoidally modulated line is written

$$\frac{A}{\pi}\{1 + m\cos(qt + \phi)\}\cos\left(\frac{2\pi t}{T}\right),$$

the sideband power is $A^2 m^2/4\pi^2$.

Equating these two powers gives $m^2 = 2T\delta q/\pi$.

6.25 Amplitude-to-phase conversion due to mistuning

(i) Show that the impedance at ω rad/s of a parallel resonant circuit tuned to ω_0 rad/s is approximately

$$\frac{R}{1 + j2Qx},$$

where $x = (\omega - \omega_0)/\omega_0$, provided that x (but not necessarily Qx) is small.

(ii) By comparing the expressions

$$\{1 + m\cos(qt + \phi)\}\cos\{\omega_c t + k\cos(qt + \theta)\} \qquad (0 < m \ll 1, 0 < k \ll 1)$$

and

$$\cos\omega_c t + a_1\cos\{(\omega_c - q)t + \alpha_1\} + a_2\cos\{(\omega_c + q)t + \alpha_2\},$$

show that

$$k^2 = (a_1 - a_2)^2 + 4a_1 a_2 \sin^2\left(\frac{\alpha_1 + \alpha_2}{2}\right).$$

(iii) A current

$$\{1 + m\cos qt\}\cos\omega_c t$$

is applied to the resonant circuit of (i). Show that the voltage across the circuit is angle-modulated with mean-square phase deviation

$$8m^2 Q^4 x_0^2 \xi^2 (1 + 4Q^2 \xi^2)^{-2},$$

where $x_0 = (\omega_c - \omega_0)/\omega_0 \ll 1$ and $\xi = q/\omega_0$.

(iv) Assuming that the amplitude modulation in (iii) is pattern-induced and that $m^2 = 2T\delta/\pi$, as for the waveform of Exercise 6.24, prove that the total mean-square phase deviation of the voltage is $x_0^2 \pi Q$.

Solution outline

(i) Assuming R,L,C in parallel, $\omega_0\sqrt{LC} = 1$ and $Q = \omega_0 CR = R\sqrt{C/L}$, the impedance at ω rad/s is

$$\frac{R}{1 + jR\left(\omega C - \dfrac{1}{\omega L}\right)} = \frac{R}{1 + jR\sqrt{\dfrac{C}{L}}\left(\omega\sqrt{LC} - \dfrac{1}{\omega\sqrt{LC}}\right)} \simeq \frac{R}{1 + j2Qx},$$

provided that x is small.

(ii)   Write each expression as the weighted sum of the five spectral components $\cos \omega_c t$, $\cos (\omega_c \pm q)t$, $\sin (\omega_c \pm q)t$. Compare coefficients and solve resulting equations for k^2.

(iii)   Subtracting the time delay $(1/\omega_c) \tan^{-1} (2Qx_0)$ of the carrier from that of each of the sidebands leads to

$$\alpha_1 = -(\omega_c - q)\left[\frac{1}{\omega_c - q} \tan^{-1}\{2Q(x_0 - \xi)\} - \frac{1}{\omega_c}\tan^{-1}(2Qx_0)\right]$$

$$\simeq \tan^{-1}\left\{\frac{2Q\xi}{1 - 4Q^2 x_0\xi}\right\} - 2Qx_0\xi \qquad (Qx_0 \ll 1)$$

and $\alpha_2 \simeq \tan^{-1}\left\{\dfrac{-2Q\xi}{1 + 4Q^2 x_0\xi}\right\} + 2Qx_0\xi.$

Also (neglecting $Q^2 x_0^2$ compared with unity)

$$a_1 \simeq (m/2)\{1 + 4Q^2(x_0 - \xi)^2\}^{-1/2}$$

$$a_2 \simeq (m/2)\{1 + 4Q^2(x_0 + \xi)^2\}^{-1/2}$$

Hence

$$k^2 \simeq 16m^2 Q^4 x_0^2 \xi^2 (1 + 4Q^2\xi^2)^{-2}$$

and the mean-square phase deviation is $k^2/2$.

(iv)   Since $\delta\xi = \delta q/\omega_0$, the total mean-square phase deviation is

$$\int_0^\infty \frac{8Q^4 x_0^2 \xi^2}{(1 + 4Q^2\xi^2)^2}\left(\frac{2T}{\pi}\right)\omega_0 d\xi = \int_0^\infty \frac{32Q^4 x_0^2 \xi^2}{(1 + 4Q^2\xi^2)^2}\,d\xi.$$

To integrate, put $2Q\xi = \tan y$. The result is $x_0^2 \pi Q$ (Bennett's formula).

6.26   Accumulation of jitter due to mistuning

Use results obtained in Exercise 6.22 (for large N) and Exercise 6.25 to show that after a large number of regenerations the peak value of the power spectral density of the systematic jitter due to mis-tuning is approximately 16 times its value after one regeneration. [Hint: assume that the peak value after N regenerations occurs when $N\omega/2a = \pi/2$.]

Show also that for the waveform of Exercise 6.24 the accumulated jitter power is less than four times Bennett's result for a single regeneration.

Solution outline

From Exercise 6.25, the jitter power due to modulation components at $\xi\omega_0$ rad/s is

$$8m^2 Q^4 x_0^2 \xi^2 (1 + 4Q^2\xi^2)^{-2}.$$

By differentiation, this may be shown to have a maximum value $m^2 Q^2 x_0^2/2$ when $\xi = 1/2Q$.

Using the solution and notation of Exercise 6.22, the jitter power due to modulation at $\xi\omega_0$ rad/s may be shown to accumulate after N regenerations to

$$8m^2 Q^4 x_0^2 \xi^2 (1 + 4Q^2\xi^2)^{-2} (4a^2/\omega^2)\sin^2 (N\omega/2a).$$

Assuming that the peak value of this occurs when $N\omega/2a = \pi/2$ and noting that $\omega = \xi\omega_0$ and $2Qa = \omega_0$, the peak value may be shown to be approximately $8m^2 Q^2 x_0^2$, as is required.

Since $\sin^2 (N\omega/2a)$ cannot exceed unity, proceeding as in Exercise 6.25 (iv) shows that the accumulated jitter power cannot exceed

$$\int_0^\infty 32Q^2 x_0^2 (1 + 4Q^2\xi^2)^{-2} \, d\xi.$$

To integrate, put $2Q\xi = \tan y$, giving $4Qx_0^2\pi$ as required.

6.27 Amplitude-to-phase conversion due to threshold misalignment

The wave $A \sin \omega t$ is applied to a decision circuit with threshold level b ($b \ll A$). Show that a small change δA in A alters the phase angle of the threshold crossing by $b\delta A/A^2$.
   Hence show that if the amplitude-modulated wave

$$A(1 + m \cos qt)\cos \omega_c t$$

is applied to the decision circuit, by way of a resonant circuit tuned to ω_c, the peak phase deviation of the threshold crossing is

$$\left| \frac{bm}{A(1 + 4Q^2\xi^2)^{1/2}} \right|,$$

where $\xi = q/\omega_c$.

Solution outline
If increasing A moves the threshold crossing from $\omega t = \theta$ to $\omega t = \theta + \delta\theta$,

$$b = A \sin \theta = (A + \delta A)\sin (\theta + \delta\theta).$$

Hence

$$A \sin \theta \simeq A \sin \theta + \delta A \sin \theta + A \, \delta\theta \cos \theta$$

i.e. $\delta\theta \simeq -\delta A (\tan \theta)/A.$

But since $b \ll A$, $\tan \theta \simeq \sin \theta = b/A$. Therefore

$$\delta\theta \simeq -b\delta A/A^2.$$

   The resonant circuit reduces the modulation index to $m(1 + 4Q^2\xi^2)^{-1/2}$ (See Exercise 6.25). Substituting this for $\delta A/A$ gives the required result.

6.28 Accumulation of jitter due to threshold misalignment

(i)   Use a result derived in Exercise 6.27 to show that at low frequencies the mean-square jitter in a 1 rad/s bandwidth due to threshold misalignment in one repeater is Tk^2/π, where $1/T$ is the frequency of the timing wave and k the threshold misalignment expressed as a fraction of the amplitude of the timing wave.
(ii)   Assuming that the injected threshold-misalignment jitter is the same at each repeater, show that after a large number N of regenerations it has accumulated to a mean-square value $\pi k^2 N/Q$.
   [In (ii), assume that the jitter due to each repeater has a uniform power spectrum and use the result derived in Exercise 6.22 that the effect of N regenerations is to multiply that power spectrum by $(4a^2/\omega^2)\sin^2 (N\omega/2a)$.]

Solution outline
(i)   As in Exercise 6.24, regard pattern-induced amplitude modulation in a bandwidth $\delta\omega$ at ω rad/s as sinusoidal modulation with index m, where $m^2 = 2T\delta\omega/\pi$.

From the result proved in Exercise 6.27, the mean-square phase deviation due to the effect of threshold misalignment on sinusoidal amplitude modulation with index m is $m^2b^2/2A^2$. Putting k for b/A and $2T\delta\omega/\pi$ for m^2, the mean-square jitter in one repeater in a bandwidth $\delta\omega$ at ω rad/s is therefore $T\delta\omega k^2/\pi$, from which the required result follows.

(ii)   After N regenerations, the total mean-square jitter has accumulated to

$$\int_0^\infty \frac{Tk^2}{\pi} \cdot \frac{4a^2}{\omega^2} \sin^2\left(\frac{N\omega}{2a}\right) d\omega.$$

Putting $N\omega/2a = \pi x$, this may be written

$$\frac{1}{2}\int_{-\infty}^\infty 2aTk^2N \operatorname{sinc}^2(x)\,dx,$$

which is equal to aTk^2N (Appendix 3). Since $aT = \pi/Q$, the required result follows.

6.29   Accumulated response of timing wave to step in pattern

In a long route using regenerative repeaters a change in the information pattern causes a step of 0.1 radians in jitter to be introduced by each repeater simultaneously. Describe the output jitter as a function of time. Show that if the resonant circuits used for timing extraction are tuned to 2.048 MHz and have Q-factor 100 there is a frequency offset of about 1 kHz lasting for about 15 ms in the timing wave in the 1000th repeater.

[Assume that the transform of the output jitter is the product of the transform of the jitter introduced at a repeater and the 'transfer function'

$$\frac{a}{s}\left\{1 - \exp\left(-\frac{Ns}{a}\right)\right\}$$

derived in Exercise 6.22.]

Solution outline
The exponential term represents a delay N/a and the factor $1/s$ represents integration. Hence the step in phase gives rise to a ramp with gradient $0.1a$ which continues until cancelled by a ramp with a complementary slope beginning after a delay N/a.

Since $a = (2\pi \times 2.048 \times 10^6)/2Q$, the frequency offset $(0.1a)$ represented by the ramp is $(2\pi \times 1024)$ rad/s ($\simeq 1$ kHz) and since $N = 1000$ the duration of the ramp is $1000/(2\pi \times 10\,240)s(\simeq 15$ ms).

6.30   HDB3 encoding

The rules for HDB3 encoding may be stated as follows, where B denotes a ternary mark obeying the bipolar rule (having opposite polarity to the previous ternary mark) and V denotes a ternary mark violating the bipolar rule.
(1)   Each binary 1 is encoded as B.
(2)   Each binary 0 in a sequence of three or less binary 0s is encoded as a ternary 0.
(3)   Each sequence of four binary 0s is encoded as either (a)   B00V, or (b)   000V, the choice of (a) or (b) being such that successive violations V are of opposite polarity.
(4)   Each sequence of five or more binary 0s is encoded by dividing it into blocks of four binary 0s, starting from the beginning. Complete blocks are encoded by rule (3) and any remaining 0s by rule (2).
  (i)   Decode the following ternary sequence

$$+0-000-+-\ +00\ +0-000-\ +-0\ +0-.$$

(ii)   Calculate the average rate at which bipolar violations occur in an HDB3 sequence derived from a random equiprobable binary message sequence. Assuming that patterns (a) and (b) occur equally often, estimate the percentage by which the power in an HDB3 signal exceeds that in an AMI signal using the same pulse waveform.

(iii) Explain briefly how a received HDB3 sequence indicates the presence of a single error.

Solution outline

(i) Locate the bipolar violations. Each denotes the end of a block of four binary 0s, thus:

$$+ 0 - \quad 000 - \quad + - + 00 + \quad 0 - \quad 000 - \quad + - 0 + 0 -$$
$$V V V$$
$$1\ 0\ 1 \quad 000\ 0 \quad 1\ 1\ 0\ 00\ 0 \quad 0\ 1 \quad 000\ 0 \quad 1\ 1\ 0\ 1\ 0\ 1$$

(ii) A binary message digit is encoded as a bipolar violation if, and only if,

it is 0 and just three   preceding binary digits are 0
or it is 0 and just seven   preceding binary digits are 0
or it is 0 and just eleven preceding binary digits are 0
or ... etc.

Hence the probability of a bipolar violation occurring is

$$2^{-5} + 2^{-9} + 2^{-13} + \ldots = 1/30$$

and therefore one ternary digit in 30 is an introduced bipolar violation. Assuming that half of these violations are also accompanied by an additional 'B' mark, the number of marks and therefore the power in an HDB3 waveform is ten per cent greater than that in the corresponding AMI waveform. In fact this assumption slightly underestimates the increase in power.

(iii) A single error either destroys a bipolar violation or creates one. In either case the alternation of the parity of the violations is violated.

6.31 B6ZS encoding

In bipolar six-zero substitution (B6ZS) encoding, an AMI signal is modified to have no all-zero sequences longer than five by replacing groups of six consecutive zeros by 0VB0VB, where B and V are defined as in Exercise 6.30.

(i) Decode the following ternary sequence:
$$+ 0 + - 0 - + 0 - + - + 0 + - 0 - + 0 + - 0 - + 00 - 0 + -$$

(ii) Calculate the average rate at which bipolar violations occur in a B6ZS signal derived from a random equiprobable binary message sequence and hence estimate the percentage by which the power in a B6ZS signal exceeds that in the AMI signal from which it is derived.

Solution outline

(i) Locate bipolar violations. Each is the second of three binary 0s:

$$+ 0 + - 0 - + 0 - + - + 0 + - 0 - + 0 + - 0 - + 00 - 0 + -$$
$$V V V V V V$$
$$1\ 0\ 0\ 0\ 0\ 0\ 0\ 0\ 1\ 1\ 1\ 1\ 0\ 0\ 0\ 0\ 0\ 0\ 0\ 0\ 0\ 0\ 0\ 0\ 0\ 0\ 0\ 1\ 0\ 1\ 1$$

(ii) By an argument analogous to that in the Solution to Exercise 6.30(ii), the probability that a given binary message digit is the last of a group of six undergoing substitution is

$$2^{-7} + 2^{-13} + 2^{-19} + \ldots = 1/126.$$

Each substituted group introduces four additional pulses and hence the power is increased by 6.3%.

6.32 Efficiency of line codes

Calculate the efficiency (i.e. the rate of transmission of binary message digits expressed as a percentage of the available information capacity) of each of the following codes:
(i) AMI, (ii) HDB3, (iii) PST, (iv) 4B3T, (v) 6B4T.
[Note that PST stands for 'pair-selected ternary' in which pairs of binary digits are encoded as pairs of ternary digits in such a way as to reduce the low-frequency content of the spectrum.]

Solution outline
(i) Each ternary digit carries one bit but has an available capacity of $\log_2 3$ bits. The efficiency is therefore $100/(\log_2 3)\% = 63.1\%$.
(ii) As for (i).
(iii) Each pair of ternary digits carries two bits but has an available capacity of $2 \log_2 3$ bits. The efficiency is again 63.1%.
(iv) Each group of three ternary digits carries four bits but has an available capacity of $3 \log_2 3$ bits. The efficiency is therefore $400/(3 \log_2 3)\% = 84.1\%$.
(v) $600/(4 \log_2 3)\% = 94.6\%$.

6.33 Frame alignment, state-transition diagram

The state-transition diagram for a simple frame-alignment system is shown in Fig. 6.19. Transitions considered impossible or highly unlikely have been omitted. After examining the n digits in the time slot(s) in which the n digit binary frame-alignment word (FAW) would occur if the receiver were in alignment, either the receiver is in the 'search' mode (having not received a correct FAW) and makes a small change to its timing or it is in the 'locked' mode (having received a correct FAW) and maintains its timing unchanged until

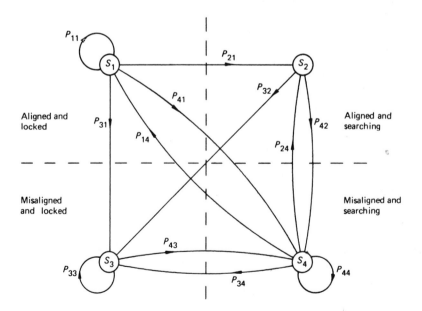

Fig. 6.19 State-transition diagram of Exercise 6.33

the next FAW time slot is examined one frame later. When a FAW time slot is examined, the receiver must be in one of the following four states:

    state 1: aligned and locked
    state 2: aligned and searching
    state 3: misaligned and locked
    state 4: misaligned and searching.

Denote by S_i the probability that the receiver is in state i.

The transition probability P_{ij} is the probability that if the receiver is in state j when a FAW time slot is examined it will go into state i during the next frame period. Clearly $\sum_i P_{ij} = 1$. The principal contributions to the transition probabilities are the following quantities:

  (i) the probability (a) that misalignment occurs during one frame;
 (ii) the probability (nb) that one or more of the FAW digits is received in error;
(iii) the probability (c) that the search procedure will have brought about alignment during one frame;
(iv) the probability (2^{-n}) that a correct FAW appears in the FAW time slot when the receiver is misaligned.

Express in terms of a, b, c and n all the P_{ij} appearing in Fig. 6.19.

Assuming that the system settles down to a condition of statistical equilibrium, write down the equations relating the S_i. Solve them to find the probability that the receiver is misaligned given that $a = 10^{-8}$, $b = 10^{-6}$, $c = 10^{-3}$ and $n = 8$. Determine whether n should be increased or decreased to reduce the probability of misalignment.

Solution outline

$$P_{11} = (1-a)(1-nb) \qquad P_{21} = (1-a)nb$$
$$P_{31} = a2^{-n} \qquad\qquad P_{41} = a(1-2^{-n})$$
$$P_{32} = 2^{-n} \qquad\qquad P_{42} = 1-2^{-n}$$
$$P_{33} = 2^{-n} \qquad\qquad P_{43} = 1-2^{-n}$$
$$P_{14} = c(1-nb) \qquad\quad P_{24} = cnb$$
$$P_{34} = (1-c)2^{-n} \qquad P_{44} = (1-c)(1-2^{-n})$$

The S_i are related by the linearly dependent equations (one of which can be disregarded)

$$S_i = \sum_j P_{ij}S_j$$

and $\sum_i S_i = 1$.

Hence

$$S_1 = (1-a)(1-nb)S_1 + c(1-nb)S_4$$
$$S_2 = (1-a)nbS_1 + cnbS_4$$
$$S_3 = a2^{-n}S_1 + 2^{-n}S_2 + 2^{-n}S_3 + (1-c)2^{-n}S_4$$
$$S_1 + S_2 + S_3 + S_4 = 1.$$

Solving these equations with the values given leads to

$$S_1 \simeq 0.992, \quad S_2 \simeq 7.9 \times 10^{-6}, \quad S_3 \simeq 3 \times 10^{-5}, \quad S_4 = 0.0079.$$

The probability that the receiver is misaligned is

$$S_3 + S_4 \simeq 0.0079.$$

In this case

$$S_3 + S_4 \simeq \frac{nb}{c}$$

and hence n should be decreased.

6.34   Frame alignment, probability of false synchronisation

In a 2048 kbit/s PCM transmission the frame-alignment word (FAW) 0011011 is inserted at fixed intervals in a sequence of equiprobable binary data digits. An 'overlap group' is a sequence of seven consecutive received digits containing some synchronising digits from the FAW and some data digits. Verify that with error-free reception no overlap group is identical to the FAW.

Assuming now that the received sequence is subject to random errors with digit error rate 0.1, calculate the probability that a received overlap group is identical to the FAW in each of the cases $m = 1, 2, 3, 4, 5$ and 6, where m is the number of synchronising digits in the overlap group.

Show that the sum of those probabilities would be less if the Barker sequence 0001101 were used as the FAW.

Solution outline
For each value of m compare the overlap group with the FAW. For example, with $m = 4$, the overlap group

    1 0 1 1 X X X

(where X denotes a data digit) is compared with

    0 0 1 1 0 1 1.

Clearly these 7-tuples always differ by at least one digit (the first) whatever the values of the data digits.

Denote by c the number of synchronising digits in an overlap group which differ from the corresponding digits in the FAW. Table 6.2 may be constructed.

For the sequence 0001101, Table 6.2 becomes Table 6.3.

Table 6.2

m	c	Probability that overlap group is identical to the FAW			
6	3	$(0.5)\,(0.9)^3(0.1)^3$	=	3.64	$\times 10^{-4}$
5	4	$(0.5)^2(0.9)\,(0.1)^4$	=	2.25	$\times 10^{-5}$
4	1	$(0.5)^3(0.9)^3(0.1)$	=	9.11	$\times 10^{-3}$
3	1	$(0.5)^4(0.9)^2(0.1)$	=	5.06	$\times 10^{-3}$
2	2	$(0.5)^5(0.1)^2$	=	3.125	$\times 10^{-4}$
1	1	$(0.5)^6(0.1)$	=	1.562	$\times 10^{-3}$
		Sum =		0.0164	

6.35   Choice of 4-digit FAW

List all the possible 4 digit FAWs for which all overlap groups (see Exercise 6.34) differ from the FAW in at least one digit. Determine which FAWs give the lowest total

Table 6.3

m	c	Probability that overlap group is identical to the FAW			
6	3	$(0.5)(0.9)^3(0.1)^3$	=	3.64	$\times 10^{-4}$
5	3	$(0.5)^2(0.9)^2(0.1)^3$	=	2.02	$\times 10^{-4}$
4	2	$(0.5)^3(0.9)^2(0.1)^2$	=	1.012	$\times 10^{-3}$
3	2	$(0.5)^4(0.9)(0.1)^2$	=	5.62	$\times 10^{-4}$
2	1	$(0.5)^5(0.9)(0.1)$	=	2.81	$\times 10^{-3}$
1	1	$(0.5)^6(0.1)$	=	1.562	$\times 10^{-3}$
			Sum =	0.0065	

probability of false synchronisation when inserted in an equiprobable binary data sequence and subject to a digit error rate of 0.1.

Solution outline
Consider an overlap group containing just one synchronising digit ($m = 1$). That digit must differ from the corresponding digit in the FAW and hence the FAW must begin and end with different digits. Thus possible FAWs are of the form 0XX1 or 1XX0. Since complementary sequences and time-inverse sequences have the same overlap properties, we consider only 0XX1.

Now consider an overlap group containing just two synchronising digits ($m = 2$). Comparison with the FAW 0XX1 shows that 0X11 and 00X1 are possible FAWs. Since one may be derived from the other by taking the complement and time inverse, we consider only 0X11.

Now consider an overlap group containing just three synchronising digits ($m = 3$). Comparison with the FAW 0X11 shows that 0111 and 0011 are possible FAWs. Taking time-inverses and complements of these gives the following list of possible FAWs:

0111, 1110, 1000, 0001, 0011, 1100.

No other 4-tuples satisfy the required condition.
For the FAW 0111, Table 6.4 may be compiled (see Exercise 6.34).

Table 6.4

m	c	Probability that overlap group is identical to FAW	
3	1	$(0.5)(0.9)^2(0.1)$	= 0.0405
2	1	$(0.5)^2(0.9)(0.1)$	= 0.0225
1	1	$(0.5)^3(0.1)$	= 0.0125

The total of three probabilities is 0.0755. For the FAW 0011, Table 6.4 becomes Table 6.5.

Table 6.5

m	c	Probability that overlap group is identical to FAW	
3	1	$(0.5)(0.9)^2(0.1)$	= 0.0405
2	2	$(0.5)^2(0.1)^2$	= 0.0025
1	1	$(0.5)^3(0.1)$	= 0.0125

In the latter case the sum of the probabilities is 0.0555 and thus the total probability of false synchronisation is lower for the FAWs 0011 and 1100 than for the other FAWs listed.

6.36 Possible 5-digit FAWs

List the 12 possible 5 digit FAWs for which all overlap groups (see Exercise 6.34) differ from the FAW in at least one digit.

Solution outline
Proceed as in Exercise 6.35 by considering $m = 1, 2, 3$ and 4. Possible FAWs must be of the form:

$m = 1$ 0XXX1

or $m = 2, 3, 4$ 00XX1.

Hence possible FAWs are:

     00001   00011   00101   00111

and their complements

     11110   11100   11010   11000

and the time inverses

     10000   10100   01111   01011.

6.37 Justification in second-order TDM

Each frame of an 8448 kbit/s second-order-multiplex PCM transmission consists of 848 digits allocated as follows:
(a)   820 message digits, digitally interleaved from four 2048 kbit/s tributaries;
(b)   four digits (one for each tributary) which may be either message digits or justified, i.e. not forming part of the tributary sequence;
(c)   12 justification control digits (three for each tributary) indicating whether or not each tributary is justified in that frame;
(d)   ten digits for frame-alignment;
(e)   two digits for other purposes.
    Calculate:
(i)   the percentage of frames in which a given tributary is justified when the digit rates of that tributary and the multiplex transmission have their nominal values;
(ii)   the maximum and minimum acceptable values for the digit rates of the tributaries when the digit rate of the multiplex transmission has its nominal value.

Solution outline
(i)   Each frame occupies 848/8448 ms and hence includes on average

$$\frac{848}{8448} \times 2048 = 205.576$$

digits from each tributary. Justification reduces the message digits allocated to a tributary in one frame from 206 to 205. The required average is obtained when each tributary is justified in 42.4% of the frames.

(ii)   The maximum digit rate for a tributary corresponds to the allocation of 206 message digits to that tributary in each frame and is therefore

$$206 \times \frac{8448}{848} = 2052.2 \text{ kbit/s.}$$

Similarly, the minimum digit rate for a tributary is

$$205 \times \frac{8448}{848} = 2042.3 \text{ kbit/s.}$$

7
Radio techniques

7.1 The superhet receiver

Radio receivers are usually of the superhet type, in which the signal is shifted in frequency to an intermediate frequency (IF) before being demodulated. The shifting is carried out by mixing the signal with a tone generated by a local oscillator in the receiver. This arrangement enables the receiver to be retuned to different signals by adjustment of the frequency of the local oscillator without any adjustment of the IF circuitry, which may therefore be set up to have a sharply defined passband and hence good rejection of signals in adjacent channels. In practice, rejection of image signals (Exercise 7.1) usually makes it necessary to precede the mixer with a radio-frequency (RF) amplifier having some selectivity. A block diagram of a simple superhet receiver is shown in Fig. 7.1. More complicated receivers involving two or three shifts of frequency are required in some applications, particularly when the required tuning range is large (Example 7.4). The superhet receiver is discussed in most books on radio engineering.

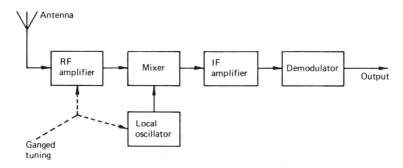

Fig. 7.1 Simple superhet receiver

7.2 Threshold extension

7.2.1 Threshold in frequency-modulation receivers

The theory summarised in Appendix 7 shows that when the signal-to-noise ratio at the input of a discriminator is high, the operation of the discriminator is dominated by the signal and the effect of the noise on the output is disproportionately small. On the other hand, when the input signal-to-noise ratio is low, it is found that the operation is dominated by the noise and that instead of improving the signal-to-noise ratio the discriminator degrades it. The transition between those two regions is quite rapid and there is therefore said to be a 'threshold' value of input signal-to-noise ratio above which the improvement is essentially as predicted in Appendix 7 and below which it deteriorates rapidly. The threshold value is found to occur at about 10 dB.

The rapid deterioration may be thought of as due to phase jumps caused by additional zero-crossings introduced by the noise. The phase jumps are equivalent to spikes of instantaneous frequency and add a flat contribution to the output noise power spectrum in addition to the parabolic contribution predicted by Appendix 7.

The existence of the threshold imposes an upper limit on the pre-detection bandwidth of a receiver and hence on the FM improvement obtainable when the received signal is weak. If the threshold signal-to-noise ratio can be lowered by any means, the frequency deviation of the transmission can be increased and a larger signal-to-noise ratio may be obtained at the receiver output for the same received signal power. Techniques for lowering the threshold are called 'threshold extension' techniques. Phase-locked loops and frequency-modulated feedback (FMFB) are both used for threshold extension and are described in Sections 7.2.2 and 7.2.3 below.

7.2.2   The phase-locked loop

The structure of the basic phase-locked loop is shown in Fig. 7.2. $F(s)$ is the transfer function of the loop filter. The difference $\theta_1(t) - \theta_2(t)$ is the phase error $\phi_1(t)$, which is small when the loop is locked. If an output is taken as shown from the input to the voltage-controlled oscillator, the phase-locked loop acts as a frequency demodulator and may be shown (Exercise 7.8) to give, when locked, an FM improvement (Appendix 7) equal to that of a discriminator.

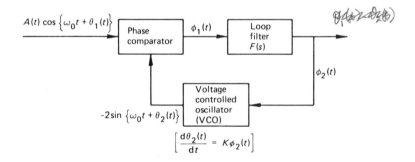

Fig. 7.2 Phase-locked loop

Noise accompanying the input may cause temporary loss of lock and the appearance at the output of voltage spikes associated with rapid changes of oscillator frequency. If the noise level is sufficiently low for $\phi_1^2(t)$ to be less than about 0.25 rad², the spikes are insufficiently frequent to contribute significantly to output noise; but for higher noise levels the rate of occurrence of the spikes increases rapidly with noise level, giving rise to a threshold effect. A phase-locked loop may be designed so that in the threshold region the effect of noise spikes may be to some extent suppressed[1], giving a threshold value of signal-to-noise ratio about 3 dB less than for a discriminator.

A linear model of a phase-locked loop is shown in Fig. 7.3. It assumes that $A(t)$ is constant and equal to A_0. If the phase comparator is a multiplier, $\phi_1(t)$ is correct for small values of $\theta_1(t) - \theta_2(t)$. If the phase comparator operates with square-wave inputs and has a linear characteristic, the interpretation of A_0 should be modified. In both cases the phase comparator is assumed to suppress ω_0 and its harmonics.

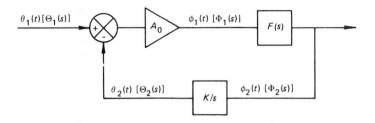

Fig. 7.3 Linear model of phase-locked loop

A thorough discussion of the choice of $F(s)$, A_0 and K for optimum performance will not be attempted here. We shall confine the discussion to a few comments. A clear introduction to phase-locked loops has been given by Gardner[2] and further analysis has been given by Viterbi[3] and Lindsey[4]. First-order loops (loops having $F(s) = 1$) are convenient for tutorial purposes (Exercises 7.7, 7.9 and 7.10), but they give very little flexibility in design and a very restricted performance. More often used are second-order loops with $F(s)$ of the form

$$F(s) = \frac{1 + s\tau_2}{1 + s\tau_1}.$$

Such second-order loops permit independent choice of bandwidth and damping and present no stability problems. A typical choice of loop parameters is indicated in Exercise 7.11.

7.2.3   FM feedback

FM feedback (FMFB) may be regarded as modulating the frequency of the local oscillator in a superhet receiver so that the frequency deviation of the signal is reduced by the mixer, thus enabling a smaller IF bandwidth to be used. The smaller IF bandwidth reduces the noise reaching the discriminator and hence lowers the threshold. A block diagram of an FMFB threshold-extension receiver is shown in Fig. 7.4. A simple approach to calculating the threshold reduction is used in Exercise 7.14. More precise calculations have suggested that reductions of 5 dB or more should be possible. For a quantitative discussion see Roberts[5].

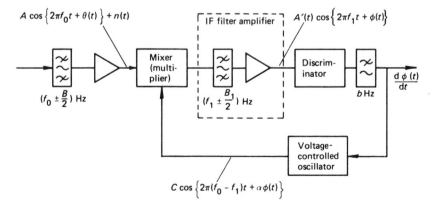

Fig. 7.4   FMFB threshold-extension receiver

7.3 **Diversity**

7.3.1 Rayleigh fading

In many radio systems the amplitude of the received signal is subject to random variations known as 'fading'. Sufficiently deep fading disrupts communication and the proportion of time for which the received signal amplitude is unacceptably small is called the 'outage rate'.

The variable r is said to be Rayleigh distributed when its probability density function $p(r)$ is given by

$$p(r) = \begin{cases} \dfrac{r}{a}\exp\left\{-\dfrac{r^2}{2a}\right\} & r \geqslant 0 \\ 0 & r < 0. \end{cases} \tag{7.1}$$

It may be shown that $\overline{r^2} = 2a$ and $(\bar{r})^2 = \pi a/2$. Fading such that a constant-amplitude transmitted signal is received with a Rayleigh-distributed amplitude is known as Rayleigh fading.

The fact that the envelope of a narrow-band gaussian process has the probability density function of Eq. 7.1 suggests that such a process might describe a sinewave subjected to Rayleigh fading. It can be shown[6] that for a gaussian process the joint probability density function $p(r, \dot{r})$ of the amplitude r and its rate of change $\dot{r}$ is of the form

$$p(r, \dot{r}) = \frac{r}{a}\exp\left\{-\frac{r^2}{2a}\right\}\frac{1}{\sqrt{(2\pi b)}}\exp\left\{-\frac{\dot{r}^2}{2b}\right\}. \tag{7.2}$$

It follows that r and $\dot{r}$ are statistically independent and that $\dot{r}$ is gaussian distributed with variance b. a is the average power of the process. It may further be shown that b is the second moment about the mid-band frequency f_c of the power spectrum of the process, i.e.

$$b = (2\pi)^2 \int_0^\infty w(f)(f-f_c)^2\,\mathrm{d}f = (2\pi)^2 \int_{-\infty}^\infty w(f+f_c)f^2\,\mathrm{d}f,$$

where $w(f)$ is the one-sided power spectrum of the process and is assumed symmetrical about f_c. From the second integral it follows that $b = -\ddot{\psi}(0)$, where $\psi(\tau)$ is the autocorrelation function of a process with power spectrum $w(f+f_c)$. Note that the power spectrum of r consists of a DC component and a continuous component. The latter is approximately triangular when $w(f)$ is rectangular, falling from a maximum at zero frequency to almost zero at twice the highest frequency component of the complex envelope[7,8].

Equation 7.2 may be used to calculate the average duration of fades in the following way. Suppose that the signal is regarded as faded when $r < R$. The number N_R of downward transitions of r through the level R in unit time is given by:

$$N_R = \int_0^\infty \dot{r}p(R, \dot{r})\,\mathrm{d}\dot{r}.$$

The average duration of fades may be calculated by dividing the outage rate by N_R.

In some systems, fading is Rayleigh when observed over a short period of time, but deviates from Rayleigh if observations extend over long periods. In such cases one can distinguish between 'fast' Rayleigh fading and 'slow' fading, the latter being relatively slow variations of the parameter a of the fast Rayleigh fading. An analogous distinction was made in Section 5.1.3 between fast (gaussian) and slow (log normal) variations of voltage in connexion with carrier-telephony waveforms.

A fading signal containing a strong non-fading component may be a more accurately described by the rician distribution[9] than by the Rayleigh distribution, but in view of its greater mathematical complexity the rician distribution is not used in the book.

7.3.2 Diversity techniques

The effects of fading can be significantly reduced if a radio receiver can make use of two or more versions of the transmitted signal which have travelled over paths sufficiently different for them to fade independently. The independently fading paths are usually referred to as 'branches'.

Techniques for providing independently fading branches include the following.

(a) **Space diversity** This technique[10] requires the use of two antennas separated typically by 100 wavelengths at either the transmitter or the receiver or both. Transmissions from different transmitting antennas require distinguishing 'markings' such as different planes of polarisation or slightly different frequencies. Space diversity is widely used in troposcatter systems and in systems involving moving vehicles.

(b) **Frequency diversity** Signals travelling over the same path on different carrier frequencies may fade independently if the carrier frequencies are sufficiently well separated. In troposcatter systems the necessary separation is typically one per cent of the carrier frequency, more for short paths[11]. Frequency diversity is widely used in HF radio telegraphy (Exercise 9.4).

(c) **Angle diversity** A useful alternative to space diversity in troposcatter systems is to use antennas having two vertically spaced feeds and therefore having two vertically separated beams[11].

7.3.3 Diversity combining

The most straightforward method of combining signals arriving in independently fading branches is to continuously monitor all the branches and to select the strongest signal by switching. This method is called 'selection combining'. Unnecessary switching may be avoided by continuing to use a signal so long as it remains acceptable, whether or not it remains the strongest.

In rejecting all but one of the branches, selection combining rejects useful signal power. A slightly better final signal-to-noise ratio can be obtained by forming a weighted sum of the branch signals, giving large weightings to branches with large signal-to-noise ratios. The optimum final signal-to-noise ratio is obtained by maximal-ratio combining, in which the weightings are chosen in the following way.

Let there be M branches. Denote the gain of the k-th branch by g_k and the transmitted signal by $\mathrm{Re}[u(t)\exp(j\omega_c t)]$. During periods of time for which g_k remains constant, the signal received from the k-th branch is $\mathrm{Re}[\{g_k u(t) + n_k(t)\}\exp(j\omega_c t)]$, where $n_k(t)$ represents added noise. When the weighting factor applied to the k-th branch is α_k, the weighted sum comprises the signal

$$s(t) = \mathrm{Re}\left[\left\{u(t)\sum_{k=1}^{M}\alpha_k g_k\right\}\exp(j\omega_c t)\right]$$

and the noise

$$n(t) = \mathrm{Re}\left[\left\{\sum_{k=1}^{M}\alpha_k n_k(t)\right\}\exp(j\omega_c t)\right].$$

The choice of weighting coefficients giving the maximum ratio of $\overline{s^2(t)}$ to $\overline{n^2(t)}$ may be shown[12] to be

$$\alpha_k = Kg_k^*/N_k,$$

where K is a constant, $N_k = \frac{1}{2}\overline{n_k^2(t)}$ and g_k^* is the complex conjugate of g_k. This choice of weighting coefficients constitutes maximal-ratio combining.

An approximation to maximal-ratio combining which is considerably simpler to implement is equal-gain combining, in which the weighting factors are chosen to have the correct arguments but equal magnitudes. Provided that the mean noise powers in the branches are equal, the performance of equal-gain combining is only slightly inferior to that of maximal-ratio combining[12].

The foregoing has assumed that the weighted sum is formed before demodulation. Unless the demodulator is linear, combining after demodulation is inferior.

7.4   Spread-spectrum techniques

7.4.1   The spread-spectrum concept

For a transmission to be described as 'spread spectrum', two criteria should be satisfied.

(i)   The transmitted signal should occupy a bandwidth much greater than that of the message signal it conveys.
(ii)   The bandwidth occupied by the transmitted signal should be determined by a prescribed waveform and not by the message signal.

Wideband frequency modulation satisfies criterion (i) but not criterion (ii) and is therefore usually considered not to be 'spread spectrum'.

In the generation of spread-spectrum signals, information is usually incorporated before the prescribed spreading waveform modulates the carrier, either by modulation of the carrier or by modulation of the spreading waveform.

The following three basic methods are used to generate the wide bandwidth in spread-spectrum systems and are described in subsequent sections.

(a)   Pseudonoise modulation, based on PSK by a suitable digital sequence (Section 7.4.2).
(b)   Frequency hopping, in which a large number of frequencies are available and the transmitter jumps rapidly from one to another in accordance with a prescribed sequence (Section 7.4.3).
(c)   Chirp, in which the transmitter carrier is swept in frequency during the course of each transmitted pulse (Section 7.4.4).

Reasons for the increasing interest in spread-spectrum techniques include the following.

A   They aid the concealment of transmissions, since the spectral density of a spread-spectrum transmission may be less than the noise spectral density in a distant receiver.
B   The despreading process in the receiver will spread the spectra of unwanted narrow-band signals, thus improving interference rejection.
C   The effect of a spread-spectrum transmission on a receiver designed to receive transmissions occupying the same frequency band but using a different spectrum-spreading pattern approximates to the effect of noise. Hence the spectrum-spreading pattern can be used to address the transmission. Pseudonoise spreading is used for this purpose in satellite systems in which centralised control of terrestrial transmitters is undesirable.
D   Wideband spread-spectrum transmissions can provide accurate timing or ranging without the inconvenience of using a very short pulse of very high power. Chirp pulses are used for this purpose in radar.

The improvement in signal-to-noise ratio due to the despreading process is known as 'process gain'.

For more information than can be included in this chapter, see Dixon[13] or special issues of IEEE journals[14,15].

7.4.2 Pseudonoise spectrum spreading

The principle of pseudonoise spread-spectrum transmission based on binary PSK (phase-shift keying, see Section 2.2.3) is shown in Fig. 7.5. The code sequence generated in the receiver is a synchronised replica of that in the transmitter so that in the receiver the signal is restored to its original narrow-band form. Since $f_s \gg b$, the general shape of the spectrum of the transmitted signal is that of the spectrum of the code sequence and there are deep minima at $f_c \pm f_s$, where f_c is the centre frequency of the narrow-band waveform. For code sequences generally used, $B = 2f_s$.

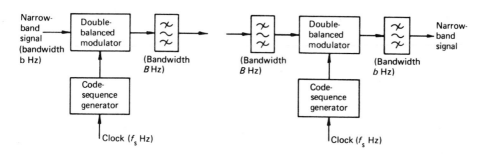

Fig. 7.5 Spectrum-spreading by phase-reversal keying ($2f_s \simeq B \gg b$)

While 'despreading' the wanted signal to occupy its original narrow bandwidth b, the receiver double-balanced modulator in Fig. 7.5 spreads the spectrum of narrow-band interference to occupy bandwidth B. Assuming that for concealment of the transmission a code sequence has been chosen which has a reasonably uniform spectrum, only a fraction b/B of the received narrowband interference accompanies the despread wanted signal. The process gain of the system is therefore approximately $10 \log (B/b)$ dB.

In Fig. 7.5, the message is assumed to be present as modulation of the narrow-band waveform. Such modulation should be constant-envelope, since considerations of concealment and efficient amplification make amplitude modulation of the transmitted signal undesirable. An alternative method of incorporating messages which are in digital form is to add the message digits to the code sequence and to use the sum to control the phase-reversal keying in the transmitter. This alternative method is sometimes referred to as 'direct-sequence spread spectrum'.

The quality of the system of Fig. 7.5 is critically dependent on the performance of its double-balanced modulators. Any direct transmission of the narrow-band waveform will impair concealment, as will the transmission of spectral lines due to asymmetry. Direct transmission through the double-balanced modulator in the receiver impairs the rejection of narrow-band interference. The effects of the latter degradation can be mitigated by using a heterodyne arrangement, in which the local oscillator is phase-modulated by the code sequence so that the incoming signal is shifted in frequency as it is despread[13].

Effective concealment of the transmitted signal requires that its spectrum should appear continuous and reasonably uniform over a wide frequency band. The appearance of spectral continuity can be ensured by choosing a code sequence with a sufficiently long repetition period. Spectral uniformity requires the choice of a code sequence having an autocorrelation consisting of very narrow pulses separated by the repetition period of the sequence and is therefore consistent with the requirement that the code sequence should be easy to synchronise. A very convenient class of code sequences is that of maximal-length sequences[16,17] generated by linear feedback shift-registers and the generation and properties of those sequences will now be outlined.

Two equivalent configurations of the n-stage linear feedback shift-register sequence generator are shown in Fig. 7.6. D_1, D_2, . . . , D_n are bi-stable delay units and c_1, c_2, . . . , c_{n-1} are multipliers which may be 0 or 1. Addition is modulo-2. Application of a clock pulse causes the register to change its state in a pre-determined way and since it has a finite number (2^n) of possible states, the register must go through a periodic sequence of states with period not greater than 2^n (clock periods). Furthermore, that sequence cannot contain the all-zero state (since the register would remain locked in that state) and therefore the period cannot exceed $2^n - 1$ clock periods. The sequence of states of one bi-stable in a register designed to go through a sequence of $2^n - 1$ different states is called a maximal-length sequence. Note that a practical generator of maximal-length sequences requires logic circuitry to prevent it from locking in the all-zero state when switched on.

Determination of the values of c_1, c_2, . . . , c_{n-1} for the generation of maximal-length sequences is a laborious process, but the results of computations have been tabulated[13,18]. The basis of the computations is to find c_1, c_2, . . . , c_{n-1} such that the characteristic polynomial

$$C(X) = 1 + c_1 X + c_2 X^2 + \ldots + c_{n-1} X^{n-1} + X^n$$

(in which addition is modulo-2) is a divisor of $1 + X^p$ for $p = 2^n - 1$ but not for any smaller value of p[16]. It may be shown[16] that suitable characteristic polynomials must be irreducible (i.e. have no factors) and therefore that reducible polynomials may be omitted

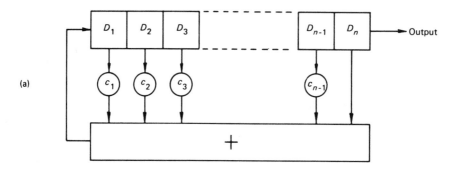

(a)

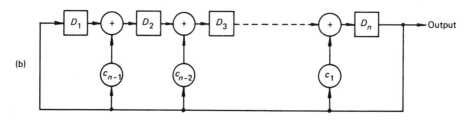

(b)

Fig. 7.6 Equivalent alternative linear feedback shift-register sequence generators (a) configuration A (b) configuration B

from the search. In fact if $2^n - 1$ is prime, irreducibility of $C(X)$ is sufficient to ensure that the sequence generated is maximal-length.

An important consequence of the linear nature of the sequence generators of Fig. 7.6 is that a maximal-length sequence added modulo-2 to a delayed replica of itself gives another delayed replica of itself. In other words, all the delayed versions of a maximal-length sequence form with the all-zero sequence a group under modulo-2 addition. This is the 'shift-and-add' property, from which it follows that the autocorrelation of the output of the sequence generator is uniformly small, except for large triangular peaks separated by the sequence period, and that a maximal-length sequence is therefore a suitable code sequence for pseudonoise spread-spectrum systems.

When different maximal-length sequences are used to give distinctive addresses to different transmissions sharing the same frequency band and the same transmission medium (code-division multiple access or CDMA, see Exercise 9.10), they should be chosen so that their cross-correlations do not contain large peaks likely to cause false synchronisation and crosstalk.

7.4.3 Frequency hopping

The transmitter of a frequency-hopping system is based upon a frequency synthesiser controlled by a digital code. Ideally the transmitted waveform is of constant amplitude and divided into constant-frequency 'chips' of short duration. The number of chips transmitted per second is called the 'chip rate'. The frequency band necessary for the transmission of a chip is called a 'channel' and its width is approximately twice the chip rate.

Concealment considerations suggest that all available chip frequencies should be used equally often in an order which appears random. An appropriate basis for the digital code controlling the synthesiser is therefore the sequence of the states of a feedback shift-register generating a maximal-length sequence. Information in digital form may be incorporated by controlling the synthesiser with the modulo-2 sum of the information and the shift-register state. The chip rate is usually either equal to the information bit rate or an integral multiple of it.

The frequency-hopping system described above is binary in the sense that the frequency of each chip has two possible values. The receiver must decide for each chip which frequency was transmitted. A possible arrangement for a receiver is shown in Fig. 7.7. The mark synthesiser and the space synthesiser are identical. They respond to the controlling code in the same way as the transmitter synthesiser except that the output is offset by the intermediate frequency f_{IF}. The code generator in the receiver is identical to that in the transmitter and synchronised to it. Coherent detection is not practicable when the chips are short.

Narrowband interference falling within a channel is spread by the mixers in the receiver of Fig. 7.7. Provided that the channels are used equally often and do not overlap, the signal-to-interference ratio at the input of one of the envelope detectors is greater than that at the mixer inputs by a process gain equal to the number of available channels. If the channels are contiguous in the frequency domain, that process gain is equal to the ratio of the RF bandwidth of the transmission to the IF bandwidth of the receiver.

7.4.4 Chirp

Modulation of the carrier frequency during the transmission of a radio-frequency pulse is principally a radar technique to improve range resolution at long range. It enables a high-energy pulse to occupy a larger bandwidth without increasing the overloading requirements of the transmitter.

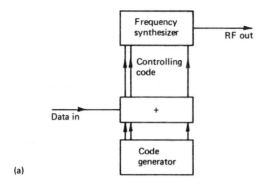

(a)

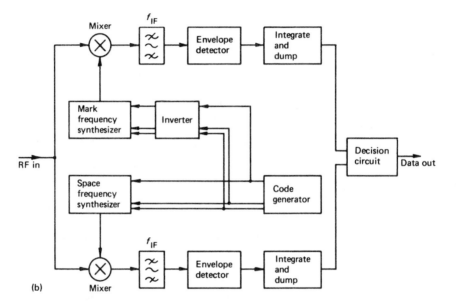

(b)

Fig. 7.7 Binary frequency-hopping system (a) transmitter (b) receiver

The simplest form of chirp consists of a linear frequency sweep resulting in transmitted pulses of the form:

$$v(t) = \begin{cases} A \cos \left\{ 2\pi f_c t \pm \pi \Delta f t^2 / T \right\} & |t| \leqslant T/2 \\ 0 & |t| > T/2. \end{cases}$$

When $T\Delta f$ is large, the energy spectrum of such a transmitted pulse is approximately rectangular. It follows that after passing through a matched filter the envelope of the pulse is approximately proportional to $\text{sinc}\{\Delta f(t - t_0)\}$, where t_0 is introduced to allow for the delay necessary to make the matched filter realisable. If the duration of this output pulse is regarded as approximately $1/\Delta f$, the matched filter may be said to have compressed the duration of the chirp pulse by the factor $T\Delta f$, which was assumed large. Radar systems using chirp pulses are usually referred to as 'pulse compression' radars.

More exact analysis shows that the spectra of chirp pulses may be expressed in terms of Fresnel integrals[19,20].

7.5 References

1 Taub, H., and Schilling, D. L., *Principles of Communication Systems*, McGraw-Hill, 1971
2 Gardner, F. M., *Phaselock Techniques*, Wiley, 1976
3 Viterbi, A. J., *Principles of Coherent Communication*, McGraw-Hill, 1966
4 Lindsey, W. C., *Synchronization Systems in Communication and Control*, Prentice-Hall, 1972
5 Roberts, J. H., *Angle Modulation*, IEE Telecommunications Series 5, Peter Peregrinus, 1977
6 Rice, S. O., The mathematical analysis of random noise', *Bell System Technical Journal*, **23**, pp. 282–332, July 1944, and **24**, pp. 46–156, January 1945
7 Lawson, J. L., and Uhlenbeck, G. E., *Threshold Signals*, MIT Radiation Laboratory Series No. 24, McGraw-Hill, 1950
8 Price, R., 'A note on the envelope and phase-modulated components of narrow-band gaussian noise', *IRE Trans*, IT-1, pp. 9–13, September 1955
9 Rice, S. O., 'Statistical properties of a sine wave plus random noise', *Bell System Technical Journal*, **27**, pp. 109–57, 1948
10 Panter, P. F., *Communication Systems Design: Line-of-Sight and Tropo-scatter Systems*, McGraw-Hill, 1972
11 Hall, M. P. M., *Effects of the Troposphere on Radio Communication*, IEE Electromagnetic Waves Series 8, Peter Peregrinus, 1979
12 Schwartz, M., Bennett, W. R., and Stein, S., *Communications Systems and Techniques*, McGraw-Hill, 1966
13 Dixon, R. C., *Spread Spectrum Systems*, Wiley, 1976
14 *IEEE Trans*, COM-25, No. 8, August 1977, Special Issue on Spread Spectrum Communications
15 *IEEE Trans*, COM-30, No. 5, Part 1, May 1982, Special Issue on Spread Spectrum Communications
16 Golomb, S., *Shift-register Sequences*, Holden Day, 1967
17 Stremler, F. G., *Introduction to Communication Systems*, Addison-Wesley, Second Edition 1982
18 Peterson, W. W., and Weldon, E. J., *Error-correcting Codes*, MIT Press, Second Edition 1972
19 Cook, C. E., 'Pulse compression—key to more efficient radar transmission', *Proc IRE*, **48**, pp. 310–16, March 1960
20 Klauder, J. R., Price, A. C., Darlington, S., and Albersheim, W. J., 'The theory and design of chirp radars', *Bell System Technical Journal*, **39**, pp. 745–809, July 1960

Exercise topics

7.1   Image-frequency rejection
7.2   Spurious responses in superhet receiver
7.3   Spurious responses in double superhet receiver
7.4   HF receiver design
7.5   Noise in FM receiver
7.6   Equivalent noise in linear model of PLL
7.7   PLL FM demodulator: frequency response
7.8   First-order PLL FM demodulator: output noise
7.9   First-order PLL: output jitter
7.10 First-order PLL: locking range
7.11 Second-order PLL FM demodulator: frequency response
7.12 FMFB receiver
7.13 FMFB receiver: output noise
7.14 FMFB receiver: threshold reduction
7.15 Rayleigh fading: frequency of fades
7.16 Rayleigh fading: outage rate
7.17 Error probability in non-coherent FSK with Rayleigh fading and selection combining
7.18 Rayleigh fading: power saving with selection combining (equal SNRs)
7.19 Rayleigh fading: reduction of outage by selection combining (unequal SNRs)
7.20 Maximal-ratio combining
7.21 Equal-gain combining
7.22 Linear feedback shift register: recurrence relations
7.23 Generation of maximal-length sequences of period 15
7.24 Maximal-length sequences: occurrence of 'all–1' and 'all–0' subsequences
7.25 Maximal-length sequences: power spectrum
7.26 Maximal-length sequences: decimation
7.27 Pseudonoise spread-spectrum radio system
7.28 Frequency-hopping radio system with interfering carrier
7.29 Frequency-hopping radio system with comb jamming
7.30 *m*-ary frequency hopping
7.31 Frequency hopping using two channels simultaneously
7.32 Energy spectrum of chirp pulse
7.33 Chirp binary data transmission

Exercises

7.1 Image-frequency rejection

The receiver of Fig. 7.1 has its local oscillator tuned to f_0 and intermediate frequency f_i ($f_0 > f_i$). Assuming that the frequency of the mixer output is the difference of the frequencies of its inputs, show that with its RF amplifier by-passed the receiver would be sensitive to two signal frequencies separated by $2f_i$. [The RF amplifier selects one of these: the other is called the image frequency.]

The receiver is required to be tunable to any signal with a frequency between 80 MHz and 100 MHz. The selectivity of the RF amplifier is provided by a single tuned circuit with a Q-factor of 40. Calculate the minimum value of f_i if the RF amplifier should have 25 dB less gain at the image frequency than it has at the signal frequency. [Note that some analysis of a tuned circuit is used in Exercise 6.25.]

f_0 is controlled by a tuned circuit with a variable capacitor. The ratio of the maximum value of the total capacitance of the tuned circuit to its minimum value is 1.5. Calculate the minimum intermediate frequency if the required frequency range is to be covered.

Solution outline
Possible values of the signal frequency f_s are $f_0 + f_i$ and $f_0 - f_i$.

Assuming that the RF amplifier is tuned to f_s, its gain at frequency f is less than its gain at f_s by $10 \log \{1 + 4Q^2 x^2\}$ dB, where $x = (f - f_s)/f_s$. This exceeds 25 dB if $|x| > 0.222$. The value of $|x|$ is least at $f = f_s \pm 2f_i$ when f_s is at the upper end of the tuning range. Hence $2f_i/100 > 0.222$, i.e. $f_i > 11.1$ MHz.

The upper end of the local-oscillator frequency range must satisfy $f_0 - (f_0/\sqrt{1.5}) > 20$, i.e. $f_0 > 108.99$ MHz. Hence f_i must exceed 8.99 MHz.

7.2 Spurious responses in superhet receiver

The receiver of Fig. 7.1 is set up to receive 100 MHz when its local-oscillator frequency is 112 MHz. Assuming that when the inputs to the mixer are at frequencies f_1 and f_2 the mixer generates all frequencies $|mf_1 \pm nf_2|$, where the integers m and n are 0, 1, 2 or 3 and $0 < m + n \leqslant 3$, list the frequencies of all spurious responses. [Spurious responses are unwanted signals which generate the receiver IF in the mixer.]

Solution outline
The frequency f MHz of a spurious response satisfies $|mf \pm 112n| = 12$. Systematically trying all permitted values of m and n gives Table 7.1.

Table 7.1

m	n	Responses
1	0	$f = 12$
1	1	$f = 100$ (wanted), $f = 124$ (image)
1	2	$f = 212, 236$
2	0	$f = 6$
2	1	$f = 50, 62$
3	0	$f = 4$

7.3   Spurious responses in double superhet receiver

A double superhet receiver is shown in Fig. 7.8. Calculate the input spurious response frequencies, assuming that
(a)   the response of each amplifier is zero outside the frequency band indicated,
(b)   mixers generate frequencies of the form $|mf_1 \pm nf_2|$ where $m, n = 0, 1, 2$ or 3.

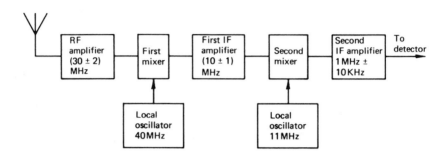

Fig. 7.8   Double superhet receiver of Exercise 7.3

Solution outline
By trying $m, n = 0, 1, 2, 3$ in the equation $|mf \pm 11n| = 1$, response frequencies of the second superhet lying within the response band of the first superhet may be shown to be 10 MHz, 10.5 MHz, 10.6 MHz. Consider each of these responses in turn.
  (i)   In $|mf \pm 40n| = 10$, $m = n = 1$ gives the wanted response at 30 MHz and $m = 3$, $n = 2$ gives a way for a 30 MHz signal to contribute distortion to the output.
  (ii)   In $|mf \pm 40n| = 10.5$, $m = n = 1$ gives a spurious response at 29.5 MHz and $m = 3$, $n = 2$ gives another at 30.16 MHz.
(iii)   In $|mf \pm 40n| = 10.6$, $m = n = 1$ gives a spurious response at 29.6 MHz and $m = 3$, $n = 2$ gives another at 30.2 MHz.

7.4   HF receiver design

Determine suitable local-oscillator and intermediate frequencies for a superhet arrangement capable of receiving single-sideband telephony at any frequency in the range 0.5 MHz to 30 MHz. Assume that
  (i)   the first intermediate frequency should be as low as possible and outside the frequency range 0.4 MHz to 32 MHz,
  (ii)   at any mixer the image frequency should be separated from the wanted frequency by at least 10%,
(iii)   rejection of adjacent channels should be carried out at 100 kHz.

Solution outline
Assumptions (i) and (ii) require that the first IF should be at 32 MHz. The range is covered by a local oscillator variable from 32.5 MHz to 62 MHz. The first IF cannot satisfy (ii) if it is below 0.4 MHz.
  Assumption (ii) requires the second IF to be greater than 1.6 MHz and also requires the final IF of 100 kHz to be preceded by an IF not exceeding 2 MHz. Hence the receiver requires three shifts of frequency and the second IF (f_2 MHz) must satisfy $1.6 \leqslant f_2 \leqslant 2$.
  Appropriate frequencies for the second and third local oscillators are $(32 \pm f_2)$ MHz and $(f_2 \pm 0.1)$ MHz.

7.5 Noise in FM receiver

A radio-telephony link transmits speech frequencies up to 3 kHz by frequency modulation. The pre-detection bandwidth of the receiver is 30 kHz. Calculate the maximum test-tone/noise ratio at the receiver audio output at maximum range, assuming that the signal/noise ratio at the discriminator input has its threshold value of 12 dB.

Re-calculate the maximum output test-tone/noise ratio when the system is modified to incorporate a threshold-extension device requiring an input signal/noise ratio of 8 dB and to have the same maximum range. Assume the same received signal power and receiver noise temperature.

Calculate the maximum output test-tone/noise ratio if single-sideband amplitude modulation is used, assuming the same received signal power and receiver noise temperature.

Solution outline
By Carson's rule (Section 5.4.2), the maximum permissible frequency deviation Δf kHz is given by $2(3 + \Delta f) = 30$. Hence the FM improvement (Appendix 7) is

$$\frac{3}{2}\left(\frac{12}{3}\right)^2\left(\frac{30}{3}\right) \quad \text{i.e. 23.8 dB.}$$

Hence the output test-tone/noise ratio cannot exceed $(12 + 23.8)$ dB, i.e. 35.8 dB.

The threshold extension device enables the pre-detection bandwidth to be increased to $(30 \times 10^{0.4})$ kHz, i.e. 75.36 kHz. This permits a peak frequency deviation of 34.68 kHz and hence an FM improvement of 37.0 dB, giving a possible output test-tone/noise ratio of 45.0 dB.

With SSB the pre-detection bandwidth is 3 kHz and hence the test-tone/noise ratio before and after demodulation is $(12 + 10)$ dB, i.e. 22 dB.

7.6 Equivalent noise in linear model of PLL

The phase comparator in the phase-locked loop of Fig. 7.2 consists of a multiplier followed by a filter stopping $2\omega_0$. By writing the input in the form

$$A_0 \cos \omega_0 t + x(t) \cos \omega_0 t - y(t) \sin \omega_0 t,$$

show that for low noise levels the effect of input noise with power spectral density N_0 is equivalent to that of noise with power spectral density $2N_0/A_0^2$ at the input to the linear model of Fig. 7.3.

Solution outline
Multiplying the input waveform by $-2 \sin \{\omega_0 t + \theta_2(t)\}$ and removing components at frequencies near to $2\omega_0$ rad/s gives

$$\phi_1(t) = -A_0 \sin \theta_2(t) - x(t) \sin \theta_2(t) + y(t) \cos \theta_2(t).$$

Since $\theta_2(t)$ is small,

$$\phi_1(t) \simeq -A_0\theta_2(t) + y(t) = A_0\{y(t)/A_0 - \theta_2(t)\},$$

which is equivalent to the effect of $y(t)/A_0$ at the input of the model.

Now the noise power at the input is $\frac{1}{2}\overline{x^2(t)} + \frac{1}{2}\overline{y^2(t)} = \overline{y^2(t)}$ and occupies twice the bandwidth of $y(t)$. Hence the power spectral density of $y(t)$ is $2N_0$. Hence the power spectral density of $y(t)/A_0$ is $2N_0/A_0^2$.

7.7   PLL FM demodulator: frequency response

Obtain an expression for the frequency response of the phase-locked loop of Fig. 7.2 when it is used to demodulate FM. (Regard the input as the instantaneous frequency deviation $d\theta_1(t)/dt$.)

Derive an expression for the cut-off frequency of the response when the loop is first-order $[F(s) \equiv 1]$.

Solution outline

The required frequency response is the ratio $\Phi_2(s)/s\Theta_1(s)$. By noting that

$$\Theta_2(s) = K\Phi_2(s)/s$$

and $\Phi_2(s) = A_0 F(s)\{\Theta_1(s) - \Theta_2(s)\},$

we may show that

$$\frac{\Phi_2(s)}{s\,\Theta_1(s)} = \frac{A_0 F(s)}{s + K A_0 F(s)}.$$

When $F(s) \equiv 1$ the response is low-pass with 3 dB cut-off at $K A_0$ rad/s.

7.8   First-order PLL FM demodulator: output noise

A phase-locked loop of the form of Fig. 7.2 and having $F(s) \equiv 1$ is used to demodulate FM. Obtain an expression for the power spectral density of the output noise when noise with one-sided power spectral density N_0 accompanies the input.

Show that at modulating frequencies well below $K A_0$ rad/s the expression agrees with that quoted in Appendix 7 for a discriminator.

Solution outline

Using the results of Exercises 7.6 and 7.7, the output noise power spectral density is

$$\frac{2N_0}{A_0^2}\left|\frac{\Phi_2(j\omega)}{\Theta_1(j\omega)}\right|^2 = \frac{2N_0\omega^2}{\omega^2 + K^2 A_0^2}.$$

When $|\omega| \ll K A_0$, this expression is approximately $8\pi^2 N_0 f^2/K^2 A_0^2$. At low frequencies the frequency response approaches $1/K$, which is therefore equivalent to K in Appendix 7.

7.9   First-order PLL: output jitter

Obtain an expression for the power spectral density of the phase at the output of the VCO in Fig. 7.2 when $F(s) \equiv 1$ and noise with one-sided power spectral density N_0 accompanies an unmodulated carrier at the input to the system. Hence show that when a first-order phase-locked loop is used for timing extraction its linear model has a noise bandwidth $K A_0/4$.

Solution outline

Using the results of Exercises 7.6 and 7.8 and noting that $K\Phi_2(j\omega) = j\omega\,\Theta_2(j\omega)$, the output noise power spectral density is

$$\frac{2N_0}{A_0^2}\left|\frac{\Theta_2(j\omega)}{\Theta_1(j\omega)}\right|^2 = \frac{2N_0}{A_0^2}\left|\frac{K\Phi_2(j\omega)}{j\omega\,\Theta_1(j\omega)}\right|^2 = \frac{2N_0 K^2}{\omega^2 + K^2 A_0^2}.$$

Integrated from $\omega = 0$ to $\omega = \infty$, this gives for the total output noise power:

$$\frac{1}{2\pi} \int_0^\infty \frac{2N_0 K^2}{\omega^2 + K^2 A_0^2} \, d\omega = \frac{N_0 K}{2A_0}.$$

This may be written $(2N_0/A_0^2)(KA_0/4)$, showing that the output noise power is equal to that which would be obtained if the phase-locked loop were a filter with bandwidth $KA_0/4$ Hz.

7.10 First-order PLL: locking range

Assuming that the phase-locked loop of Fig. 7.2 is first-order $[F(s) \equiv 1]$ and uses a multiplier as a phase comparator, show that when the input is the unmodulated wave $A_0 \cos \omega_c t$, the phase error $\phi \, (= \theta_1 - \theta_2)$ satisfies the equation

$$\dot{\phi} = (\omega_c - \omega_0) - KA_0 \sin \phi.$$

Sketch a graph of the variation of $\dot{\phi}$ with ϕ and use it to show that the loop cannot lock if

$$|\omega_c - \omega_0| > KA_0.$$

Solution outline
The VCO is described by

$$\dot{\theta}_2 = K\phi_1$$

and the comparator by

$$\phi_1 = A_0 \sin(\theta_1 - \theta_2) = A_0 \sin \phi.$$

Hence

$$\dot{\theta}_2 = KA_0 \sin \phi.$$

In this case

$$\dot{\theta}_1 = \omega_c - \omega_0.$$

Hence

$$\dot{\phi} = \dot{\theta}_1 - \dot{\theta}_2 = (\omega_c - \omega_0) - KA_0 \sin \phi,$$

which is sketched in Fig. 7.9.

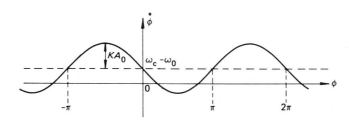

Fig. 7.9 Solution to Exercise 7.10

  Any point on the curve represents a possible state of the system, but only points at which $\dot{\phi} = 0$ correspond to equilibrium conditions. If $|\omega_c - \omega_0| > KA_0$, there are no such points. It can be further argued that stable states correspond to those $\dot{\phi} = 0$ points at which the curve has a negative gradient. This is a simple example of a system trajectory in the phase plane[3].

7.11 Second-order PLL FM demodulator: frequency response

A phase-locked loop as shown in Fig. 7.2 has $F(s) \equiv (1 + s\tau_2)/(1 + s\tau_1)$ and is to be used to demodulate FM. Find values for KA_0, τ_1, and τ_2 so that the following conditions are satisfied.

(i) The response to high modulating frequencies behaves as that of a single-pole low-pass filter with 3 dB cut-off at 10^4 rad/s.

(ii) The poles of the frequency response are of the form $r(-1 \pm j)$ (i.e. the damping factor of the denominator is $1/\sqrt{2}$).

(iii) $\tau_2 \gg 1/KA_0$. Take $\tau_2 = 100/KA_0$.

Find the frequency at which the magnitude of the frequency response has a maximum value and express that maximum value in dB relative to the low-frequency value.

Solution outline

$$\frac{\Phi_2(s)}{s\Theta_1(s)} = \frac{A_0(1 + s\tau_2)}{KA_0 + s(1 + KA_0\tau_2) + s^2\tau_1} = H(s), \text{ say.}$$

Condition (i) requires that $KA_0\tau_2/\tau_1 = 10^4$. Since $KA_0\tau_2 = 100$, the denominator has the required form if $2KA_0\tau_1 = (101)^2$. Hence

$$\tau_1 = 0.01, \ \tau_2 = 0.000\,196 \text{ and } KA_0 = 510\,050.$$

At about $\omega = 5600$ rad/s $|H(j\omega)|$ has a maximum value about 2 dB above its low-frequency value.

7.12 · FMFB receiver

Show that if the input to the FMFB receiver of Fig. 7.4 is sinusoidally modulated with peak frequency deviation Δf Hz at a frequency less than b and $n(t) = 0$ then the output power is $(2\pi\Delta f)^2/2 (1 + \alpha)^2$.

Solution outline

Since the IF filter amplifier selects frequencies near to f_1,

$$A' \cos\{2\pi f_1 t + \phi(t)\} = \frac{1}{2} AC \cos\{2\pi f_1 t + \theta(t) - \alpha\phi(t)\}.$$

Therefore

$$\phi(t) = \theta(t) - \alpha\phi(t)$$

and hence

$$\dot{\phi}(t) = \dot{\theta}(t)/(1 + \alpha).$$

Since the output power is $\overline{\dot{\phi}^2(t)}$ and $\overline{\dot{\theta}^2(t)} = (2\pi\Delta f)^2/2$, the result follows.

7.13 FMFB receiver: output noise

Show that if in the FMFB receiver of Fig. 7.4 $\theta(t) = 0$ and $n(t)$ is gaussian with one-sided power spectral density N_0 $(2A^2 \gg BN_0)$ then the one-sided power spectral density of the output is

$$\frac{8\pi^2 f^2 N_0}{A^2(1 + \alpha)^2}.$$

Hence show that the FM improvement is equal to that quoted in Appendix 7 for a discriminator.

Solution outline

$n(t)$ may be written $x(t)\cos(2\pi f_0 t) - y(t)\sin(2\pi f_0 t)$, where the power spectral densities of $x(t)$ and $y(t)$ are both $2N_0$. Hence

$$A'(t)\cos\{2\pi f_1 t + \phi(t)\} = \frac{C}{2}\{A + x(t)\}\cos\{2\pi f_1 t - \alpha\phi(t)\} - \frac{C}{2}y(t)\sin\{2\pi f_1 t - \alpha\phi(t)\}$$
$$= A'(t)\cos\{2\pi f_1 t - \alpha\phi(t) + \psi(t)\},$$

where

$$\psi(t) = \tan^{-1}[y(t)/\{A + x(t)\}] \simeq y(t)/A.$$

Hence

$$\phi(t) = -\alpha\phi(t) + y(t)/A$$

and since the mean square value of the output is $\overline{\phi^2(t)}$, the result follows.

The total output noise power is (by integration) $8\pi^2 N_0 b^3/3A^2(1+\alpha)^2$. By using the expression proved in Exercise 7.12 and by assuming that the effects of noise and angle modulation are independent, the output signal/noise ratio may be shown to be $3A^2(\Delta f)^2/4N_0 b^3$, an improvement by the factor $3(\Delta f)^2 B/2b^3$ on the input signal/noise ratio $A^2/2N_0 B$.

7.14 FMFB receiver: threshold reduction

Given that in the FMFB receiver of Fig. 7.4, $f_0 = 100$ MHz, $f_1 = 20$ MHz, $B = 88$ kHz (the Carson bandwidth), $b = 4$ kHz (the maximum modulating frequency) and $\alpha = 5$, estimate B_1 and the amount in dB by which the threshold signal-to-noise ratio is reduced.

Solution outline

Since the Carson bandwidth of the input is 88 kHz, the peak frequency deviation at the input is 40 kHz. Hence the peak frequency deviation associated with $\phi(t)$ is 6.7 kHz $[\phi(t) = \theta(t)/(1+\alpha)]$. Hence B_1 should be $2(4+6.7)$ kHz $= 21.4$ kHz.

Assuming $\theta(t) = 0$ and $n(t) \neq 0$, the VCO output is a constant-amplitude wave with random narrow-band angle modulation (bandwidth $\ll B$). Therefore the carrier/noise-density ratio is approximately unaltered by the mixer and so the carrier/noise ratio at the discriminator is greater by about B/B_1 ($\simeq 6$ dB) than it would be in a simple receiver.

7.15 Rayleigh fading: frequency of fades

A carrier wave transmitted over a fading radio link is received with a gaussian power spectrum centred on 10 MHz and having standard deviation 0.025 Hz. The received signal is regarded as faded when its envelope has fallen more than 20 dB below its median value. Assuming the received waveform to be a gaussian process, calculate the outage rate, the average number of fades occurring per minute and the average duration of the fades.

Solution outline

Denote the mean received power by a (as in Section 7.3.1). The median value R_0 of the amplitude r is given by:

$$\int_0^{R_0} \frac{r}{a}\exp\left\{-\frac{r^2}{2a}\right\}dr = 0.5.$$

Hence $R_0^2 = 2a\ln 2$. The required outage rate is

$$\int_0^{R_0/10} \frac{r}{a}\exp\left\{-\frac{r^2}{2a}\right\}dr = 0.0069.$$

The one-sided power spectral density $w(f)$ of the received signal is given by:

$$w(f) = \frac{a}{\sigma\sqrt{(2\pi)}} \exp\left\{-\frac{(f-10^7)^2}{2\sigma^2}\right\}$$

where $\sigma = 0.025$. Hence the inverse transform $\psi(\tau)$ of $w(f+10^7)$ is given by:

$$\psi(\tau) = a\exp(-2\pi^2\sigma^2\tau^2).$$

Hence

$$\ddot{\psi}(0) = -4\pi^2\sigma^2 a.$$

The average number N of fades occurring per second is given by:

$$N = \int_0^\infty \dot{r}\frac{\sqrt{(2a\ln 2)}}{10a}\exp\left\{-\frac{\ln 2}{100}\right\}\frac{1}{\sqrt{(2\pi b)}}\exp\left\{-\frac{\dot{r}^2}{2b}\right\}d\dot{r},$$

where $b = -\ddot{\psi}(0) = 4\pi^2\sigma^2 a$ (see Section 7.3.1). Hence $N = 0.0073$, i.e. 0.440 fades per minute.

The average duration of the fades is $0.0069/0.0073 = 0.95$ s.

7.16   Rayleigh fading: outage rate

A carrier wave is received subject to Rayleigh fading and accompanied by noise. Show that the probability $P(x)$ that the ratio of the received carrier power to the average received noise power is less than x is given by:

$$P(x) = 1 - \exp(-x/\Gamma),$$

where Γ is the ratio of the average carrier power to the average noise power.

Calculate by how much (in dB) the average received carrier power should exceed the received carrier power at the threshold of acceptability if the outage rate should not exceed 0.1%.

Solution outline
The average noise power is a/Γ, where a is the average received carrier power. From Section 7.3.1,

$$P(x) = \int_0^{\sqrt{(2ax/\Gamma)}} \frac{r}{a}\exp\left\{-\frac{r^2}{2a}\right\}dr = 1 - \exp\left\{-\frac{x}{\Gamma}\right\}.$$

Show that

$$1 - \exp(-x/\Gamma) < 10^{-3} \text{ if } 10\log(\Gamma/x) > 30 \text{ dB}.$$

7.17   Error probability in non-coherent FSK with Rayleigh fading and selection combining

The digit error probability of a radio system using non-coherent binary frequency-shift keying is

$$\tfrac{1}{2}\exp\left(-\frac{\gamma}{2}\right),$$

where γ is the receiver signal-to-noise ratio at the output of the filter containing the signal. Calculate the digit error probability when the system is subject to Rayleigh fading, given that it is 10^{-4} when γ is equal to its average value.

Calculate the digit error probability if the system is modified to have two identical receivers and selection combining.

Solution outline

The average value Γ of γ is given by $(1/2)\exp(-\Gamma/2) = 10^{-4}$. Therefore $\Gamma = 17.03$.

The probability $P(x)$ that γ is less than x is given by

$$P(x) = 1 - \exp(-x/\Gamma)$$

(Exercise 7.16) and hence the probability density function dP/dx is $(1/\Gamma)\exp(-x/\Gamma)$. The required probability is therefore

$$\int_0^\infty \left\{\frac{1}{2}\exp\left(-\frac{x}{2}\right)\right\}\left\{\frac{1}{\Gamma}\exp\left(-\frac{x}{\Gamma}\right)\right\}dx = \frac{1}{\Gamma+2} = 0.0525.$$

With selection combining the required probability is

$$\int_0^\infty \left[\frac{d}{dx}\left\{1-\exp\left(-\frac{x}{\Gamma}\right)\right\}^2\right]\left\{\frac{1}{2}\exp\left(-\frac{x}{2}\right)\right\}dx = \frac{4}{(\Gamma+2)(\Gamma+4)} = 0.01.$$

7.18 Rayleigh fading: power saving with selection combining (equal SNRs)

A radio system combines by selection two independently Rayleigh-fading branches in which the ratios of average signal power to noise are equal. After combination the outage rate is 10^{-4}. Calculate the reduction in transmitted power resulting from the use of four branches to provide the same outage rate.

Solution outline

Let the minimum acceptable signal-to-noise ratio be Γ_0. Let the required ratio of average signal power to noise in each branch be Γ_1 when there are two branches and Γ_2 when there are four. From the result proved in Exercise 7.16

$$\{1-\exp(-\Gamma_0/\Gamma_1)\}^2 = \{1-\exp(-\Gamma_0/\Gamma_2)\}^4 = 10^{-4}.$$

Hence $\Gamma_1/\Gamma_2 = 10.48$ i.e. 10.2 dB. Hence there is a saving of 7.2 dB in transmitted power.

7.19 Rayleigh fading: reduction of outage by selection combining (unequal SNRs)

Four independently Rayleigh-fading branches with average signal-to-noise ratios 18 dB, 22 dB, 25 dB and 27 dB are combined by selection. Calculate the outage rate if the signal-to-noise ratio at the threshold of acceptability is 12 dB.

Solution outline

In the result proved in Exercise 7.16, substitute in turn $10^{0.6}$, 10, $10^{1.3}$, $10^{1.5}$ for the ratio Γ/x and multiply the probabilities. The result is

$$0.22 \times 0.095 \times 0.049 \times 0.031 = 3.2 \times 10^{-5}.$$

7.20 Maximal-ratio combining

The transmitted carrier-wave $\cos\omega_c t$ is received by two independently Rayleigh-fading branches as $\text{Re}[g_1\exp(j\omega_c t)]$ and $\text{Re}[g_2\exp(j\omega_c t)]$ accompanied by narrow-band noise processes $\text{Re}[n_1(t)\exp(j\omega_c t)]$ and $\text{Re}[n_2(t)\exp(j\omega_c t)]$ respectively. The fading is sufficiently slow for g_1 and g_2 to be regarded as constant for a time long enough for the calculation of the average noise power $\overline{n_1^2(t)}/2$ and $\overline{n_2^2(t)}/2$ to give values close to the long-term averages N_1 and N_2. The two branches are combined in a maximal-ratio combiner.

(i)   Obtain an expression for the signal-to-noise ratio after combining and show that it is equal to the sum of the signal-to-noise ratios in the branches.

(ii)   By using the Laplace transform of a probability density function as its characteristic function, show that when the mean signal-to-noise ratios in the branches are equal and equal to Γ, the probability density function of the signal-to-noise ratio x at the combined output is

$$\frac{x}{\Gamma^2}\exp\left\{-\frac{x}{\Gamma}\right\}.$$

(iii)   Show that if Γ is 23.5 dB above the threshold of acceptability for the combined output the outage rate is about 10^{-5} and calculate what the outage rate would be were selection combining used.

   [Note that the Laplace transforms of $\exp(kx)$ and $x\exp(kx)$ are $(p-k)^{-1}$ and $(p-k)^{-2}$ respectively).]

Solution outline

(i)   The weighting coefficients in the combiner are Kg_1^*/N_1 and Kg_2^*/N_2 [Section 7.3.3]. The combined signal is therefore

$$K\,\mathrm{Re}\left[\left\{\frac{|g_1|^2}{N_1}+\frac{|g_2|^2}{N_2}\right\}\exp(j\omega_c t)\right]$$

which has mean power

$$\frac{1}{2}K^2\left\{\frac{|g_1|^2}{N_1}+\frac{|g_2|^2}{N_2}\right\}^2.$$

   Since the noise processes are independent with average powers N_1 and N_2, the noise power after combining is

$$N_1\left|\frac{Kg_1^*}{N_1}\right|^2+N_2\left|\frac{Kg_2^*}{N_2}\right|^2,$$

which simplifies to

$$K^2\left\{\frac{|g_1|^2}{N_1}+\frac{|g_2|^2}{N_2}\right\}.$$

Hence the signal-to-noise ratio after combining is

$$\frac{|g_1|^2}{2N_1}+\frac{|g_2|^2}{2N_2},$$

which is the sum of the signal-to-noise ratios in the branches.

(ii)   The probability density function of the signal-to-noise ratio after combining is the convolution of the probability density functions of the signal-to-noise ratios in the branches. Hence it may be obtained as the inverse Laplace transform of the product of the Laplace transforms of the probability density functions for the branches. Each of those probability density functions is $(1/\Gamma)\exp(-x/\Gamma)$ (by differentiation of the cumulative probability derived in Exercise 7.16), which has Laplace transform $(1+p\Gamma)^{-1}$. The product is therefore $(1+p\Gamma)^{-2}$, the inverse transform of which is the required result.

(iii)   The cumulative probability for the signal-to-noise ratio after combining is

$$\int_0^x (y/\Gamma^2)\exp(-y/\Gamma)\,dy.$$

This may be integrated by parts to give

$$1 - \left(1 + \frac{x}{\Gamma}\right)\exp\left(-\frac{x}{\Gamma}\right),$$

which is approximately 10^{-5} when $10 \log (\Gamma/x) = 23.5\,\mathrm{dB}$.

When selection combining is used, the outage rate is $\{1 - \exp(-x/\Gamma)\}^2$, which is approximately 2×10^{-5} when $10 \log (\Gamma/x) = 23.5\,\mathrm{dB}$.

7.21 Equal-gain combining

Obtain an expression for the signal-to-noise ratio after combining when the two branches described in Exercise 7.20 are combined by equal-gain combining.

Solution outline
The weighting coefficients in the combiner are $Kg_1^*/|g_1|$ and $Kg_2^*/|g_2|$. The combined signal is therefore

$$K\,\mathrm{Re}\left[\left\{\frac{g_1 g_1^*}{|g_1|} + \frac{g_2 g_2^*}{|g_2|}\right\}\exp(j\omega_c t)\right],$$

which has mean power

$$\tfrac{1}{2}K^2(|g_1| + |g_2|)^2.$$

The noise power after combining is

$$N_1\left|\frac{Kg_1^*}{|g_1|}\right|^2 + N_2\left|\frac{Kg_2^*}{|g_2|}\right|^2,$$

which simplifies to $K^2(N_1 + N_2)$.

Hence the signal-to-noise ratio after combining is

$$\frac{(|g_1| + |g_2|)^2}{2(N_1 + N_2)}.$$

7.22 Linear feedback shift register: recurrence relations

For each of the maximal-length-sequence-generator configurations shown in Fig. 7.6, obtain a recurrence relation relating the r-th state (a_r) of the n-th stage to previous states $(a_{r-1}, a_{r-2}, $ etc.), thus showing that the configurations are interchangeable.

Solution outline
Configuration A We may consider instead the sequence of states of the first stage, since it is the required sequence advanced by $n-1$ clock periods.

When the first stage is in state a_{r-1}, the later stages are in states $a_{r-2}, a_{r-3}, \ldots, a_{r-n}$. Hence

$$a_r = c_1 a_{r-1} + c_2 a_{r-2} + \ldots + c_{n-1} a_{r-n+1} + a_{r-n}.$$

Configuration B In this case the state of the n-th stage may be regarded as the weighted sum of its previous states thus:

$$a_r = c_1 a_{r-1} + c_2 a_{r-2} + \ldots + c_{n-1} a_{r-n+1} + a_{r-n}.$$

7.23   Generation of maximal-length sequences of period 15

$1 + X^{15}$ may be expressed as the product

$$(1 + X)(1 + X + X^2)(1 + X + X^4)(1 + X^3 + X^4)(1 + X + X^2 + X^3 + X^4)$$

in which addition is modulo-2 and the factors are irreducible. Use this result to find all the generators of the form of Fig. 7.6(a) which generate a maximal length sequence of period 15. For each generator, write down the sequence of states and hence the output binary sequence. State the relationship between the binary sequences you obtain.

Solution outline

$1 + X + X^4$ and $1 + X^3 + X^4$ are suitable characteristic polynomials: $1 + X + X^2 + X^3 + X^4$ is not, because it is a factor of $1 + X^5$. The only possible generators are those shown in Fig. 7.10. The sequences of states are given in Table 7.2.

A column from (a) or (b) gives one period of the output sequence for generator (a) or (b) respectively. The sequence generated by (a) is the time-inverse of that generated by (b).

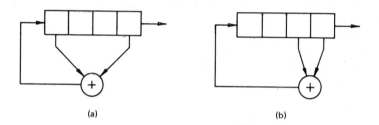

Fig. 7.10   Maximal-length-sequence generators with period 15 (solution to Exercise 7.23)

Table 7.2

(a)	(b)
1 0 0 0	1 0 0 0
1 1 0 0	0 1 0 0
1 1 1 0	0 0 1 0
1 1 1 1	1 0 0 1
0 1 1 1	1 1 0 0
1 0 1 1	0 1 1 0
0 1 0 1	1 0 1 1
1 0 1 0	0 1 0 1
1 1 0 1	1 0 1 0
0 1 1 0	1 1 0 1
0 0 1 1	1 1 1 0
1 0 0 1	1 1 1 1
0 1 0 0	0 1 1 1
0 0 1 0	0 0 1 1
0 0 0 1	0 0 0 1
(1 0 0 0)	(1 0 0 0)

7.24   Maximal-length sequences: occurrence of 'all-1' and 'all-0' subsequences

Prove that a maximal-length sequence generated by the linear-feedback shift-register of Fig. 7.6(a) has the following properties.

(i) The longest unbroken sequence of 0s is of length $n - 1$.
(ii) The longest unbroken sequence of 1s is of length n.
(iii) There is no unbroken sequence of 1s of length just $n - 1$.
(iv) The ratio of the number of 0s to the number of 1s is $1 - 2^{1-n}$.

Solution outline
(i) There can be no sequence of n 0s since the all-zero state is not assumed. A sequence of $(n - 1)$ 0s begins as the state changes from $00 \ldots 01$ to $10 \ldots 00$.
(ii) The state $11 \ldots 11$ must be followed by the state $01 \ldots 11$.
(iii) A sequence of just $(n - 1)$ 1s could begin only when the state changed from $11 \ldots 10$ to $01 \ldots 11$. Such a change would require feedback to be taken from an odd number of stages (including the last) and is therefore inconsistent with an irreducible characteristic polynomial.
(iv) In the $2^n - 1$ possible states there are $n2^{n-1}$ 1s and $(n2^{n-1} - n)$ 0s.

7.25 Maximal-length sequences: power spectrum

(i) Obtain an expression for the envelope of the power spectrum of a polar switching waveform derived from a maximal length binary sequence of period $p = 2^n - 1$ by representing 0 and 1 by NRZ rectangular pulses of $+1$ V and -1 V respectively. Express the result as a function of p and the digit rate f_c.
(ii) Repeat (i) but in this case assume that the pulses are $+1$ V and 0 V.
(iii) Repeat (ii), but in this case assume that the duration of each pulse is half the clock period.

Solution outline
(i) By using the facts that
   (a) shifting and multiplying the derived waveform is equivalent to shifting and adding (modulo-2) the maximal-length binary sequence, and
   (b) there are $(p - 1)/2$ positive pulses and $(p + 1)/2$ negative pulses in one period of the derived waveform,
the autocorrelation of the derived waveform may be shown to be as described in Fig. 7.11(a).
The power spectrum consists therefore of DC power $1/p^2$ and the line spectrum

$$U(f)\frac{1}{T} \sum_{n \neq 0} \delta\left(f - \frac{n}{T}\right),$$

(Appendix 1.5), where $T = p/f_c$ and $U(f)$ (the Fourier transform of a triangular pulse) is given by

$$U(f) = \frac{1+p}{pf_c} \text{sinc}^2\left(\frac{f}{f_c}\right)$$

(Appendix 1.1). The envelope of the two-sided power spectrum (disregarding the $f = 0$ line) is therefore

$$\frac{1+p}{p^2} \text{sinc}^2\left(\frac{f}{f_c}\right).$$

(ii) Halving the amplitude changes the envelope in (i) to

$$\frac{1+p}{4p^2} \text{sinc}^2\left(\frac{f}{f_c}\right) \qquad (f \neq 0).$$

The DC power is now $(p - 1)^2/4p^2$.

(iii)   The autocorrelation may be shown to be as described in Fig. 7.11(b). It may be regarded as the sum of two functions, one with period $1/f_c$ and one with period p/f_c. Hence the power spectrum may be regarded as the sum of two contributions:

(a) a line spectrum with lines at f_c and its harmonics and with (two-sided) envelope (including $f = 0$)

$$\frac{p-3}{16p}\,\text{sinc}^2\!\left(\frac{f}{2f_c}\right),$$

(b) a line spectrum with lines at f_c/p and its harmonics and with (two-sided) envelope (including $f = 0$)

$$\frac{p+1}{16p^2}\,\text{sinc}^2\!\left(\frac{f}{2f_c}\right).$$

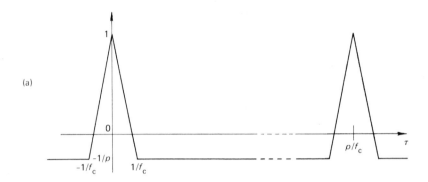

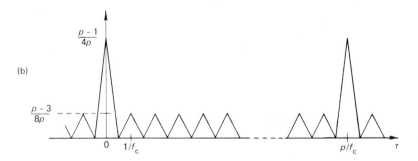

Fig. 7.11   Autocorrelation functions of waveforms derived from maximal-length sequences (Exercise 7.25) (a) NRZ polar waveform (b) RZ unipolar waveform

7.26   Maximal-length sequences: decimation

Use the maximal-length sequence generated by the shift-register of Fig. 7.6(a) with $n = 3$, $c_1 = 1$ and $c_2 = 0$ to verify the following properties of maximal-length sequences.

(i)   The sequence obtained by selecting every q-th term of a maximal-length sequence is the same sequence if $q = 2^k$, where k is a positive integer.

(ii)   When the period p of a maximal-length sequence is a prime number, the sequence obtained by selecting every q-th term of that sequence is also maximal length with period p for any integer q.

Solution outline
(i) The output sequence of the shift register is

   0 0 1 1 1 0 1 0 0 1 1 1 0 1 0 0 1 1 1 0 1....

Selecting every second term gives

   0 1 1 1 0 1 0 0 1 1 1....

Selecting every fourth term gives

   0 1 0 0 1 1 1 0 1 0 0....

Selecting every eighth is equivalent to selecting every term. Selecting every sixteenth is equivalent to selecting every second term etc.
(ii) Selecting every third, fifth or sixth term gives the inverse sequence, which is also maximal length.
   Selecting every q-th term is known as 'decimation'.

7.27 Pseudonoise spread-spectrum radio system

In the transmitter of a pseudonoise spread-spectrum radio link, a 100 MHz carrier is frequency-modulated by the message to occupy a 25 kHz bandwidth and then subjected to phase-reversal keying in accordance with a maximal-length binary sequence so as to occupy a 1 MHz bandwidth.
(i) Estimate the process gain and the minimum acceptable ratio of signal to interfering carrier at the input to the receiver if the ratio of signal to interference at the discriminator input should not be less than 10 dB.
(ii) An eavesdropping conventional receiver of bandwidth 1 kHz is searching for the spread-spectrum transmission. Calculate the minimum length of the shift register generating the maximal-length sequence if in the absence of message modulation the eavesdropping receiver is always receiving at least ten spectral lines when tuned to 100 MHz.
(iii) Interference is received from a second transmitter, identical to that described above except that it uses a different maximal-length sequence. Estimate the minimum value of the signal-to-interference ratio at the receiver input if it should not be less than 10 dB at the discriminator input.
(iv) Show that an eavesdropping receiver which generates the square of a received pseudonoise spread-spectrum transmission can extract the message without any knowledge of the sequence used for spreading the spectrum.

Solution outline
(i) The process gain is approximately 40 (16 dB). Hence the signal-to-interference ratio at the receiver input should not be less than $(10 - 16)$ dB $= -6$ dB. [See Section 7.4.2.]
(ii) The clock frequency of the sequence generator is approximately 500 kHz. To satisfy the requirement, the repetition frequency of the sequence must not exceed 100 Hz. Hence the length of the sequence must exceed 5000. Since $2^{12} = 4096$ and $2^{13} = 8192$, 13 stages are required.
(iii) Assume that the effect of the despreader on a transmission with the incorrect code sequence is to leave unchanged its power spectral density around 100 MHz. Hence the despreader improves the signal-to-interference ratio by about 16 dB, so that the ratio may be as low as -6 dB at the receiver input.

(iv)   The signal received is $s(t) \cos\{\omega_c t + \phi(t)\}$, where $s(t)$ is always either $+1$ or -1. After squaring, the signal is

$$\tfrac{1}{2}[1 + \cos\{2\omega_c t + 2\phi(t)\}],$$

which has the original angle modulation with twice the original frequency deviation.

7.28   Frequency-hopping radio system with interfering carrier

The binary frequency-hopping system of Fig. 7.7 transmits data at 100 kbit/s and has contiguous channels occupying a radio-frequency bandwidth of approximately 20 MHz. A strong interfering carrier within the frequency band of the transmitted signal causes a chip error when it falls within the channel available but not used during that chip. Calculate the digit error probability for the data output in each of the following cases.
   (i)   The chip rate is equal to the data rate.
   (ii)   The chip rate is equal to five times the data rate and the data output is a majority decision based on five consecutive chip decisions.
   (iii)   The chip rate is equal to half the data rate. In this case the system is modified to be quaternary. Two consecutive data digits determine which of four available frequencies is transmitted during each chip and the receiver chooses the largest output from four integrate-and-dump circuits.
   Repeat the calculation in cases (i) and (ii) when there are five interfering carriers evenly distributed across the frequency band of the transmitted signal.

Solution outline
With one interfering carrier:
   (i)   there are 100 channels and the digit error probability is therefore 0.01;
   (ii)   there are 20 channels and the digit error probability is therefore

$$\binom{5}{3}\left(\frac{1}{20}\right)^3\left(\frac{19}{20}\right)^2 + \binom{5}{4}\left(\frac{1}{20}\right)^4\left(\frac{19}{20}\right) + \left(\frac{1}{20}\right)^5 = 0.001\,16;$$

   (iii)   there are 200 channels; during any chip there is one channel in which interference produces two data errors and two channels in which it produces one; the average number of data errors in one chip is therefore

$$2\left(\frac{1}{200}\right) + \left(\frac{1}{200}\right) + \left(\frac{1}{200}\right) = 0.02,$$

giving a digit error probability of 0.01.
   With five interfering carriers the digit error probability is

   (i)   0.05,

   (ii) $\binom{5}{3}\left(\frac{5}{20}\right)^3\left(\frac{15}{20}\right)^2 + \binom{5}{4}\left(\frac{5}{20}\right)^4\left(\frac{15}{20}\right) + \left(\frac{5}{20}\right)^5 = 0.104.$

7.29   Frequency-hopping radio system with comb jamming

In a binary frequency-hopping system (Fig. 7.7) the chip rate is $(2k+1)$ times the data rate (where k is a non-negative integer) and each output data digit is a majority decision based on $(2k+1)$ consecutive chip decisions. The radio-frequency bandwidth is 20 MHz. An interfering 'comb' jammer produces interfering tones in 100 of the channels.
   Calculate the maximum data rate which can be achieved by suitable choice of k if the digit error probability for the data output must not exceed 0.01. Assume that the channels

are contiguous and that each has bandwidth twice the chip rate. Assume also that an interfering tone causes a chip error when it falls within the channel available but not used during that chip.

Solution outline
Consider in turn the cases $k = 0, 1, 2, \ldots$, continuing until the maximum data rate consistent with the required error rate begins to decrease. If there are n channels, the chip error probability p is approximately $100/n$ and the data rate r cannot exceed $10^7/(2k+1)n$ bit/s.
  (i) $k = 0$. $p < 0.01$ requires $n > 10^4$ and $r < 1000$ bit/s.
  (ii) $k = 1$.

$$\binom{3}{2}p^2(1-p) + p^3 < 0.01$$

requires $p < 0.0589$. Hence $n > 1698$ and $r < 1963$ bit/s.
  (iii) $k = 2$.

$$\binom{5}{3}p^3(1-p)^2 + \binom{5}{4}p^4(1-p) + p^5 < 0.01$$

requires $p < 0.105\,64$. Hence $n > 946$ and $r < 2114$ bit/s.
  (iv) $k = 3$. $p < 0.1423$, $n > 702$, $r < 2035$ bit/s.
  (v) $k = 4$. $p < 0.1710$, $n > 584$, $r < 1903$ bit/s.
Hence the best choice of k is $k = 2$, giving a data rate of about 2.11 kbit/s.

7.30   *m*-ary frequency hopping

In an *m*-ary frequency-hopping radio system with $m = 8$, the chip rate is 10 kHz and the data rate is 30 kbit/s. In each chip three consecutive binary data digits determine which of the eight available frequencies $f_c \pm 10$ kHz, $f_c \pm 30$ kHz, $f_c \pm 50$ kHz, $f_c \pm 70$ kHz is transmitted. The value of f_c changes from chip to chip in a pseudorandom manner, assuming with equal probability 500 different values uniformly spaced over a 20 MHz bandwidth. In the receiver the incoming signal is shifted to a 1 MHz intermediate frequency by mixing it with a carrier hopping in synchronism with f_c and is then passed to eight band-pass filters, each followed by an envelope detector and sample-and-hold circuit. At the end of each chip, the three output binary data digits are determined by selecting the integrator with the largest output. Calculate the digit error rate at the data output due to a strong interfering carrier.

For comparison, calculate the digit error rate at the output when the 20 MHz transmission bandwidth is occupied by binary frequency hopping with chip rate 30 kHz.

Solution outline
The probability that the interfering carrier lies within a given channel is 0.001.

Therefore during any chip there is a probability 0.001 that three errors are introduced into the data, a probability 0.003 that two errors are introduced and a probability 0.003 that one error is introduced. Hence the average number of data errors per chip is

$$(3 \times 0.001) + (2 \times 0.003) + (1 \times 0.003) = 0.012$$

i.e. 0.004 per data digit.

In the binary case there are 20 000/60 channels and therefore the digit error rate is 0.003.

7.31   Frequency hopping using two channels simultaneously

In a frequency-hopping radio system with 1000 contiguous channels, five adjacent channels are available in each chip. Three bits of information are transmitted per chip by transmitting simultaneously on just two of the five available channels. Calculate the probability that the information derived from a chip is in error when a strong interfering carrier is present.

Solution outline
Errors are caused when the interfering carrier falls in one of the three available but unoccupied channels. The probability of that occurring is 0.003.

7.32   Energy spectrum of chirp pulse

Using the approach outlined in Section 7.4.4., derive an expression for the energy spectrum of the chirp pulse $v(t)$ defined by

$$v(t) = \begin{cases} A\cos\left[2\pi f_c t + 2t\Delta f\sin^{-1}\left(\dfrac{2t}{T}\right) - T\Delta f\left\{1 - \sqrt{\left(1 - \dfrac{4t^2}{T^2}\right)}\right\}\right] & \text{when } |t| \leqslant \dfrac{T}{2} \\ 0 & \text{otherwise} \end{cases}$$

Show that after passing through a matched filter the pulse has an envelope with null-to-null duration $3/\Delta f$.

Solution outline
The instantaneous frequency f is given by

$$f = f_c + \frac{\Delta f}{\pi}\sin^{-1}\left(\frac{2t}{T}\right) \qquad |t| \leqslant \frac{T}{2}.$$

Assuming that the (two-sided) energy spectrum $|V(f)|^2$ varies inversely with df/dt, at positive frequencies,

$$|V(f)|^2 = K\cos\left\{\frac{\pi(f-f_c)}{\Delta f}\right\} \qquad |f-f_c| \leqslant \frac{\Delta f}{2},$$

for some constant K. Since the energy in the pulse is $A^2 T/2$, $K = \pi A^2 T/8\Delta f$.

The Fourier transform of the envelope of the pulse at the output of a matched filter is therefore (Section 2.2.1 and Appendix 1.4)

$$2|V(f+f_c)|^2 = 2K\cos\left(\frac{\pi f}{\Delta f}\right) \qquad |f| \leqslant \frac{\Delta f}{2}.$$

The null-to-null duration of the inverse transform of this is $3/\Delta f$ (Exercise 6.5).

7.33   Chirp binary data transmission

In a 'chirp' binary data transmission system the received signal elements are bursts of amplitude A and duration T during which the frequency is swept linearly from $f_0 - \frac{1}{2}\Delta f$ to $f_0 + \frac{1}{2}\Delta f$ for a mark and from $f_0 + \frac{1}{2}\Delta f$ to $f_0 - \frac{1}{2}\Delta f$ for a space. In the receiver two matched filters are followed by two envelope detectors, the outputs of which are compared at suitable sampling instants.

(i) Assuming the spectrum of each signal element to be rectangular, obtain an expression for the waveform of a signal element at the output of the filter matched to it.

(ii)   Show that in the absence of noise the ratio of the detector output voltages at the sampling instants is approximately $\sqrt{(2T\Delta f)}$.

(iii)   For a 200 baud system in which $\Delta f = 3$ kHz, $T = 5$ ms and the received signal (of power 10^{-13} W) is accompanied by noise with power density 10^{-18} W/Hz, estimate:

(a)   the signal-to-noise ratio at the sampling instant at the output of a matched filter and the change in that signal-to-noise ratio due to a Doppler shift of 1 kHz in the received signal,

(b)   the timing error which would cause a 3 dB reduction in sample values,

(c)   the signal-to-interference ratio at the receiver input when an interfering carrier at a frequency within the band of the received signal produces at the mark detector output a voltage 20 dB less than the peak produced by a mark signalling element.

Solution outline

(i)   The energy of a signal element is $A^2T/2$ and therefore its (two-sided) energy density is $A^2T/4\Delta f$. The power gain of a matched filter is therefore $A^2T/4\Delta f$. The required waveform is the inverse Fourier transform of a real spectrum of magnitude $A^2T/4\Delta f$ restricted to rectangles about $+f_0$ and $-f_0$ and is therefore

$$(A^2T/2)\operatorname{sinc}(t\Delta f)\cos 2\pi f_0 t$$

(Section 2.2.1 and Appendices 1.4, 3).

(ii)   At the output of a matched filter both wanted and unwanted signal elements have energy $A^4T^2/8\Delta f$. The unwanted signal element has an approximately constant amplitude and extends for a duration approximately $2T$. Therefore the amplitude of the unwanted signal element is $A^2T/(2\sqrt{2T\Delta f})$. Since from (i) the maximum value of the amplitude of the wanted signal is $A^2T/2$, the required result follows.

(iii)   (a)   Either use the general result for a matched filter ($2E/N_0$, Section 2.2.1) or use (i), noting that the (one-sided) noise bandwidth of the matched filter is Δf. The required ratio is $A^2T/2N_0 = 30$ dB. Doppler shift reduces it by about $20 \log 1.5$ dB $= 3.5$ dB.

(b) $\operatorname{sinc}(t\Delta f) = 1/\sqrt{2}$ approximately when $t\Delta f = 0.44$. Hence the allowable timing error is $\pm 0.44/\Delta f = 0.15$ ms.

(c)   Denote by C the amplitude of the incoming interfering carrier. At the mark detector output the sampled signal voltage is $A^2T/2$ and the voltage due to the interference is $C\sqrt{(A^2T/4\Delta f)}$. For the ratio of these to be 10, $A/C = 2.58$ (8.2 dB). Alternatively argue that compression of the mark by the factor $T\Delta f (= 15)$ gives it an advantage of 11.8 dB over interference.

8
Radio paths

8.1 Propagation in the troposphere

8.1.1 Refraction in the troposphere

The troposphere is the lowest part of the atmosphere. It is characterised by its temperature decreasing with increasing height and it extends to a height of approximately 10 km.

The refractive index $n(h)$ in the troposphere usually varies with height (h) according to the law:

$$n(h) = 1 + a \exp(-bh),$$

where a and b are quantities dependent upon weather conditions. The values $a = 315 \times 10^{-6}$ and $b = 0.136$ (which assume that h is in km) define the CCIR 'reference atmosphere'[1].

When $n(h)$ is a decreasing function of h, there is a tendency for rays to bend so as to extend the radio horizon. Under typical atmospheric conditions the bending may be taken into account by assuming that propagation is rectilinear and that the earth has a radius of about 8800 km.

8.1.2 Reflexion by the surface of the earth

In many situations received signal strength is affected by reflexions from the surface of the earth.

The reflexion coefficient varies with polarisation, frequency, angle of incidence, surface roughness and the electrical and magnetic properties of the reflecting medium. A few useful generalisations are given below; for more information the reader should consult more specialised texts[2,3,4].

The magnitude of the reflexion coefficient is greater for horizontally polarised waves than for vertically polarised waves, and is greater for reflexion by sea water than for reflexion by soil.

For angles of incidence approaching 90° (near-grazing incidence) vertically polarised waves suffer a phase reversal on reflexion, whereas for larger angles of incidence they do not. The angle of incidence at which the reflexion coefficient for vertical polarisation changes sign is typically 70° for soil and 85° for sea water. For horizontally polarised waves there is a phase reversal on reflexion for any angle of incidence.

8.1.3 Diffraction

Diffraction is the spreading of a wave caused by the partial obstruction of it. As a consequence of diffraction, shadows have blurred edges, the extent of the blurring increasing with wavelength. Thus the strength of a received radio signal may be influenced by obstacles which do not actually obstruct the line of sight between transmitter and receiver and useful signals may sometimes be received even when the line of sight is obstructed.

The extent to which a line-of-sight path is clear of obstacles is usually expressed in terms of Fresnel zones. Fresnel zones are the regions between adjacent ellipsoids of revolution in a confocal family with the foci located at the transmitting and receiving antennas. The outer radius r_n of the n-th Fresnel zone at distances a and b from the antennas is given by

$$r_n = \sqrt{\frac{n \lambda ab}{a+b}},$$

where λ (the wavelength), a, b and r_n are all expressed in the same units. The ellipsoid forming the outer boundary of the n-th Fresnel zone has the geometrical property that the path of a ray travelling from one antenna to the other by way of reflexion at that boundary is longer than the direct path between the antennas by $n\lambda/2$ (see Exercise 8.6). A widely used diffraction parameter v is defined by:

$$v = -h\sqrt{2}/r_1,$$

where h is the distance by which the line of sight misses an obstacle under consideration and r_1 is the radius of the first Fresnel zone in the region of that obstacle[5]. v is positive when the obstacle obstructs the line of sight and negative when it does not.

For an obstacle in the form of a semi-infinite plate terminating in a straight knife-edge, the diffraction loss (i.e. the increase in path loss due to the presence of the obstacle) is as shown in Fig. 8.1. If the edge of the obstacle is not a sharp knife-edge but rounded with a radius of curvature not small compared with wavelength, the diffraction loss in the shadow

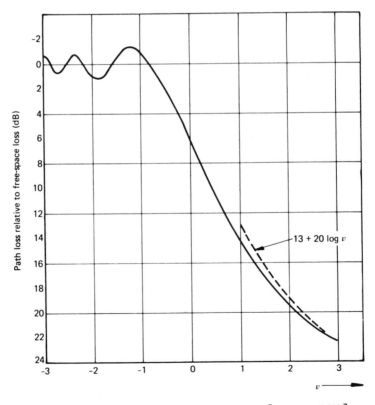

Fig. 8.1 Path loss due to knife edge diffraction [based on CCIR]

region ($v > 0$) is much greater than is predicted by Fig. 8.1[5]. Figure 8.1 gives rise to the rule of thumb that diffraction effects may be disregarded if the first Fresnel zone is unobstructed, since the semi-infinite plate does not intrude into the first Fresnel zone if $v < -\sqrt{2}$.

Several methods have been proposed for the estimation of diffraction loss when there are several knife-edges near the line of sight. Deygout's method[7] is outlined in Exercise 8.11.

Diffraction theory has been applied[5] to the propagation of radio waves near to a smooth spherical earth at frequencies too high for surface-wave propagation (Section 8.1.6) to be of significance. The strength of signals received over the horizon in this way is sometimes increased by the presence of sharp-edged obstacles, the decrease in path loss being referred to as 'obstacle gain'.

8.1.4 Attenuation

In this context the word 'attenuation' refers to those contributions to path loss which are due to the scattering of a radio wave by small objects in its path or to the absorption of energy by media through which the wave passes. In this sense, therefore, attenuation is additional to the inverse-square-law spreading covered by the free-space-loss formula (Appendix 8) and to diffraction loss.

Objects commonly causing scattering are raindrops and ice particles. At frequencies low enough for the objects to be small compared with wavelength, the scattered energy is proportional to the fourth power of frequency (Rayleigh's law), but at higher frequencies the scattered energy is more-or-less independent of frequency. In temperate climates, attenuation due to scattering is rarely significant at frequencies below 10 GHz, but it may be considerable at frequencies above 10 GHz when there is heavy rain.

Typical values of the attenuation per kilometre due to scattering by raindrops are given in Fig. 8.2 and may be multiplied by an 'equivalent path length' to give the total attenuation due to scattering for a given radio path. Equivalent path length is a measure of the distance the radio wave travels through the rain. For the case of earth-to-satellite paths, typical values of the equivalent path length are given in Fig. 8.3. These are empirically derived values which take into account the fact that heavy rain is usually more localised than light rain. For a more detailed discussion of scattering the reader is referred to more specialised texts[5,8].

Attenuation due to absorption in the atmosphere is dominated by the effects of oxygen and water vapour. Typical values of the attenuation per kilometre due to absorption are given in Fig. 8.4. They may be used for different concentrations of water vapour by assuming that the attenuation in dB/km due to absorption by water vapour is proportional to the density of the water vapour present. Frequencies near to absorption peaks are usually avoided. In an earth-to-satellite path, attenuation due to absorption may be estimated by assuming a path length of $3.1 \operatorname{cosec} \theta$ km, where θ is the angle of elevation of the path at the earth terminal and is assumed to be not less than five degrees.

Attenuation due to absorption also occurs when radio waves pass through walls or vegetation[5].

8.1.5 Troposcatter systems

A troposcatter radio system is one in which over-the-horizon communication is achieved at microwave frequencies by directing high-gain transmitting and receiving antennas towards a region of the troposphere (the scatter volume) located above the straight line between them. Frequencies from about 200 MHz to about 5 GHz have been used for this purpose and useful ranges up to about 1000 km have been obtained. Various theories have

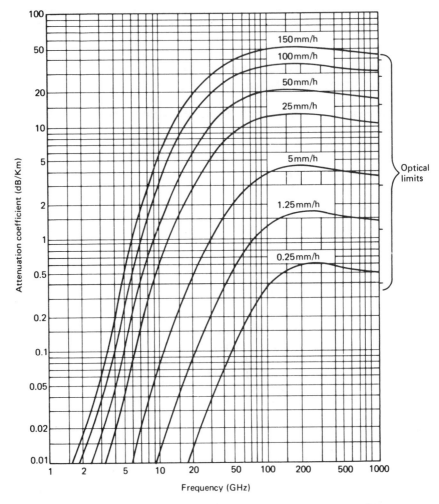

Fig. 8.2 Attenuation coefficient due to rain [based on CCIR]

been proposed to account for the strength of the received signal, which may be much greater than that predicted by diffraction over the intervening terrain[4].

The path loss of a troposcatter radio link may be estimated by adding two quantities, forward scattering loss and aperture-to-medium coupling loss, to the path loss which would be obtained if the link were line-of-sight.

Forward scattering loss takes account of the fact that very little of the power reaching the scatter volume is redirected so as to reach the receiver. It is a fluctuating quantity with a median value depending on meteorological conditions, frequency and scattering angle (the angle through which the beam is diverted at the scatter volume). Table 8.1 shows typical median values at 900 MHz. The median value for a given scattering angle increases by about 3 dB for each octave increase in frequency. Methods have been established[9] for incorporating climatic conditions into the prediction of forward scattering loss.

Aperture-to-medium coupling loss (L_d) takes account of the reduction in the gain of highly directional antennas due to the distribution of the scattering process. It may be

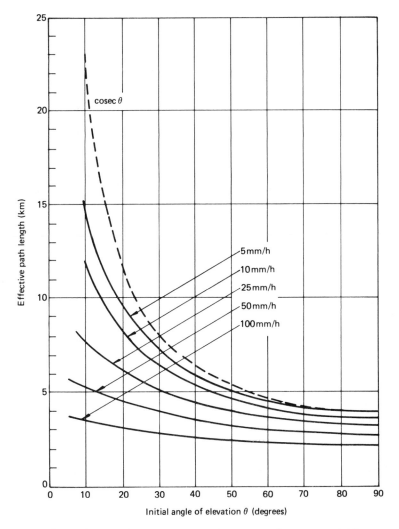

Fig. 8.3 Equivalent path lengths of slant paths [based on CCIR]
--- Cosecant θ law for 5 mm/h

estimated from the semi-empirical formula[9]

$$L_d = 0.07 \exp\left\{0.055\,(G_T + G_R)\right\} \text{ dB},$$

where G_T and G_R are the gains in dB of the transmitting and receiving antennas when operating under free-space conditions, each assumed to be less than 50 dB.

Observations over periods of a few minutes show that the fluctuations of the received signal approximate to Rayleigh fading (Section 7.3.1). The fading rate (which is sometimes defined as the average number of times that the envelope of the received signal falls through its median value in unit time) is found to increase from a few fades per minute at VHF to a few fades per second at SHF[5]. Over periods longer than a few minutes, changes in meteorological conditions, cause slow changes in the variance of the Rayleigh distribution.

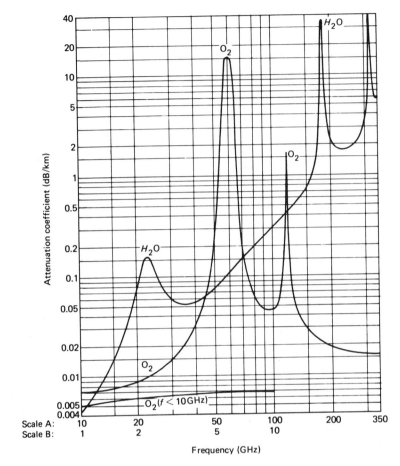

Fig. 8.4 Specific attenuation by atmospheric gases (use scale B for oxygen absorption below 10 GHz) [based on CCIR]
Pressure: 1 atm (1013.6 mb)
Temperature: 20°C
Water vapour density: 7.5 g/m^3

Table 8.1 Typical median values of forward scattering loss at 900 MHz.

Scattering angle (m rad)	Median forward scattering loss at 900 MHz (dB)
7.5	42
10	46
20	58
30	66
40	74
50	80
70	91
90	103

8.1.6 Surface wave

A vertically polarised radio wave launched by an antenna close to the ground travels over the ground as a surface wave, the attenuation of which is influenced by the electrical constants of the ground over which it passes.

Theoretical study of propagation over a smooth spherical earth suggests that it should be possible to distinguish the following three regions.

Region 1 At short ranges the electrical properties of the ground have little effect on attenuation and therefore the field strength falls approximately inversely with distance as for free-space propagation.

Region 2 At ranges greater than about $20\lambda^{1/3}$ km (where the wavelength is in metres) attenuation may be regarded as due solely to diffraction and field strength falls exponentially with distance.

Region 3 At intermediate ranges field strength falls approximately inversely with the square of distance. In this region attenuation is influenced at low frequencies primarily by the conductivity σ of the ground and at high frequencies primarily by the relative permittivity ε of the ground.

The range at which the transition from Region 1 to Region 3 occurs depends upon the electrical constants of the ground and when attenuation is low (as over sea water) Region 3 may not be distinguishable.

An indication of ranges available by propagation in the surface-wave mode is given by the theoretical predictions of Table 8.2. For further discussion of the principal predictions of the theory of surface waves see for example Picquenard[4] and for charts of the predictions see CCIR publications[10].

Table 8.2 Distance in km at which the field strength of a wave launched by a short vertical dipole close to the surface of a smooth spherical earth is 40 dB below its value over a plane, perfectly conducting earth.

Frequency	Soil				Sea water
	$\varepsilon = 4$ $\sigma = 10^{-5} \text{S/m}$	$\varepsilon = 4$ $\sigma = 10^{-4} \text{S/m}$	$\varepsilon = 4$ $\sigma = 10^{-3} \text{S/m}$	$\varepsilon = 4$ $\sigma = 10^{-2} \text{S/m}$	$\varepsilon = 80$ $\sigma = 4 \text{S/m}$
10 kHz	2200	4800	5300	5200	5100
100 kHz	240	500	1400	2300	2400
1 MHz	27	27	85	310	1100
10 MHz	2.6	2.4	3.0	9.5	380

8.2 Propagation in the ionosphere

8.2.1 The structure of the ionosphere

Ionisation of the atmosphere by solar radiation tends to be concentrated in layers.

The most intensely ionised layers occur between about 150 km and about 500 km above the surface of the earth, in what is called the F–region. The height of the ionised layers depends on the temperature of the atmosphere and hence varies considerably between day and night, particularly in summer. It is usually found that F–region ionisation is concentrated in a single F-layer at night, but during the day two layers (F1 and F2) can usually be distinguished.

Between about 50 km and about 100 km above the surface of the earth, a diffuse, weakly ionised layer called the D–layer occurs during the day.

Intermediate between the D–layer and the F–layers in height and in ionisation intensity is the E–layer, which is much more weakly ionised at night than during the day. Occasionally a thin but relatively strongly ionised layer called the sporadic E–layer appears at a height between about 90 km and 130 km.

In addition to diurnal and seasonal variations in ionisation intensity, there are variations with solar activity (sunspots) with a period of about 11 years.

A lucid introduction to propagation in the ionosphere is given by Glazier and Lamont[2]. For more details consult Davies[11].

8.2.2   Maximum usable frequency

The usefulness of the ionosphere in telecommunications lies in its ability to provide sky-wave, over-the-horizon communications by redirecting radio waves so that they are returned to the earth's surface.

In ionised regions, the presence of free electrons increases the phase velocity of radio waves and hence gives rise to refraction. The refractive index n at frequency f Hz when there are N free electrons per cubic metre is given approximately by

$$n \simeq \left(1 - \frac{81N}{f^2}\right)^{1/2}.$$

By the optical laws of refraction, a ray entering an ionised layer at an angle of incidence θ penetrates the layer until it reaches a height at which $\sin \theta = n$. At that height it is horizontal. It follows that for a given value of θ, reflexion occurs provided that $f < f_{\text{m.u.f.}}$, where $f_{\text{m.u.f.}}$ (the maximum usable frequency) is given approximately by

$$f_{\text{m.u.f.}} \simeq 9 \sqrt{N_{\text{max}}}/\cos \theta,$$

N_{max} being the maximum value of N in the ionised layer.

The maximum usable frequency for vertical incidence ($\theta = 0$) is called the 'critical frequency' and denoted by f_c. Values of f_c associated with the F–layers usually lie between 3 MHz and 10 MHz. Typical day-time values of f_c for the E–layer lie between 3 MHz and 4 MHz.

For given values of f and N_{max}, reflexion occurs if $\theta > \theta_c$, where θ_c (the 'critical angle') is given by

$$\cos \theta_c = f_c/f \simeq 9 \sqrt{N_{\text{max}}}/f.$$

Thus when $f > f_c$ there is a distance (the skip distance) below which sky-wave reception is not possible.

Ranges up to about 4000 km are obtainable by means of a single reflexion in the F1–layer or F2–layer (single-hop transmission). Longer ranges are obtainable by means of a series of reflexions in the ionosphere and by the ground alternately (multi-hop transmission) (see Exercise 8.16).

8.2.3   Attenuation in the ionosphere

Collisions between ionisation electrons and gas molecules take energy from radio waves in the ionosphere. When expressed in dB/km, the resulting attenuation is inversely proportional to the square of frequency. This form of attenuation is therefore minimised by using the highest frequency which can be relied upon to be reflected and is most serious in the relatively dense D–layer.

Considerable additional attenuation occurs at frequencies near to the gyro frequency for an electron in the earth's magnetic field. The gyro frequency is the period of revolution of an electron in its circular orbit in a steady magnetic field and for the earth's magnetic field has an average value of about 1.4 MHz.

Signals received by way of ionospheric reflexion suffer from severe fading. It is usual to distinguish between fast fading, in which the time between fades is typically between 0.1 s and 10 s, and slow fading, in which the time between fades is measured in minutes, hours or days.

The two principal types of fast fading are interference fading and polarisation fading. Interference fading arises because the received signal is effectively the sum of many components each having travelled by a different path of fluctuating length. Polarisation fading arises from fluctuations in the polarisation presented to the receiving antenna and is caused by the splitting of the ray by the earth's magnetic field into two elliptically polarised rays following different paths. Sometimes fast fading is well described by the Rayleigh distribution, but when there is a strong specular component present the rician distribution is more appropriate. When an unmodulated carrier is transmitted, the power spectrum of the received signal is often approximately gaussian in shape[12], the standard deviation being referred to as the 'coherence bandwidth' of the fading channel and used as a measure of the rapidity of the fading.

In radio systems in which ionospheric reflexion occurs, it is often found that the received signal is the sum of two independently fading components which have travelled by paths with different propagation times. In such cases the fading of the total received signal varies with frequency (the fading is 'frequency-selective') and the two components may cancel to give deep fading at a 'comb' of frequencies separated by the reciprocal of the difference in propagation times. At long ranges, that time difference may be as much as 10 ms, but it is usually much less. Components which have travelled by paths with different delays can be added in phase by the technique known as 'rake'[13] which thus makes the fading less frequency-selective and reduces intersymbol interference.

8.3 Noise in radio paths

8.3.1 Antenna noise temperature and system noise temperature

The noise temperature of a terrestrial receiving antenna may be regarded as the sum of three components, representing noise generated (i) in the antenna itself, (ii) in the earth's atmosphere and (iii) outside the earth's atmosphere.

Noise generated in the antenna itself is associated with losses in the antenna. It may be taken into account by assuming that there is an attenuator at the output of the antenna and will not be discussed further.

Generation of noise in the atmosphere is associated with the absorption of radio waves. Its contribution to the antenna noise temperature may be shown[14] to be $T_2 (1 - \alpha)$, where T_2 is the temperature of the atmosphere and α the fraction to which power from a source outside the atmosphere is reduced by absorption in the atmosphere. The derivation of this result is analogous to that in Exercise 8.20 (ii).

If the contribution to antenna noise temperature of a source outside the atmosphere would be T_1 were the atmosphere absent, absorption in the atmosphere reduces that contribution to αT_1.

Antenna noise originating in the atmosphere or beyond it is often referred to as 'sky noise'.

The total noise accompanying the signal in a radio receiver may be regarded as being due to a single source of noise at some convenient point, for example at the input to the first amplifier in the receiver. The noise temperature of that source is called the 'system noise temperature' and is the sum of contributions representing antenna noise, feeder noise and noise generated within the amplifiers of the receiver.

8.3.2   Extra-terrestrial sources of noise

At frequencies below about 1 GHz, noise from sources outside the solar system may be significant, particularly when a narrow-beam antenna is directed towards the centre of the galaxy. This 'galactic' or 'cosmic' noise decreases with increase of frequency (f) and is roughly proportional to $f^{-2.4}$. It varies widely with direction, but for a wide-beam antenna at 100 MHz contributes typically about 1000 K to the antenna noise temperature[5,2].

When a narrow-beam receiving antenna is directed towards the sun, solar noise may make a large contribution to the antenna noise temperature. Solar noise varies widely with sun-spot activity, but when the sun is quiet the contribution falls with increasing frequency, being typically 4.8 K at 100 MHz, 1.7 K at 1 GHz and 0.053 K at 10 GHz for an isotropic antenna (assuming no atmospheric absorption).

8.3.3   Atmospheric noise

Estimates of the contribution to antenna noise temperature from atmospheric noise due to absorption may be made from Fig. 8.5, which is drawn from calculations based on a theoretical model. The model assumes that the earth has an effective radius of 8500 km, that the water vapour content of the atmosphere decreases exponentially with height and that the antenna beam is infinitely narrow.

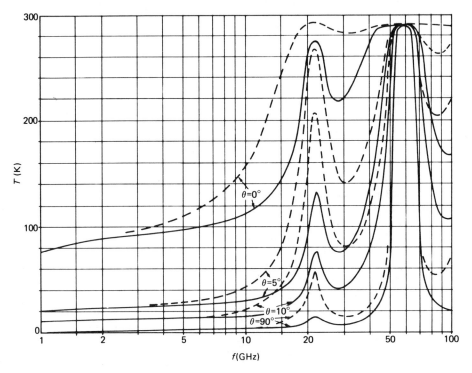

Fig. 8.5 Sky noise temperature (clear air) [based on CCIR]
Surface pressure 1 atm
Surface temperature 20°C
———surface water-vapour concentration 3 g/m³ (17 % humidity)
θ = elevation angle
– – –surface water-vapour concentration 17 g/m³ (98 % humidity)

Additional noise is generated by lightning discharges in the atmosphere. There are on average about 100 flashes every second, mostly occurring in tropical regions, and the spectrum of the radiated energy is at a maximum in the region of 10 kHz.

8.4 References

1 *CCIR*, Vol. 5, Propagation in non-ionized media, Recommendation 369-3, Geneva, 1982
2 Glazier, E. V. D., and Lamont, H. R. L., *The Services Textbook of Radio*, Vol. 5, 'Transmission and propagation', HMSO, 1958
3 Matthews, P. A., *Radio Wave Propagation*, Chapman and Hall, 1965
4 Picquenard, A., *Radio Wave Propagation*, Macmillan (Philips Technical Library), 1974
5 Hall, M. P. M., *Effects of the Troposphere on Radio Communication*, IEE Electromagnetic Waves Series 8, Peter Peregrinus, 1979
6 Carl, H., *Radio Relay Systems*, Macdonald, 1966
7 Deygout, J., Multiple knife-edge diffraction of microwaves, *IEEE Trans*, AP-14, pp. 480–9, July 1966
8 Brodage, H., and Helmut, W., *Planning and Engineering of Radio Relay Links*, Siemens AG and Heyden and Son Ltd, 1977
9 *CCIR*, Vol. 5, Propagation in non-ionized media, Report 238–4, Geneva, 1982
10 *CCIR*, Vol. 5, Propagation in non-ionized media, Recommendation 368-4, Geneva, 1982
11 Davies, K., *Ionospheric Radio Propagation*, Dover, 1966
12 *CCIR*, Vol. 6, Propagation in ionized media, Report 266–5, Geneva, 1982
13 Price, R., and Green, P. E., 'A communication technique for multipath channels', *Proc. IRE*, **46**, pp. 555–70, March 1958
14 Pierce, J. R., and Posner, E. C., *Introduction to Communication Science and Systems*, Plenum Press, 1980
15 Millington, G., 'Ground-wave propagation over an inhomogeneous smooth earth', *Proc. IEE*, **96**, Part III, pp. 56–64, January 1949

Exercise topics

8.1   Effective radius of earth
8.2   Modified refractive index
8.3   Reflexion due to abrupt change in permittivity
8.4   Reflexion due to linear change in permittivity
8.5   Plane-earth transmission formula
8.6   Reflexion points giving cancellation
8.7   Signal reduction due to surface reflexion
8.8   Frequency of fading due to reflexions
8.9   Effective height of obstacles
8.10 Knife-edge diffraction
8.11 Diffraction at two ridges
8.12 Diffraction at three ridges
8.13 Optimum frequency for troposcatter link
8.14 Path loss of troposcatter link
8.15 Surface wave: Millington's method and recovery effect
8.16 Ionospheric reflexion: available ranges
8.17 Group velocity in the ionosphere
8.18 Virtual path and virtual height
8.19 HF soundings of ionosphere
8.20 Noise due to networks in cascade
8.21 Receiver signal-to-noise ratio
8.22 Noise in terrestrial receiver directed towards satellite
8.23 Formula for estimating galactic noise
8.24 Principle of radiometry

Exercises

8.1 Effective radius of earth

Show that the path of a radio wave travelling horizontally in the troposphere curves downwards with a radius of curvature $-n\{dn/dh\}^{-1}$, where n is the refractive index at height h. Verify that near the ground in a basic reference atmosphere that radius is approximately 23 300 km. Taking the earth to be a sphere of radius 6370 km, verify that the position of the path relative to the earth is approximately the same as for rectilinear propagation above a sphere of radius 8800 km (the effective radius of the earth) and calculate the distance of the radio horizon from an antenna 100 m above the ground.

Solution outline
If a wavefront moves from AA′ to BB′ in the short time interval Δt, it appears to rotate about P, the intersection of AA′ and BB′ (Fig. 8.6(a)). Since $AB = c\Delta t/n(h)$ and $A'B' = c\Delta t/n(h + dh)$, it follows that $PA \to -n(dh/dn)$ as $dh \to 0$.

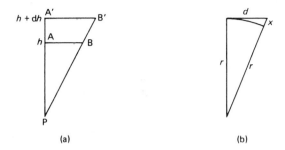

Fig. 8.6   Diagram of solution to Exercise 8.1

For a basic reference atmosphere

$$n(h) = 1 + (315 \times 10^{-6}) \exp(-0.136h) \qquad \text{(Section 8.1.1)}.$$

Hence at ground level the radius is

$$-n\left\{\frac{dn}{dh}\Big|_{h=0}\right\}^{-1} \simeq 23\,300 \text{ km}.$$

In Fig. 8.6(b), $2xr \simeq d^2$ when $d \ll r$. Hence the required radius R' is given by

$$\frac{d^2}{2R'} = \frac{d^2}{2 \times 6370} - \frac{d^2}{2 \times 23\,300}.$$

Note that the reciprocal of radius of curvature is known as curvature and the curvature of the hypothetical sphere is the difference between the curvature of the earth and the curvature of the radio path.

The result $2xr = d^2$ also shows that the distance of the radio horizon of an antenna at a height x is $\sqrt{2xR'}$. Putting $x = 0.1$ and $R' = 8800$ gives a distance 42 km.

8.2 Modified refractive index

Show that the path of a radio wave relative to the surface of the earth may be calculated by assuming the earth to be flat and the refractive index of the atmosphere at height h to be

$n'(h)$ (the 'modified refractive index') given by

$$n'(h) = n(h) + h/R$$

where $n(h)$ is the actual refractive index of the atmosphere and R the radius of the earth.

Under certain weather conditions, $n'(h)$ is found to increase linearly with height, being equal to 1.003 at a height of 50 m and 1.005 at a height of 650 m. Deduce the angle of elevation of a ray as it leaves a transmitting antenna at a height 50 m if that ray meets the ground at grazing incidence.

Solution outline
Since $n(h) \simeq 1$, adding h/R to the refractive index of the atmosphere decreases the downward curvature of a horizontally travelling wave by $1/R$ (See Exercise 8.1) and hence gives correct predictions of height over a flat earth. Alternatively use an argument based on Snell's law[2, 3].

With the figures given, $n'(h) = n(0) + h/300$. Hence with respect to a plane earth the path curves upwards with radius of curvature 300 km. Using the fact that in Fig. 8.6(b) $2xr \simeq d^2$ when $d \ll r$, the required elevation may be shown to be about $-1.0°$.

8.3   Reflexion due to abrupt change in permittivity

A vertically polarised plane radio wave travelling in air with relative permittivity (ε) equal to unity meets a horizontal layer of air with relative permittivity $1 + \Delta\varepsilon$. Show that for sufficiently small $\Delta\varepsilon$ the reflexion coefficient is equal to

$$\Delta\varepsilon \left\{ \frac{1}{2} - \frac{1}{4\cos^2\theta_i} \right\},$$

where θ_i is the angle of incidence at the layer. [Assume that the wave impedance η is real and inversely proportional to $\sqrt{\varepsilon}$.]

A 10 kW, 300 MHz transmission is received at a distance of 100 km after reflexion due to an abrupt change in atmospheric permittivity. Calculate the received power assuming that
(a) each antenna is at sea level, has a 3 dB gain and is correctly aligned,
(b) the atmospheric permittivity increases abruptly by 0.01 % at a height of 5 km but is uniformly unity below that height, and
(c) the earth is flat.
Calculate also the electric field at the receiving antenna.

Solution outline
With the notation of Fig. 8.7, since the tangential components of electric and magnetic fields are continuous at the interface,

$$E_i \cos\theta_i - E_r \cos\theta_i = -E_t \cos\theta_t$$

and

$$\frac{E_i}{\eta_1} + \frac{E_r}{\eta_1} = \frac{E_t}{\eta_2}.$$

Since the normal component of the electric displacement is continuous at the interface,

$$\varepsilon_1 E_i \sin\theta_i + \varepsilon_1 E_r \sin\theta_i = \varepsilon_2 E_t \sin\theta_t.$$

By eliminating θ_t, putting $\varepsilon_1^2/\varepsilon_2^2 = 1 - 2\Delta\varepsilon$ and putting $\eta_1^2/\eta_2^2 = 1 + \Delta\varepsilon$, the reflexion

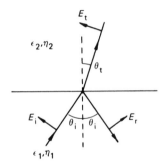

Fig. 8.7 Diagram of solution to Exercise 8.3

coefficient $\rho(= E_r/E_i)$ may be shown to satisfy

$$\rho^2 + 2\rho \left\{ 1 - \frac{2\cos^2\theta_i}{\Delta\varepsilon\cos 2\theta_i} \right\} + 1 = 0,$$

from which the required result follows for small $\Delta\varepsilon$.

When $\tan\theta_i = 10$ and $\Delta\varepsilon = 10^{-4}$, $\rho \simeq 2.47 \times 10^{-3}$, a loss on reflexion of 52.1 dB. The free-space loss for the path length $2\sqrt{(5)^2 + (50)^2}$ km is 122.0 dB (Appendix 8). The received power is therefore $(40 + 3 - 122.0 - 52.1 + 3)$ dBW $= 1.55 \times 10^{-13}$ W. The effective area of the receiving antenna is $2.0/4\pi\,\mathrm{m}^2$ and hence the power density at the receiver is

$$(1.55 \times 10^{-13})/(2.0/4\pi) = 9.7 \times 10^{-13}\ \mathrm{W/m}^2.$$

Hence the r.m.s. electric field strength is 1.9×10^{-5} V/m.

8.4   Reflexion due to linear change in permittivity

A vertically polarised radio wave with wavelength λ travelling in the atmosphere at a small angle ϕ (radians) to the horizontal meets a horizontal plane at which the refractive index of the air decreases abruptly by Δn. Prove that the reflexion coefficient is approximately $\Delta n/2\phi^2$.

Show further that if the decrease Δn takes place linearly with height throughout a layer of thickness Δh, the magnitude of the reflexion coefficient is approximately

$$\frac{\Delta n}{2\phi^2} \left\{ \frac{\sin(2\pi\phi\Delta h/\lambda)}{2\pi\phi\Delta h/\lambda} \right\}.$$

A vertically polarised 300 MHz signal from a powerful transmitter at sea level is received 300 km away at sea level after reflexion from a layer 5 m thick at a height of 10 km in which the refractive index falls linearly by 0.0001. Estimate the loss associated with the reflexion. (Assume the earth to be a sphere of radius 6370 km and the refractive index of the atmosphere to be uniformly unity below the reflecting layer.)

Solution outline
The reflexion coefficient for an abrupt change in refractive index follows from the expression derived in Exercise 8.3 with $\cos\theta_i \simeq \phi \ll 1$ and $\Delta\varepsilon \simeq -2\Delta n$.

By regarding the linear change as a succession of small abrupt changes, the reflexion coefficient in the second case may be written

$$\int_{h_0-(\Delta h/2)}^{h_0+(\Delta h/2)} \frac{1}{2\phi^2} \exp\left\{-j\frac{4\pi h \sin\phi}{\lambda}\right\}\frac{\Delta n}{\Delta h}\,dh.$$

From the data given, $\phi = 0.0783$ rad and therefore $20\log|\rho| = -53.6$ dB.

8.5   Plane-earth transmission formula

Use the formula given in Appendix 8 for the power P_R available at the terminals of a receiving antenna under free-space conditions to show that when the antennas are situated at heights h_1 and h_2 above a perfectly reflecting horizontal plane, P_R is given at low frequencies by:

$$P_R = P_T G_T G_R (h_1 h_2/r^2)^2$$

(the 'plane-earth' transmission formula). Assume $h_1 \ll r, h_2 \ll r$, phase reversal on reflexion and antenna gains the same for the reflected ray as for the direct ray.

Solution outline
The path of the reflected ray is longer than that of the direct ray by

$$\{r^2+(h_1+h_2)^2\}^{1/2} - \{r^2+(h_1-h_2)^2\}^{1/2} \simeq 2h_1h_2/r.$$

Hence

$$P_R = P_T G_T G_R \left(\frac{\lambda}{4\pi r}\right)^2 \left|1-\exp\left\{-j\left(\frac{2\pi}{\lambda}\right)\left(\frac{2h_1h_2}{r}\right)\right\}\right|^2,$$

which reduces to the required result when $h_1 h_2 \ll \lambda r$.

8.6   Reflexion points giving cancellation

A radio signal transmitted from A is received at B both directly and after reflexion (with phase reversal) at C. Show that the reflected signal is in anti-phase at B with the direct signal when

$$r = \sqrt{\frac{2k\lambda ab}{a+b}},$$

where a, b and r are defined in Fig. 8.8, λ is the wavelength and k is a positive integer. Assume that $a \gg r$ and $b \gg r$.

Solution outline
The direct and reflected signals are in anti-phase when

$$\sqrt{a^2+r^2} + \sqrt{b^2+r^2} - (a+b) = k\lambda$$

for any integer k.

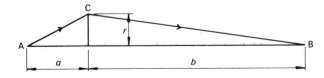

Fig. 8.8   Diagram of Exercise 8.6

8.7 Signal reduction due to surface reflexion

Two ships 5 km apart are in radio communication at 3.3 GHz over the sea. The antenna on one ship is 10 m above the surface of the water and that on the other 40 m. Calculate the change in received signal level due to the existence of a ray reflected in the sea. Assume that the magnitude of the reflexion coefficient is 0.9, that phase reversal occurs on reflexion, that the ocean is smooth and flat and that propagation is rectilinear.

Solution outline
The path of the reflected ray is longer than that of the direct ray by $2h_1 h_2/r$ (Exercise 8.5). Hence, with the data given, the phase difference at the receiver is 93.6° and the effect of the reflected ray is to increase the received power by a factor $|1 + 0.9\underline{/93.6°}|^2$, i.e. an increase of 2.3 dB.

8.8 Frequency of fading due to reflexions

A 600 MHz signal from a transmitter 200 m above sea level on the coast is received over the sea by an aircraft flying at 1000 ft above sea level towards the transmitter. Calculate the rate at which fades occur when the aircraft is flying at 700 m/s and is 15 km from the transmitter [1 ft = 0.305 m].

Solution outline
With the notation of Exercise 8.5, R fades occur per second, where

$$R = \frac{1}{2\pi} \frac{d}{dr} \left\{ \left(\frac{2\pi}{\lambda} \right) \left(\frac{2h_1 h_2}{r} \right) \right\} \frac{dr}{dt}.$$

With the data given, $R = 0.76$.

8.9 Effective height of obstacles

A transmitter and a receiver are at a distance y apart on the surface of the earth. Show that the curvature of the earth increases the effective height of an obstacle at a distance x from the transmitter by approximately $x(y - x)/2R$, where R is the radius of the earth.

A 300 MHz signal is transmitted from an antenna 5 m above a summit which is 300 m above sea level. A receiver is to be located on another summit which is 450 m above sea level and 30 km from the transmitter. Between the summits and 20 km from the transmitter is a third summit 325 m above sea level. Calculate the height h of the mast necessary to support the receiving antenna if the third summit is to lie outside the first Fresnel zone. Take the effective radius of the earth to be 8800 km.

Solution outline
The required result may be obtained directly by equating the products of the segments into which two chords of a circle divide each other.

At the third summit the radius of the first Fresnel zone is 81.65 m (Section 8.1.3) and the increase in effective height due to the curvature of the earth is 11.36 m (formula above). Hence

$$\frac{(450 + h) - 305}{(325 + 11.36 + 81.65) - 305} = \frac{3}{2} \qquad \text{i.e. } h = 24.5 \text{ m.}$$

8.10 Knife-edge diffraction

A 3 GHz radio signal is transmitted between two antennas 1 km apart and 10 m above non-reflecting level ground. Midway between the antennas the line of sight is obstructed by a large building, its roof acting as a long horizontal knife-edge at right angles to the line of sight and 20 m above the ground. Estimate the received signal power when the transmitted power is 1.5 W and the gain of each antenna is 25 dB.

Estimate also the power in a signal received by way of reflexion in a lake at ground level on the same side of the building as the transmitter, assuming the modulus of the (E–field) reflexion coefficient to be 0.95.

Solution outline

At the obstacle the radius of the first Fresnel zone is 5 m (Section 8.1.3). Hence $v = 2\sqrt{2}$ and the additional path loss $(13 + 20\log_{10}v)$ is 22.0 dB. The free-space loss is 102.0 dB (Appendix 8) and hence the received power is

$$[10\log 1.5 + 25 - 102.0 - 22.0 + 25] \, \text{dBW} = -72.2 \, \text{dBW}.$$

The reflected ray appears to come from an image 10 m below ground level. Hence $v = 4\sqrt{2}$. The signal received by way of reflexion is therefore

$$[10\log 1.5 + 25 - 102.0 - 28.1 + 25 - 0.4] \, \text{dBW} = -78.7 \, \text{dBW}.$$

8.11   Diffraction at two ridges

A 600 MHz radio signal with equivalent isotropic radiated power 1 W is transmitted from an antenna 150 m above sea level. Use each of the three methods outlined below to estimate the field strength at a receiving antenna 15 km distant and 150 m above sea level when the ground between the antennas is approximately at sea level, apart from sharp ridges of heights 190 m and 200 m at 3 km and 6 km respectively from the transmitter. Neglect the effects of refraction and assume the earth to be flat. Calculate path loss by adding diffraction losses to the free-space loss.

Method A (Bullington, 1947)   Assume diffraction at a single, hypothetical knife-edge located so that it is just visible from both antennas.

Method B (Epstein and Peterson, 1953)   Calculate the loss due to the presence of the ridges as the sum of two diffraction losses (expressed in dB). The first diffraction loss is regarded as the effect of the first ridge on a transmission path from the transmitting antenna to the peak of the second ridge, and the second diffraction loss as the effect of the second ridge on a transmission path from the peak of the first ridge to the receiving antenna.

Method C (Deygout, 1966)   Calculate v (Section 8.1.3) for each ridge with the presence of the other ridge disregarded and call the ridge with the larger value of v the 'main obstacle'. Calculate the loss due to the presence of the ridges as the sum of two diffraction losses, the first diffraction loss being that due to the main obstacle with the other ridge disregarded, and the second diffraction loss being the effect of the other ridge on a transmission path between the peak of the main obstacle and one of the antennas.

Solution outline
Calculation of diffraction loss:
Method A   The radius r_0 of the first Fresnel zone at the hypothetical ridge is given by

$$r_0 = \sqrt{\frac{\lambda a_0 b_0}{a_0 + b_0}}, \qquad \text{(Fig. 8.9(a) and Section 8.1.3)}$$

where $a_0 = 4.41$ km and $b_0 = 10.59$ km. Hence $r_0 = 39.46$ m. Since $h_0 = -58.8$ m, $v = h_0\sqrt{2}/r_0 = 2.11$ (Section 8.1.3) and hence the diffraction loss is approximately 19.6 dB (Fig. 8.1).

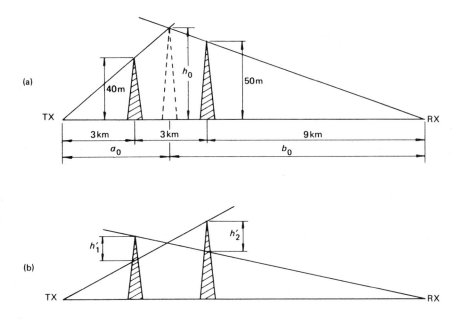

(a)

(b)

(c)

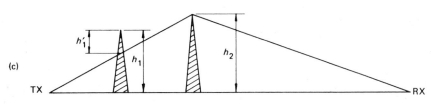

Fig. 8.9 Methods of estimating diffraction losses due to multiple knife-edges (solution to Exercise 8.11)

Method B In Fig. 8.9(b), $h'_1 = 15$ m and $h'_2 = 20$ m. At the first ridge the radius of the first Fresnel zone is 27.38 m. (Put $a = b = 3$ km in the formula in Section 8.1.3.) Hence $v = 0.775$ and the diffraction loss is therefore approximately 12.5 dB. For the second ridge, $a = 3$ km and $b = 9$ km; hence the diffraction loss is approximately 13.0 dB. The total diffraction loss is therefore approximately 25.5 dB.

Method C For the first ridge, $v_1 = 40\sqrt{2}/34.64 = 1.63$; for the second ridge, $v_2 = 50\sqrt{2}/42.43 = 1.67$. Hence the second ridge is the main obstacle and contributes about 18 dB to the diffraction loss. For the effect of the first ridge on the path from the transmitter to the second ridge, $v = 15\sqrt{2}/27.38 = 0.775$, giving a further diffraction loss of approximately 12.5 dB. The total diffraction loss is therefore approximately 30.5 dB.

With no obstructions the power density at the receiver would be $1/4\pi(15\,000)^2$ W/m^2, corresponding to a field strength of $\sqrt{120\pi/4\pi(15\,000)^2} = 3.65 \times 10^{-4}$ V/m (Appendix 8). The diffraction losses calculated by Methods A, B and C reduce this to 38 μV/m, 19 μV/m and 11 μV/m respectively.

8.12   Diffraction at three ridges

A 10 W 3 GHz signal is transmitted from an antenna 20 m above sea level. Use Deygout's method (Exercise 8.11, Method C) to estimate the power received by an antenna 12 km distant and 380 m above sea level given that the antennas are correctly aligned with gains 10 dB and that the only obstructions which need be considered are ridges of heights 220 m, 270 m and 330 m at distances 6 km, 8 km and 10 km respectively from the transmitter. [Apply the method several times. At each application determine the main obstacle and further subdivide the path. Assume the earth to be plane.]

Solution outline

Calculating v for each ridge (Section 8.1.3) gives $20\sqrt{2}/17.32$, $10\sqrt{2}/16.33$ and $10\sqrt{2}/12.91$. The first is the largest and corresponds to a diffraction loss of approximately 18.5 dB.

Calculating v for the two ridges between the first ridge and the receiver gives $-3.33\sqrt{2}/11.54$ and $3.33\sqrt{2}/11.54$. The latter is the larger and corresponds to a diffraction loss of approximately 9.5 dB.

Regarding the second ridge as an obstacle between the first ridge and the third ridge gives $v = -5\sqrt{2}/10$, corresponding to a diffraction loss of approximately 0.5 dB.

The free-space loss for the path is 123.6 dB (Appendix 8). Hence the received power is

$$(10 + 10 - 123.6 - 18.5 - 9.5 - 0.5 + 10)\,\text{dBW} = -122.1\,\text{dBW}.$$

8.13   Optimum frequency for troposcatter link

In a troposcatter link the transmitting and receiving antennas both have an effective area of 200 m². Estimate the frequency at which the path loss is a minimum, basing your estimate on the information contained in Section 8.1.5.

Solution outline

The path loss L may be expressed as

$$L = (L_0 + L_s - G_T - G_R + L_d)\,\text{dB},$$

where L_0 is the free-space loss, L_s the forward scattering loss, G_T and G_R the antenna gains and L_d the aperture-to-medium coupling loss, all expressed in dB. These quantities may be expressed as functions of frequency f as follows (f in MHz).

$L_0 = 20 \log_{10} f + \text{constant}$	(Appendix 8),
$L_s = 10 \log_{10} f + \text{constant}$	(Section 8.1.5),
$G_T + G_R = -31.08 + 40 \log_{10} f$	(Appendix 8),
$L_d = 0.07 \exp\{0.055 \times (-31.08 + 40 \log_{10} f)\}$	(Section 8.1.5)
$\quad = 0.012\,67\,(f)^{0.9554}.$	

The value of f minimising L may be shown by differentiation to be about 472 MHz.

8.14   Path loss of troposcatter link

Estimate the path loss for a 2 GHz, 300 km tropospheric-scatter link given that each antenna has a gain of 45 dB when used under free-space conditions and that the two antennas are aligned so that their directions of maximum radiation are at elevations of $-0.25°$ and $+1°$. Assume that the earth is a sphere of effective radius 8800 km.

Solution outline

The 300 km path subtends $1.95°$ at the apparent centre. The scattering angle is therefore

$$-0.25° + 1° + 1.95° = 2.70° = 47.1 \text{ mrad}.$$

Hence from Table 8.1 the median forward scattering loss is approximately 78 dB at 900 MHz and therefore about 81.5 dB at 2 GHz.

The free-space loss is 148.0 dB (Appendix 8) and the aperture-to-medium coupling loss (Section 8.1.5) is approximately 9.9 dB.

Hence the path loss is approximately

$$(148.0 + 81.5 + 9.9 - 90) \text{ dB} = 149.4 \text{ dB}.$$

8.15 Surface wave: Millington's method and recovery effect

Figure 8.11 shows the field strength $E(d)$ in dB relative to 1 $\mu V/m$ near the surface of a smooth, homogeneous earth at distance d from a vertically polarised 10 MHz, 1 kW isotropic radiator at the surface of the earth. Millington's method [10, 15] for calculating $E(d)$ when the path consists of sections S_1, S_2, S_3 having conductivity and dielectric constant $\sigma_1, \varepsilon_1; \sigma_2, \varepsilon_2; \sigma_3, \varepsilon_3$ (Fig. 8.10) is as follows. Denote by $E_i(d)$ the field strength at distance d from a 1 kW radiator as given by the curve on Fig. 8.11 appropriate to section S_i. $E(d)$ is then calculated as

$$E(d) = \tfrac{1}{2}(E_R + E_T),$$

where

$$E_R = E_1(d_1) + \{E_2(d_1 + d_2) - E_2(d_1)\} + \{E_3(d_1 + d_2 + d_3) - E_3(d_1 + d_2)\}$$

and $\quad E_T = E_3(d_3) + \{E_2(d_3 + d_2) - E_2(d_3)\} + \{E_1(d_3 + d_2 + d_1) - E_1(d_3 + d_2)\}.$

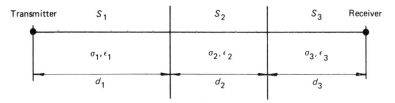

Fig. 8.10 Diagram of Millington's method of Exercise 8.15

Adaptation to paths consisting of more or less than three sections is straightforward.

For the first 100 km from a 10 MHz 10 kW ground-mounted isotropic radiator the ground is smooth with $\varepsilon = 4$ and $\sigma = 10^{-4}$ S/m. Beyond 100 km is sea water. Use Millington's method to calculate the field strength 2 km, 4 km, 6 km and 50 km out to sea and thereby show that it predicts an initial increase with distance from land, which is known as the 'recovery effect'.

Solution outline

2 km out to sea,

$$\begin{aligned} E_R &= E_1(100) + E_2(102) - E_2(100) \\ &= (-13 + 60.5 - 61) \text{ dB}(\mu V/m) \qquad \text{(from Fig. 8.11)} \\ &= -13.5 \text{ dB}(\mu V/m), \end{aligned}$$

and $E_T = E_2(2) + E_1(102) - E_1(2)$
$$= (103.5 - 14 - 65)\ dB(\mu V/m)$$
$$= +24.5\ dB(\mu V/m).$$

Hence
$$\tfrac{1}{2}(E_R + E_T) = 5.5\ dB(\mu V/m).$$

Since in this case the power transmitted is 10 kW, the field strength is

$$(5.5 + 10)\ dB(\mu V/m) \qquad \text{i.e. } 6.0\ \mu V/m.$$

Similarly at 4, 6 and 50 km the field strength may be shown to be approximately 7.7 $\mu V/m$, 8.7 $\mu V/m$ and 6.3 $\mu V/m$.

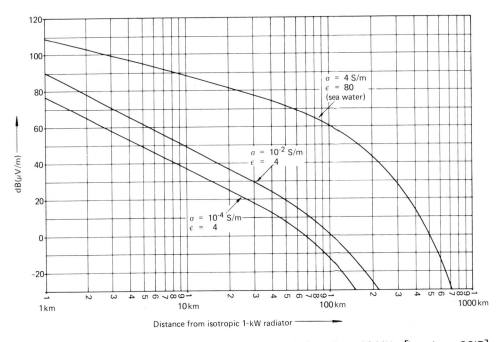

Fig. 8.11 Ground-wave propagation over smooth spherical earth at 10 MHz [based on CCIR] (Exercise 8.15)

8.16 Ionospheric reflexion: available ranges

Assuming the ionosphere to consist of a uniform layer at a height of 300 km and with critical frequency 5 MHz, calculate the minimum and maximum ranges at which a 12 MHz signal may be received by means of single-hop transmission. Calculate also the ranges at which both single-hop and double-hop transmission may contribute to the received signal. Take the radius of the earth to be 6370 km.

Solution outline
Reflexion occurs for angles of incidence greater than $\cos^{-1}(5/12)$.

By using the sine rule the minimum range for single-hop transmission may be shown to be 1507 km.

The maximum range for single-hop transmission is obtained when the rays are horizontal at both transmitter and receiver and is therefore 3836 km.

Both single-hop and double-hop transmission contribute at ranges between 2×1507 $(= 3013)$ km and 3836 km.

8.17 Group velocity in the ionosphere

Given that the group (or envelope) velocity $v_g(f)$ of a wave at frequency f is given by

$$\frac{1}{v_g(f)} = \frac{d}{df}\left\{\frac{f}{v_p(f)}\right\},$$

where $v_p(f)$ is the phase velocity, show that in the ionosphere
(i) the group velocity is $cn(f)$, where c is the free-space velocity and $n(f)$ the refractive index,
(ii) the horizontal component of the group velocity is constant provided that the ionisation occurs in horizontal layers.

Solution outline
(i) $\dfrac{1}{v_g} = \dfrac{d}{df}\left\{\dfrac{fn(f)}{c}\right\}$

$\qquad = \dfrac{d}{df}\left\{\dfrac{f}{c}\left(1 - \dfrac{81N}{f^2}\right)^{1/2}\right\}$ (Section 8.2.2)

$\qquad = \dfrac{1}{cn(f)}.$

(ii)   Consider a horizontal boundary between regions with refractive indices n_1 and n_2. Consider a ray crossing the boundary making angle θ_1 to the vertical in one region and angle θ_2 to the vertical to the other. The horizontal component of group velocity is $cn_1 \sin\theta_1$ in one region and $cn_2 \sin\theta_2$ in the other. By Snell's law, $n_1 \sin\theta_1 = n_2 \sin\theta_2$.

8.18 Virtual path and virtual height

Figure 8.12 shows a ray refracted in a flat ionised region and the 'virtual path' of the ray, assuming a reflector at P and no ionisation. Show that the time taken for a burst of radio-frequency energy to travel the length of the virtual path is equal to that taken over the actual path and hence show that the 'virtual height' a of P is given by

$$a = \int_0^b \frac{\cos\theta\,dh}{\sqrt{n^2(h) - \sin^2\theta}},$$

where $n(h)$ is the refractive index at height h.

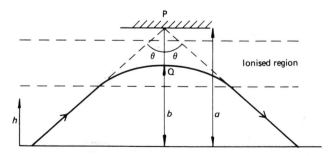

Fig. 8.12   Virtual path of Exercise 8.18

A pulse radar transmitting vertically from a position on the ground directly below P determines the apparent height of a reflecting layer by multiplying half the propagation time by the free-space velocity. Show that if the carrier frequency of the radar is $f \cos \theta$ (where f is the carrier frequency associated with ray path of Fig. 8.12) the radar pulses are reflected at Q and the apparent height of Q is a.

Solution outline

The horizontal component of group velocity remains constant at $c \sin \theta$ (Exercise 8.17). Hence the times taken for a burst to travel the actual and virtual paths are equal.

Since the group velocity at height h is $cn(h)$, its vertical component is

$$\sqrt{c^2 n^2(h) - c^2 \sin^2 \theta}.$$

Hence the time T taken for a burst to reach Q (or P on the virtual path) is given by

$$T = \int_0^b \frac{dh}{c\sqrt{n^2(h) - \sin^2 \theta}}.$$

Since $a = Tc \cos \theta$, the required result follows.

At Q the vertical component of the group velocity is zero and hence the electron density $N(h)$ at $h = b$ must satisfy

$$1 - \frac{81 N(b)}{f^2} = \sin^2 \theta.$$

For the radar, $\theta = 0$, and hence reflexion takes place at Q if

$$1 - \frac{81 N(b)}{f'^2} = 0,$$

where f' is the carrier frequency of the radar. Elimination of $N(b)$ shows that $f' = f \cos \theta$.

Putting $\theta = 0$ in the formula for a derived above, the apparent height measured by the radar is

$$\int_0^b \frac{dh}{n'(h)},$$

where $n'(h)$ is the refractive index at the frequency f' of the radar. Since $f' = f \cos \theta$,

$$\{n'(h)\}^2 = 1 - \frac{81 N(h)}{f^2 \cos^2 \theta} = \frac{1}{\cos^2 \theta} \left\{ 1 - \frac{81 N(h)}{f^2} - \sin^2 \theta \right\} = \frac{n^2(h) - \sin^2 \theta}{\cos^2 \theta}$$

and hence the apparent height is given by the same expression as a.

8.19 HF soundings of ionosphere

Sounding experiments have been carried out using a pulse-modulated HF radar radiating vertically upwards. Figure 8.13 shows the echoes received as the frequency was increased slowly from 2 MHz to 10 MHz. A communication link is to be established between two points 2500 km apart on the surface of the earth by way of F–layer reflexion vertically above the radar. Determine the maximum frequency which can be used on the link.

Assuming that 18 MHz is actually chosen, estimate the height at which reflexion takes place and the minimum range obtainable by reflexion vertically above the radar.

Take the radius of the earth to be 6370 km.

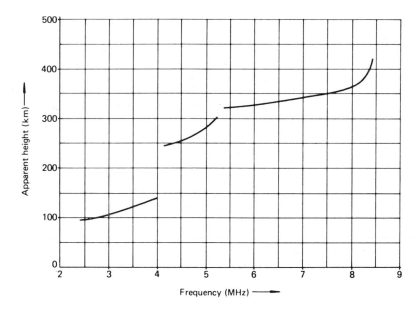

Fig 8.13 Data for Exercise 8.19

Solution outline

Assume that the F–layer accounts for radar reflexions in the frequency range from 4.1 MHz to 8.4 MHz (Section 8.2.1).

Assume an arbitrary value between 320 km and 420 km for the apparent height a of reflexion. Read from Fig. 8.13 the carrier frequency f' at which the radar pulses are reflected at the assumed apparent height. Use the cosine and sine rules to calculate the angle of incidence θ for the assumed value of a. Hence calculate the frequency $f'/\cos\theta$ for which reflexion occurs at the assumed apparent height (refer to Exercise 8.18). Repeat the calculation for different values of a to find the value of a which gives a maximum value for $f'/\cos\theta$. The maximum value of $f'/\cos\theta$ occurs when $a \simeq 365$ km and is about 22.1 MHz, which is therefore the maximum frequency usable on the link.

To find the apparent height of reflexion when 18 MHz is used, proceed as above, but find instead the apparent height for which $f'/\cos\theta = 18$ MHz. That height is approximately 329 m.

To find the minimum range when 18 MHz is used, proceed as follows. Assume an arbitrary value for a between 320 km and 420 km, find f' for that value of a (Fig. 8.13) and calculate the angle of incidence θ [$= \cos^{-1}(f'/18)$] at which 18 MHz signals are reflected at the assumed apparent height. Use the sine rule to calculate the range given by that value of θ. Repeat the calculation for different values of a until a minimum range is found. A broad minimum of roughly 1680 km occurs when $a \simeq 360$ km.

8.20 Noise due to networks in cascade

(i) A two-port network consists of a cascade of three subnetworks having noise temperatures T_1, T_2, T_3 and gains G_1, G_2, G_3 and matching at their interconnexions. Write down an expression for the noise temperature of the network. Hence show that the noise factor F of the network is given by

$$F = F_1 + \frac{F_2 - 1}{G_1} + \frac{F_3 - 1}{G_1 G_2},$$

where F_1, F_2, F_3 are the noise factors of the subnetworks defined for the same source temperature as F.

(ii)   An attenuator is at temperature T and its loss when supplying a matched load is $10 \log_{10} L$ dB. Show that its noise temperature is $T(L-1)$. Hence show that for a source temperature also equal to T its noise factor is equal to L.

(iii)   Two identical 20 dB amplifiers and a 16 dB attenuator are connected in cascade with the attenuator between the amplifiers. Calculate the overall noise factor for a source temperature of 290 K assuming
   (a)   the noise factor of each amplifier for a source temperature of 290 K is 3 dB, and
   (b)   matching at all interconnexions.

(iv)   For the arrangement described in (iii), calculate the changes (in dB) in output noise power,
   (a)   if the source temperature were increased by 100 K, and
   (b)   if the temperature of the attenuator were increased by 100 K.

Solution outline

(i)   The noise temperature T of the network is given by

$$T = T_1 + \frac{T_2}{G_1} + \frac{T_3}{G_1 G_2}.$$

Dividing by the source temperature T_s and using the relationship $T = T_s(F-1)$ (Appendix 6) gives the required result.

(ii)   The noise available from the attenuator with its source connected is equal to that from a single source with noise temperature T. The source connected to the attenuator contributes T/L to that noise temperature. Hence by its definition (Appendix 6) the noise factor of the attenuator is $T/(T/L)$.

(iii)   In the formula proved in (i) substitute $F_1 = F_3 = 1.995$, $G_1 = 100$, $F_2 = 1/G_2 = 39.81$. Hence $F = 2.78$ (4.44 dB).

(iv)   (a)   The output noise per Hz is $290kFG_1G_2G_3$ and would be increased by $100kG_1G_2G_3$, an increase of 0.51 dB.
   (b)   The output noise per Hz would be increased by $100(39.81-1)kG_2G_3$, an increase of 0.20 dB.

8.21   Receiver signal-to-noise ratio

A radio receiver of bandwidth 20 MHz and noise factor 1.5 dB (for a 290 K source) is receiving a 2 GHz signal arriving with power density 5×10^{-12} W/m² from a distant transmitter. The receiving antenna has gain 35 dB and noise temperature 50 K and is directed towards the transmitter. Calculate the signal-to-noise ratio at the demodulator input.

Solution outline

The effective area of the antenna is $(3 \times 10^8/2 \times 10^9)^2 \, 10^{3.5}/4\pi$ m² and hence the received signal power is 2.83×10^{-11} W (Appendix 8).

The noise temperature of the receiver is $290 \, (10^{0.15} - 1) = 119.6$ K. Hence the total noise power in a 20 MHz band and referred to the receiver input is (Appendix 6)

$$(119.6 + 50)(1.38 \times 10^{-23})(2 \times 10^7) = 4.68 \times 10^{-14} \text{ W}.$$

Hence the signal-to-noise ratio required is 27.8 dB.

8.22 Noise in terrestrial receiver directed towards satellite

The system noise temperature of a terrestrial 10 GHz receiver is 154 K when its antenna (of gain 40 dB) is directed towards a satellite vertically overhead. Estimate the change in the signal-to-noise ratio at the demodulator input for each of the following changes.

(i) A feeder with attenuation 1 dB and at a temperature 293 K is inserted between the antenna and the receiver.

(ii) The sun moves into the antenna beam.

(iii) The satellite and the antenna beam move to an elevation of 5°.

Assume a quiet sun and that the atmosphere is at 20°C with a 3 g/m³ concentration of water vapour. Neglect any noise transmitted by the satellite.

Solution outline

The sky-noise temperature is initially about 4 K (Fig. 8.5) and hence the receiver itself contributes about 150 K to the system noise temperature.

(i) The system noise temperature becomes

$$(4 \times 10^{-0.1}) + 293(1 - 10^{-0.1}) + 150 = 213.4 \text{ K}$$

corresponding to an increase of 1.4 dB in total noise. Since the signal power is decreased by 1 dB, the signal-to-noise ratio is decreased by 2.4 dB.

(ii) With no atmospheric attenuation and an isotropic antenna, a quiet sun contributes 0.053 K to the sky-noise temperature (Section 8.3.2). The equivalent path length through the atmosphere is 3.1 km (Section 8.1.4) and the attenuation coefficient is approximately

$$0.007 + (0.0044 \times 3/7.5) = 0.0088 \text{ dB/km} \qquad \text{(Fig. 8.4)}.$$

Hence the sun adds

$$0.053 \times 10^4 \times 10^{-(3.1 \times 0.000\,88)} = 527 \text{ K}$$

to the noise temperature, decreasing the signal-to-noise ratio by about 6.5 dB.

(iii) The sky-noise temperature is increased to about 31 K. Hence the system noise temperature is increased to 181 K, decreasing the signal-to-noise ratio by about 0.7 dB.

8.23 Formula for estimating galactic noise

Assuming that the field strength of galactic noise in bandwidth B Hz at f MHz at a terrestrial receiving station is

$$a + 10 \log B + b \log f \text{ dB} \, (\mu\text{V/m}),$$

where a and b are independent of B and f, use information given in Section 8.3.2 to estimate the values of a and b.

Solution outline

If the r.m.s. field strength in a 1 Hz bandwidth due to galactic noise is E μV/m, the power density at the receiver is $E^2/(120\pi \times 10^{12})$ W/m².

The effective area of an isotropic antenna at f MHz is $(3 \times 10^8)^2/4\pi f^2 10^{12}$ (Appendix 8).

Hence the contribution of the galactic noise to the antenna noise temperature is

$$\left\{ \frac{E^2}{120\pi \times 10^{12}} \right\} \left\{ \frac{(3 \times 10^8)^2}{4\pi f^2 \, 10^{12}} \right\} \left\{ \frac{10^{23}}{1.38} \right\} \text{K}.$$

Equating this to $1000(100/f)^{2.4}$ and taking logarithms,

$$10 \log_{10}(E^2) = -43.4 - 4 \log_{10} f.$$

Hence $a = -43.4$ and $b = -4$.

8.24 Principle of radiometry

Show that the process of averaging a signal over intervals of duration τ is equivalent to passing that signal through a low-pass filter with noise bandwidth $1/2\tau$.

It may be shown[14] that when the input to a radio receiver with RF bandwidth B_1 and post-detection bandwidth B_2 $(B_2 \ll B_1)$ is white noise, the output noise power fluctuates about its mean value P_0 with variance σ^2 given by

$$\sigma^2 \simeq P_0^2 \frac{2B_2}{B_1}.$$

Use this result to calculate the standard deviation of estimates of its antenna noise temperature made by a receiver with RF bandwidth 100 MHz and noise temperature 100 K. Assume that each estimate is made by averaging detector output noise over a 0.1 s interval and that the antenna noise temperature is 50 K. Calculate also the sensitivity of P_0 in dB/K to small changes in antenna noise temperature from 50 K.

Solution outline
Denote the signal by $v(t)$ and its Fourier transform by $V(f)$. Its average over an interval of duration τ is

$$\frac{1}{\tau} \int_{-\tau/2}^{\tau/2} v(t+t')\,dt' = \frac{1}{\tau} \int_{-\tau/2}^{\tau/2} \left[\int_{-\infty}^{\infty} V(f) \exp\{j2\pi f(t+t')\}\,df \right] dt'$$

$$= \int_{-\infty}^{\infty} V(f) \exp(j2\pi ft) \left[\frac{1}{\tau} \int_{-\tau/2}^{\tau/2} \exp(j2\pi ft')\,dt' \right] df$$

$$= \int_{-\infty}^{\infty} V(f) \left\{ \frac{\sin(\pi f\tau)}{\pi f\tau} \right\} \exp(j2\pi ft)\,df,$$

in which $\sin(\pi f\tau)/(\pi f\tau)$ is the transfer function of a low-pass filter with (one-sided) noise bandwidth $1/2\tau$ (Appendix 3).

The antenna noise temperature T_A may be calculated from measured values of P_0 by the relation:

$$P_0 = k(T_R + T_A) B_1 G,$$

where G is the receiver gain and T_R the receiver noise temperature. The standard deviation σ' of the values of T_A so calculated is therefore given by:

$$\sigma' = \sigma/k B_1 G.$$

Since $\sigma^2 = P_0^2(2B_2/B_1)$, $T_R = 100$ K, $T_A = 50$ K, $B_1 = 10^8$ Hz and $B_2 = 5$ Hz, σ' may be shown to be about 0.047 K.

The required sensitivity is

$$\frac{d}{dT} [10 \log_{10}\{k(T_R + T_A) B_1 G\}] = \frac{10}{(T_R + T_A) \ln 10} = 0.029 \text{ dB/K}.$$

9
Miscellaneous radio systems—exercises

Exercise topics

9.1   System value
9.2   Noise and interference in FM receiver
9.3   Radio-telephone nets using FM and pseudonoise spread-spectrum modulation
9.4   HF over-the-horizon radio link
9.5   Linked compandor for HF radio telephony
9.6   Multitone signalling for HF radio telegraphy
9.7   Troposcatter link
9.8   Urban mobile UHF radio
9.9   Terrestrial receivers for various satellite transmissions
9.10 Communication by satellite using CDMA

Exercises

9.1   System value

The system value S of the terminals of an angle-modulation radio link is defined as the sum of the signal-to-noise ratio (in dB) at the receiver output (with noise measured psophometrically) and the path loss (in dB) of the link. Show that when phase modulation is used

$$S = 10 \log_{10} \left\{ \frac{P_T \overline{\phi_s^2 w}}{N_0 b} \right\},$$

where
  P_T is the power available at the transmitter output,
  N_0 is the one-sided power spectral density referred to the receiver input of noise accompanying the angle-modulated signal at the demodulator input,
  b is the bandwidth of the modulating signal,
  $\overline{\phi_s^2}$ is the mean-square phase deviation due to the modulating signal,
  w is a psophometric weighting factor.
For sinusoidal modulation at 1 kHz with peak frequency deviation 5 kHz, the system value of certain phase-modulation, single-channel, radio-telephony terminals is 180 dB. Given that the carrier frequency is 60 MHz, the post-detection (audio) bandwidth is 3.1 kHz and the system noise temperature referred to the receiver input is 900 K, calculate the available transmitter output power.

The maximum range of the radio-telephony equipment when antennas of gain 2 dB are used is stated to be 10 km. Calculate the margin allowed for propagation losses additional to free-space loss if a 1 kHz tone transmitted with peak frequency deviation 5 kHz should be at least 40 dB above the noise at the receiver output.

Solution outline
Denoting path loss by A, the received signal power is P_T/A and the mean-square phase deviation due to noise $N_0 b/(P_T/A)$ (Section 5.4.3).
  From the given data, $S = 180$, $w = 10^{0.25}$, $b = 3100$, $N_0 = (1.38 \times 10^{-23}) \times 900$ and $\overline{\phi_s^2} = 0.5 \,(5000/1000)^2 = 12.5$. Hence from the formula derived, $P_T = 1.73$ W.
  The path loss for a free-space 10 km path is $[20 \log_{10} \{ (4\pi \times 10^4)/5 \} - 4]$ dB i.e. 84.0 dB. Hence under free-space conditions the output is $(180 - 84)$ dB $= 96$ dB above the noise, allowing an increase of 56 dB in path loss.

9.2   Noise and interference in FM receiver

In an FM receiver, the antenna gain is 10 dB, the antenna noise temperature is 290 K, the noise factor of the RF amplifier is 4 dB, the bandwidth of the IF amplifier is 180 kHz and the demodulator is followed by a low-pass filter cutting off at 15 kHz. The receiver receives a 90 MHz radio wave with field strength 2.5 μV/m and modulated sinusoidally in frequency by a 1 kHz signal so that the peak frequency deviation is 75 kHz. Calculate the output signal-to-noise ratio.

Calculate the output signal-to-noise ratio if a de-emphasis network with voltage gain $(1 + j\omega/\omega_0)^{-1}$ at ω rad/s (where $\omega_0 = 20\,000$ rad/s) is incorporated between the demodulator and the low-pass filter.

Assuming that there is no de-emphasis, calculate the ratio of signal to interference at the receiver output when the input signal is accompanied by an unmodulated 90.002 MHz interfering carrier 20 dB lower in level.

Solution outline

From Appendix 8, multiplying power density $(2.5 \times 10^{-6})^2/120\pi$ by effective area $(3.33)^2 \times 10/4\pi$ gives 1.466×10^{-13} W for the signal power at the receiver input. If sources of noise other than those given are ignored, noise power referred to the receiver input is

$$(1.38 \times 10^{-23}) \times 290 \times 10^{0.4} \times (1.8 \times 10^5) = 1.809 \times 10^{-15} \text{ W}.$$

The FM improvement is (Appendix 7)

$$(3/2) (75/15)^2 (180/15) = 450.$$

Hence when there is no de-emphasis the signal-to-noise ratio at the receiver output is

$$\{ (1.466 \times 10^{-13})/(1.809 \times 10^{-15}) \} \times 450 = 36\,450 \qquad (45.6 \text{ dB}).$$

The effect of de-emphasis is to reduce the signal power by

$$20 \log_{10} |1 + j(2\pi/20)| \text{ dB} = 0.4 \text{ dB}$$

and to reduce the noise power by

$$10 \log_{10} \left\{ \int_0^{15} \frac{f^2 \, df}{1 + f^2 (2\pi/20)^2} \right\} - 10 \log_{10} \left\{ \int_0^{15} f^2 \, df \right\} \text{dB} = 10.2 \text{ dB}.$$

Hence the output signal-to-noise ratio is changed to

$$(45.6 - 0.4 + 10.2) \text{ dB} = 55.4 \text{ dB}.$$

The interference produces approximately sinusoidal angle modulation with peak phase deviation approximately 0.1 rad and therefore with peak frequency deviation approximately 200 Hz. The required signal-to-interference ratio is therefore $\{20 \log_{10} (75\,000/200)\}$ dB, i.e. 51.5 dB.

9.3 Radio-telephone nets using FM and pseudonoise spread-spectrum modulation

The transceivers of an FM radio-telephone net use omnidirectional antennas (gain 0 dB), have signal and receiver bandwidth 15 kHz and have noise temperature 3000 K. An airborne transceiver at a height of 20 m has antenna gain 10 dB, receiver bandwidth 15 kHz and noise temperature 800 K. It attempts to jam the net by transmitting 10 W of unmodulated carrier when a transmission from the net produces 0 dB signal-to-noise ratio at the input of its demodulator. Propagation is assumed to be 'plane-earth' throughout (Exercise 8.5) and the net antennas 2 m above the ground. Calculate the maximum distance from a net transmitter at which the jamming is activated, assuming that the net transmitter produces a 25 dB signal-to-noise ratio at the discriminator input of a net receiver 5 km distant. Calculate also how near the airborne transceiver must approach to the net receiver for the received jamming power to equal the power received from the net transmitter 5 km distant.

Repeat the calculations assuming that transmitted powers are unaltered but that the net transceivers are modified to utilise pseudonoise spread-spectrum transmissions occupying a 2 MHz bandwidth (Section 7.4.2).

A second pseudonoise spread-spectrum net has transmitters identical to the net transmitter described above, except for the code sequence used, and its transmissions occupy the same frequency band. How near can one of its transmitters be placed to a receiver of the first net if the interference power from the transmitter should not exceed the noise power at the discriminator input of the receiver?

Solution outline

Calculations are based on the 'plane-earth' expression for path loss, namely $r^4/h_1^2h_2^2G_TG_R$ (Exercise 8.5).

If the signal-to-noise ratio is 0 dB at the input to the discriminator of the airborne receiver at distance d_1 from the net transmitter,

$$\left(\frac{20}{2}\right)^2\left(\frac{5000}{d_1}\right)^4 10 = 10^{-2.5}\left(\frac{800}{3000}\right).$$

Hence $d_1 = 1.65 \times 10^5$ m.

If the power from a jammer at distance d_2 is equal to the signal power from a transmitter 5 km distant,

$$10 \times 10 \times \left(\frac{2 \times 20}{d_2^2}\right)^2 = (1.38 \times 10^{-23}) \times 15\,000 \times 3000 \times 10^{2.5}.$$

Hence $d_2 = 3.004 \times 10^4$ m.

When the transmissions are spread-spectrum, the power received by the airborne receiver is modified by the factor 15/2000 (approximately). Hence d_1 becomes

$$(1.65 \times 10^5) \times \left(\frac{15}{2000}\right)^{1/4} = 4.86 \times 10^4 \text{ m}.$$

The jamming power reaching the discriminator of the net receiver is also modified by the same factor. Hence d_2 becomes 8.84×10^3 m.

If the required distance from a transmitter of the second net is d_3,

$$\left(\frac{5000}{d_3}\right)^4\left(\frac{15}{2000}\right) = 10^{-2.5}.$$

Hence $d_3 = 6.20 \times 10^3$ m.

9.4   HF over-the-horizon radio link

The Exercise is based upon an HF radio link between two small islands 3000 km apart as measured along a great circle on the surface of the earth. Assume reflexion at a height of 350 km by a uniform ionised F–layer with critical frequency 4 MHz. The transmitted power is 100 W, each antenna has gain 8 dB, the receiver system noise temperature is 900 K and the receiver bandwidth is 3 kHz. Experiments show that when an unmodulated 10 MHz carrier is transmitted, the average signal-to-noise ratio at the receiver output is 54 dB. Assume the earth to be a sphere of radius 6370 km.

 (i)   Determine the range of frequency over which single-hop propagation is possible but multi-hop propagation is not.

 (ii)   Calculate the average loss due to absorption at 10 MHz.

(iii)   Assuming that there is flat Rayleigh fading and that outage is deemed to have occurred when the signal-to-noise ratio at the receiver output is less than 20 dB, determine how many diversity branches must be provided to reduce the outage rate to less than 10^{-10} without increasing the total transmitted power above 100 W. Assume 10 MHz transmissions and selection combining.

 (iv)   Determine the range at which a signal-to-noise ratio of 20 dB could be obtained by ground-wave transmission at 10 MHz over the sea. (Use Fig. 8.11.)

 (v)   Estimate the signal-to-noise ratio at the receiver output when a 7 MHz signal is received by single-hop propagation.

 (vi)   Assuming that all sky-wave losses additional to free-space loss are due to absorption in a uniform D–layer at a height of 75 km, compare the strength of a 7 MHz signal received by double-hop propagation with that of a 7 MHz signal received by single-hop propagation.

(vii)   In 'two-tone telegraphy' data is transmitted as binary FSK and received in two ASK receivers (one tuned to each transmitted frequency), the outputs of which are combined by selection combining. Suggest a suitable frequency spacing for a two-tone telegraphy transmission at about 7 MHz.

Solution outline

(i)   For single-hop propagation the angle of incidence at the reflecting layer is 70.52° and the path length is 2×1576 km = 3153 km. Hence reflexion occurs at frequencies less than (4/cos 70.52°) MHz = 12.0 MHz, and the free-space loss at 10 MHz ($\lambda = 30$ m) is $20 \log \{ (4\pi \times 3.153 \times 10^6)/30\}$ dB = 122.4 dB.

For double-hop propagation reflexion occurs at frequencies less than (4/cos 62.22°) MHz = 8.6 MHz and the length of the path is 4×845.7 km = 3383 km.

Hence at frequencies between 8.6 MHz and 12.0 MHz single-hop propagation is possible but double-hop propagation is not.

(ii)   Referred to the receiver input, the system noise is $10 \log (1.38 \times 10^{-23} \times 900 \times 3000)$ dBW = -164.3 dBW.

Hence if the average loss due to absorption is L dB,

$$20 + 8 + 8 - 122.4 - L = -164.3 + 54$$

and therefore $L = 23.9$ dB.

(iii)   Without diversity, outage occurs when the received signal-to-noise ratio is more than 34 dB below its average value and the outage rate is therefore $1 - \exp(-10^{-3.4})$ (Exercise 7.16). With n branches, the outage rate is

$$\{1 - \exp(-n\,10^{-3.4})\}^n \simeq (n\,10^{-3.4})^n$$

which is less than 10^{-10} if $n \geqslant 4$.

(iv)   The effective area of the receiving antenna is $(30)^2 \times 10^{0.8}/4\pi$ and hence the field strength E at the receiving antenna is given by

$$\frac{E^2}{120\pi} \times \frac{(30)^2 \times 10^{0.8}}{4\pi} = 1.38 \times 10^{-23} \times 900 \times 3000 \times 100.$$

Hence $E = 5.575 \times 10^{-8}$ V/m = -25.1 dB (μV/m). Since Fig. 8.11 applies to a 1 kW transmitter and an isotropic transmitting antenna, the distance required is that corresponding to a field strength of $(-25.1 + 10 - 8)$ dB (μV/m) and is seen from the 'sea-water' curve to be about 660 km.

(v)   The free-space loss for single-hop propagation is 119.3 dB. The average loss due to absorption is approximately $23.9 \times (10/7)^2$ dB = 48.8 dB (Section 8.2.3). Hence the required signal-to-noise ratio is $20 + 8 + 8 - 119.3 - 48.8 + 164.3 = 32.2$ dB.

(vi)   Assume that the absorption by the D–layer is proportional to the cosecant of the angle between the ray and the layer. This angle is about 10.6° in the single-hop case and about 22.7° in the double-hop case. Hence the absorption in the double-hop case is greater by

$$48.8\left\{ 2\frac{\text{cosec } 22.7°}{\text{cosec } 10.6°} - 1 \right\} = -2.3 \text{ dB}.$$

In the double-hop case the free-space loss is 119.9 dB and hence the received signal strength is greater by about $2.3 - 0.6 = 1.7$ dB.

(vii)   If the transmitted wave is $\cos \omega t$ and the received signal $s(t)$ is the sum of two fading versions with delays τ_1 and τ_2, $s(t)$ may be written

$$s(t) = r_1(t) \cos\{\omega(t - \tau_1) - \phi_1(t)\} + r_2(t) \cos\{\omega(t - \tau_2) - \phi_2(t)\}$$
$$= r(t) \cos\{\omega t - \phi(t)\},$$

where

$$r(t) = [r_1^2(t) + r_2^2(t) + 2r_1(t)r_2(t) \cos\{\omega(\tau_1 - \tau_2) + \phi_1(t) - \phi_2(t)\}]^{1/2}.$$

Hence at frequencies separated by $(\tau_1 - \tau_2)^{-1}$ Hz the fading is identical.

In this case $(\tau_1 - \tau_2)^{-1} = 3 \times 10^5/230 = 1304$ Hz, suggesting a spacing of about 650 Hz if the frequencies used should not fade together.

9.5 Linked compandor for HF radio telephony

Figure 9.1 shows the principal components of a linked compressor-expander used to reduce the effects of fading in HF radio telephony. The speech waveform is transmitted as a constant-level signal and syllabic variations of level are re-inserted in the receiver. Information on level is conveyed by a frequency-modulated control signal accompanying the constant-level signal and 5 dB below it in level. The effects of fading are reduced by a fast-acting constant-volume amplifier in the receiver. The control signal can vary between 2840 Hz and 2960 Hz, each hertz of deviation representing a 0.5 dB change in level. Assuming that the power spectrum of the noise at the output of the SSB receiver is uniform and extends from 0 Hz to 3000 Hz, calculate

(i) the root-mean-square error (in dB) in the level of the audio output when the signal-to-noise ratio at the receiver output is 10 dB,

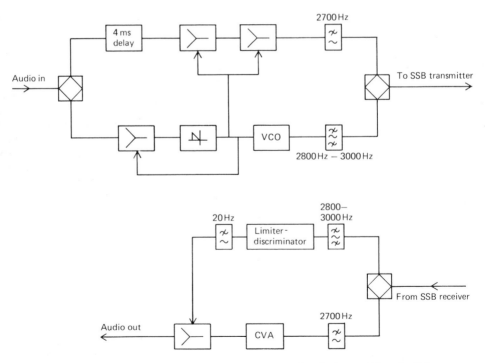

Fig. 9.1 Linked compressor and expander for HF radio telephony of Exercise 9.5

(ii)   the probability that the received control signal is below the FM threshold if the signal at the receiver output is subject to Rayleigh fading and has an average power 10 dB above the accompanying noise.

Solution outline
(i)   The received power is divided between the control signal and the constant-level speech signal in the ratio 0.240 to 0.760. Hence the signal-to-noise ratio at the discriminator input is $10 \times 0.240 \times (3000/200)$, i.e. 36.

Integration of the expression for noise power spectral density in Appendix 7 leads to $\{8\pi^2 K^2 N_0 b^3 / 3 A^2\}^{1/2}$ for the required root-mean-square error. Putting $K = 1/4\pi$ dB/(rad/s), $b = 20$ Hz and $A^2/2N_0 B = 36$ (where $B = 200$ Hz) gives 0.30 dB.
(ii)   Assume that the threshold is at a signal-to-noise ratio of 10 at the discriminator input (Section 7.2.1). The average signal-to-noise ratio is 36 (from (i)). Using the result proved in Exercise 7.16, the required probability is $1 - \exp(-10/36)$, i.e. 0.24.

9.6   Multitone signalling for HF radio telegraphy

In the 'Piccolo'* system each of the 32 teleprinter characters is represented by a 100 ms burst of one of 32 audio-frequency tones and received by a bank of 32 lossless resonator circuits each with an envelope detector. Immediately before the arrival of each burst all the resonators are heavily damped. At the end of each burst samples of the outputs of the envelope detectors are compared and the largest is selected.
(i)   Show that if the frequencies used are multiples of 10 Hz each burst contributes to the sample taken from only one detector.
(ii)   Show that at the sampling instant the ratio of the signal envelope to noise at the output of the resonator containing the signal is equal to that obtainable from a matched filter.
(iii)   Calculate the ratio of energy per bit to noise power spectral density necessary at the resonator inputs for a character error rate of 10^{-4}. Assume the transmitted characters to be independent and equiprobable.
(iv)   Re-calculate the ratio in (iii) when only two frequencies are used and the characters are transmitted synchronously by a 5 unit binary code.
   Hint: First derive the following results.
(a)   When input current $I \sin \omega t$ is connected at $t = 0$ to a lossless resonator tuned to ω_0 rad/s and storing no energy the subsequent voltage across the resonator is

$$\frac{I\omega}{C(\omega^2 - \omega_0^2)} \{\cos \omega_0 t - \cos \omega t\}.$$

(b)   When the received signal is accompanied by white gaussian noise the mean-square noise voltage across the resonator tuned to ω_0 is approximately equal at a sampling instant to that obtained by passing the input white noise through a filter with power gain

$$\frac{1}{C^2 (\omega - \omega_0)^2} \sin^2 \{(\omega - \omega_0) T/2\},$$

where T is the duration of a burst.

* Robin, H. K., Bayley, D., Murray, T. L., and Ralphs, J. D., 'Multitone signalling system employing quenched resonators for use on noisy radio-teleprinter circuits', *Proc. IEE*, **110**, pp. 1554–68, September 1963

Ralphs, J. D., *Principles and Practice of Multi-frequency Telegraphy*, IEE Telecommunications Series 11, 1985

Solution outline

(a)   The voltage across the resonator is $\int_{-\infty}^{\infty} u(\tau)h(t-\tau)\,d\tau$ (Appendix 2) where

$$u(t) = \begin{cases} I\sin\omega t & (t \geq 0) \\ 0 & (t < 0) \end{cases}, \text{ and } h(t) = \begin{cases} \dfrac{1}{C}\cos\omega_0 t & (t \geq 0) \\ 0 & (t < 0) \end{cases}.$$

(b)   At a sampling instant the noise voltage across the resonator is the convolution of input noise voltage with $f(t)$, where

$$f(t) = \begin{cases} \dfrac{1}{C}\cos\omega_0 t & 0 < t < T \\ 0 & \text{otherwise.} \end{cases}$$

The transfer function of the resonator and sampler is therefore the Fourier transform $F(\omega)$ of $f(t)$. The required expression is $|F(\omega)|^2$ with contributions due to spectral 'folding' ignored.

(i)   Since

$$\frac{I\omega}{C(\omega^2 - \omega_0^2)}\{\cos\omega_0 t - \cos\omega t\} = \frac{2I\omega}{C(\omega^2 - \omega_0^2)}\sin\left(\frac{\omega + \omega_0}{2}t\right)\sin\left(\frac{\omega - \omega_0}{2}t\right),$$

the envelope of a resonator output is proportional to $\sin\{(\omega - \omega_0)t/2\}$ and is zero at $t = T$ if $\omega - \omega_0$ is a multiple of $2\pi/T$.

(ii)   As ω approaches ω_0, the resonator output approaches the linearly increasing sinewave $(It/2C)\sin\omega_0 t$ and therefore the signal power in a sample of the envelope is $(IT/2C)^2$.

If the received burst is accompanied by white noise with one-sided power spectral density N_0, the noise power across the resonator and hence at the output of its envelope detector is at the sampling instant

$$\frac{N_0}{2\pi}\int_0^{\infty} \frac{1}{C^2(\omega - \omega_0)^2}\sin^2\{\tfrac{1}{2}(\omega - \omega_0)T\}\,d\omega,$$

which equals $N_0 T/4C^2$ (Appendix 3), provided that $\omega_0 \gg 2\pi/T$. Hence the signal-to-noise ratio for the sample is $I^2 T/N_0$.

Since the energy E of the received burst is $I^2 T/2$, the signal-to-noise ratio for the sample may be written $2E/N_0$, the result for a matched filter (Section 2.2.1).

(iii)   For a character to be received correctly, 31 binary decisions must be made correctly and hence the character error probability p_c is given by

$$p_c = 1 - (1-p)^{31} \simeq 31\,p,$$

where $p = \frac{1}{2}\exp(-E/2N_0)$ (Table 2.1). Hence

$$E/2N_0 = \ln(1/2p) = \ln(31/2p_c) = 11.95.$$

Since each character is equivalent to five bits of information the required ratio is

$$E/5N_0 = 11.95 \times (2/5) = 4.8.$$

(iv)   In this case $p_c \simeq 5p$. Hence $E/2N_0 = 10.13$ and the required ratio is $10.13 \times 2 = 20.3$.

9.7   Troposcatter link

A troposcatter radio link operates at 2 GHz and has a range of 400 km. Quadruple space diversity is used and at each terminal there are two paraboloidal antennas each of gain 45 dB under free-space conditions. At one terminal the antenna beams are horizontal and at the other terminal they are one degree above the horizontal. The system noise temperature for each receiver is 600 K. The modulation used is differential quadrature phase-shift keying for which the error-rate is unacceptably high when the E/N_0 ratio (energy per bit/noise power per hertz) after combining is below 12.8 dB.

  (i) Estimate the maximum signalling speed possible without intersymbol interference.

  (ii) Assuming information transmission at 1 Mbit/s, independent Rayleigh fading in the branches and selection combining, calculate the transmitted power necessary for an outage rate of 10^{-6}.

(iii) Recalculate (ii) assuming maximal-ratio combining.

Solution outline
(i) Assuming that α and θ are small and that the scatter volume is midway between the terminals, the difference between the propagation times for the paths TAR and TBR in Fig. 9.2 may be shown to be approximately $l\alpha\theta/2c$.

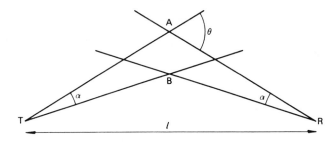

Fig. 9.2 Diagram of solution to Exercise 9.7

In this case the beamwidth α is given approximately by

$$\pi(\alpha/2)^2 \times 10^{4.5} = 4\pi,$$

whence $\alpha \simeq 22.5$ mrad, and the scattering angle θ is approximately

$$\left(\frac{400}{8800} + \frac{\pi}{180}\right) \times 10^3 = 62.9 \text{ mrad.}$$

Hence $l\alpha\theta/2c \simeq 9.4 \times 10^{-7}$, suggesting a maximum signalling speed of about 10^6 baud.
(ii) The mean value Γ of the E/N_0 ratio in each branch is related to the value x of that ratio at the threshold of acceptability by (Exercise 7.16)

$$[1 - \exp(-x/\Gamma)]^4 = 10^{-6}.$$

Hence $x/\Gamma = 0.0321$ (14.9 dB) and therefore the mean E/N_0 ratio must be at least $(12.8 + 14.9)$ dB. Since the median E/N_0 ratio is 1.6 dB below the mean (Section 7.3.1 and Exercise 7.15), the median received E/N_0 ratio for each branch must be at least $(12.8 + 14.9 - 1.6)$ dB i.e. 26.1 dB.

    From the information given,
    free space loss = 150.5 dB (Appendix 8),
    antenna-to-medium coupling loss = 9.9 dB (Section 8.1.5),
    median scattering loss = 87 dB (Section 8.1.5), and
    $N_0 = (1.38 \times 10^{-23} \times 600)$ W/Hz $= -200.8$ dBW/Hz.
Hence, noting that the bit rate is 10^6 bit/s, the minimum necessary transmitted power P_T dBW for each branch is given by

$$P_T - 60 + 45 - 150.5 - 87 - 9.9 + 45 + 200.8 = 26.1.$$

Hence $P_T = 42.7\,\text{dBW}$ (18.6 kW). Since there are two transmitters, the total transmitted power must be at least 37.2 kW.

(iii)   By extension of the argument applied to dual diversity in Exercise 7.20, the E/N_0 ratio after combining is the sum of the E/N_0 ratios in the branches and its probability density function is therefore

$$\frac{x^3}{6\Gamma^4} \exp\left(-\frac{x}{\Gamma}\right)$$

[the inverse Laplace transform of $(1 + p\Gamma)^{-4}$]. Integration by parts leads to the cumulative probability

$$1 - \tfrac{1}{6}\left\{\left(\frac{x}{\Gamma}\right)^3 + 3\left(\frac{x}{\Gamma}\right)^2 + 6\left(\frac{x}{\Gamma}\right) + 6\right\}\exp\left(-\frac{x}{\Gamma}\right)$$

and equating this to 10^{-6} gives $x/\Gamma \simeq 0.071$. Hence Γ, the mean E/N_0 ratio in each branch, must be 11.5 dB [$10 \log (1/0.071)\,\text{dB}$] above the threshold of acceptability, compared with 14.9 dB in the selection-combining case. Hence maximal-ratio combining permits a reduction of 3.4 dB in transmitted power.

9.8   Urban mobile UHF radio

In an urban mobile radio system, operating near 1 GHz, terminals in vehicles are in communication with a central base station.

Receivers in vehicles use dual space diversity with selection combining. They have pre-detection bandwidth 25 kHz and system noise temperature 3000 K and they operate satisfactorily provided that the pre-detection signal-to-noise ratio is not less than 10 dB. The outage rate must not exceed 10^{-4} for a receiver in a vehicle in an area in which the mean received signal strength is 50 dB below its free-space value at distances up to 10 km from the base station. Assuming Rayleigh fading and antenna gains of 2 dB, calculate the power required from the base-station transmitter.

Calculate the number of fades occurring per second in each branch when the vehicle is travelling at 100 km/hr, assuming that the outage rate is 10^{-4}, that the transmitted signal is a sinewave carrier and that the power spectrum of the received signal is rectangular and of width twice the maximum Doppler shift.

Solution outline
Using the expression derived in Exercise 7.16, the threshold signal-to-noise ratio x must be related to the mean signal-to-noise ratio Γ in each branch by

$$\left\{1 - \exp\left(-\frac{x}{\Gamma}\right)\right\}^2 = 10^{-4}.$$

Hence Γ must be 20.0 dB above x.

Since the noise level referred to the receiver input is $(1.38 \times 10^{-23} \times 3000 \times 25\,000)$ W, i.e. -149.9 dBW, the mean branch signal level at the receiver input is $(-149.9 + 10.0 + 20.0)\,\text{dBW} = -119.9\,\text{dBW}$. The free-space loss is 112.4 dB and hence the required transmitter power is

$$(-119.9 + 112.4 - 2 - 2 + 50)\,\text{dBW} = 38.5\,\text{dBW}\ (7.1\ \text{kW}).$$

Denoting the mean power of the received signal by a and its bandwidth by B, the necessary autocorrelation $\psi(\tau)$ (Section 7.3.1) is given by:

$$\psi(\tau) = \int_{-B/2}^{B/2} \frac{a}{B} \exp{(j2\pi f\tau)} \, df = a \, \frac{\sin{(\pi B\tau)}}{(\pi B\tau)}.$$

(Compare with Exercise 7.15.) Hence for small τ

$$\psi(\tau) \simeq a\left\{1 - \frac{1}{6}(\pi B\tau)^2\right\}$$

and therefore

$$\ddot{\psi}(0) \simeq -a\pi^2 B^2/3.$$

From Section 7.3.1, the average number of fades occurring in unit time is

$$\int_0^\infty \dot{r}\left(\frac{R}{a}\right)\exp{\left(-\frac{R^2}{2a}\right)}\frac{1}{\sqrt{(2\pi b)}}\exp{\left(-\frac{\dot{r}^2}{2b}\right)}\mathrm{d}\dot{r},$$

where $b = -\ddot{\psi}(0) = a\pi^2 B^2/3$, $R^2/2a = 0.010$ (x/Γ above), $B = 2 \times$ (maximum Doppler shift) $= 185.2$ Hz. It is therefore about 19 fades/sec.

9.9 Terrestrial receivers for various satellite transmissions

A 15 GHz down-link is available from a geosynchronous satellite (orbit radius 42 162 km) with transmitter EIRP + 34 dBW. Terrestrial receivers must be capable of operating with the satellite at elevations down to $10°$, in 25 mm/hr rain and with 17 g/m³ of water vapour in the atmosphere. The 'figure of merit' of a receiver is defined as

$$10 \log\left\{\frac{\text{Receiving-antenna gain } (G_R)}{\text{System noise temperature } (T)}\right\} \text{ dB/K}.$$

Calculate the minimum figure-of-merit necessary in a terrestrial receiver for each of the following applications. Assume the earth to be a sphere of radius 6370 km.

(i) The transmission is binary DPSK at 1200 bit/s and with differentially coherent detection. The error rate should not exceed 10^{-6}.

(ii) The transmission consists of 800 QPSK 64 kbit/s signals in FDM. The receiver has a bandwidth of 38 kHz and admits one signal only. Satisfactory reception requires that the ratio of signal to noise power spectral density at the demodulator input is 59.3 dB.Hz.

(iii) The transmission consists of 800 single-channel FM telephony signals in FDM. The receiver has a bandwidth of 38 kHz and admits one signal only. Satisfactory reception requires that the ratio at the demodulator output of the mean speech power to the noise power should not be less than 25 dB when speech is present. Assume that the mean square frequency deviation is controlled (is the same for all talkers) and that pre-emphasis is not used.

(iv) There is a single FM transmission occupying 36 MHz. The modulating signal extends from 12 kHz to 3284 kHz and consists of 792 SSB telephone channels in FDM. After demodulation the noise power should not exceed -50 dBm0p in any channel. There is no pre-emphasis.

(v) A system identical to (iv) except that pre-emphasis is used in the frequency modulation. Assume that the power gain of the pre-emphasis is proportional to the square of the frequency.

Solution outline
The maximum path length is 40 586 km and hence the maximum free-space loss is 208.1 dB (Appendix 8).

The maximum attenuation due to rain is $1.6 \times 7.6 = 12.2$ dB (Figs 8.2 and 8.3) and that due to water vapour is $(0.0076 + 0.034) \times 3.1 \operatorname{cosec} 10° = 0.7$ dB.

Hence the received power P_R in the worst case is given by
$$10 \log P_R = (+34 - 208.1 - 12.2 - 0.7 + 10 \log G_R) \text{dBW} = (10 \log G_R - 187.0) \text{ dBW}.$$

(i) In this case the lowest acceptable value of E/N_0 is given by (Section 2.2.3):
$$\tfrac{1}{2} \exp\left(-\frac{E}{N_0} \right) = 10^{-6},$$

i.e. $E/N_0 = 11.2$ dB. But $N_0 = kT$ and $P_R = ER$ (where R is the bit-rate). Hence $E/N_0 = P_R/kTR$ and
$$10 \log \left(\frac{G_R}{T} \right) \geqslant 187 + 10 \log k + 10 \log 1200 + 11.2 = 0.4 \text{ dB/K}.$$

(ii)   The requirement is that $10 \log(P_R/800 \, N_0) \geqslant 59.3$ dB.Hz. Hence
$$10 \log \left(\frac{G_R}{T} \right) \geqslant 187 + 29 + 10 \log k + 59.3 = 46.7 \text{ dB/K}.$$

(iii) From Section 1.1.4, the channel should handle a test-tone 10 dB above the mean speech power. Hence a modulating tone producing the maximum allowable frequency deviation should be 35 dB above the noise at the demodulator output.

By Carson's rule (Section 5.4.2) the maximum frequency deviation is 15.6 kHz (assuming the maximum modulating frequency to be 3.4 kHz). Therefore the FM advantage is 15.0 dB (Appendix 7).

Hence at the demodulator input the ratio of the signal power to the noise power in a 3.4 kHz bandwidth should be at least $(35 - 15)$ dB $= 20$ dB. Hence
$$(10 \log G_R - 187.0 - 10 \log 800) - 10 \log (kT3400) \geqslant 20.0$$
and hence
$$10 \log \left(\frac{G_R}{T} \right) \geqslant 42.7 \text{ dB/K}.$$

(iv)   By Carson's rule, the maximum allowable peak frequency deviation is about 14.7 MHz and hence the maximum allowable mean-square value is about $(14.7)^2/10$ (MHz)^2 (Section 5.4.2). The maximum mean-square deviation due to a test tone in one channel is $(10 \log 792 - 15)$ dB below this (assuming conventional loading, Section 5.1.2 and 5.2.3) and is therefore 8.63×10^{11} (Hz)^2. Hence in the noisiest channel (centred on 3282 kHz) the noise is less than -50 dBm0p if (Section 5.4.3)
$$\frac{kT \times 3100 \times (3.28 \times 10^6)^2}{C \times (8.63 \times 10^{11})} < 10^{(-5+0.25)},$$

where C is the power in the received FM signal and an increase of 2.5 dB has been allowed due to the measurement of the noise power on a psophometer (Section 1.1.3). Hence
$$\frac{C}{T} > 3.00 \times 10^{-14},$$
whence
$$10 \log \left(\frac{G_R}{T} \right) \geqslant 51.8 \text{ dB/K}.$$

(v)   For a pre-emphasis characteristic αf^2, the same mean-square frequency deviation as in (iv) requires that

$$\int_{f_1}^{f_2} \alpha f^2 \, df = f_2 - f_1$$

(where f_1 and f_2 are the minimum and maximum modulating frequencies). When f is expressed in Hz, this gives $\alpha = 2.78 \times 10^{-13}$.

For a channel at frequency f_m, the mean-square deviation due to a test-tone is $\alpha f_m^2 \times (8.63 \times 10^{11})$ (Hz)^2 and therefore the channel noise is less than -50 dBm0p if (Section 5.4.3)

$$\frac{kT \times 3100 \times f_m^2}{C \alpha f_m^2 \times (8.63 \times 10^{11})} < 10^{(-5+0.25)}.$$

This is satisfied for all channels if

$$\frac{C}{T} > 1.00 \times 10^{-14},$$

which requires that

$$10 \log\left(\frac{G_R}{T}\right) \geqslant 47.0 \text{ dB/K}.$$

Alternatively, argue as follows. Allowing for guard bands, the effect of square-law emphasis is to make the signal-to-noise ratios in the channels equal and approximately 1.2 dB greater than the signal-to-noise ratio of the SSB/FDM multiplex signal before the channels are separated. Assuming conventional loading, the mean value of the latter signal-to-noise ratio should therefore be not less than $(50 - 2.5 - 15 - 1.2)$ dB ($= 31.3$ dB). Since a mean-square frequency deviation of $(14.7)^2/10$ (MHz)^2 is possible (as in (iv)), an FM advantage of 7.8 dB is possible. Hence we require:

$$10 \log\left(\frac{G_R}{T}\right) > 187 + 31.3 + 10 \log k + 10 \log 3\,270\,000 - 7.8.$$

9.10 Communication by satellite using CDMA

Simultaneous digital communication between terrestrial stations is to be provided by a transponder in a satellite approximately 41 000 km distant.

The transponder receives and transmits on frequencies close to 8 GHz. It has bandwidth 20 MHz, has figure-of-merit (G_R/T, Exercise 9.9) $+ 7$ dB/K, delivers EIRP (Appendix 8) $+ 50$ dBW and contains a hard limiter (see note below). Absorption losses of 3 dB should be assumed in both up-links and down-links.

Two types of terrestrial stations are to be used:
(a) static stations with EIRP $+ 55$ dBW and figure-of-merit $+ 13$ dB/K, and
(b) portable stations with EIRP $+ 33$ dBW and figure-of-merit $+ 4$ dB/K.
Both types of terrestrial station use code-division multiple access (CDMA), in which each transmission is continuous and is spread to occupy the entire pass-band of the transponder by phase-reversal keying according to a code sequence (Section 7.4.2). Each transmission has a distinctive code sequence, associated with the receiver to which it is addressed. In a receiver, phase-reversal keying restores the wanted signal to a narrow-band binary DPSK waveform while leaving unchanged the power density of unwanted transmissions and noise. The DPSK waveform is subsequently demodulated by a differentially coherent demodulator and the digit error rate after demodulation is not to exceed 10^{-6}.

(i) Estimate the system noise temperature of a static terrestrial receiver if the diameter of its antenna is 1.5 m and the aperture efficiency is 0.7.

(ii)   Assuming a satellite system noise temperature of 1000 K, make a rough estimate of the diameter of the terrestrial region covered by the satellite antenna when the satellite is directly overhead.

(iii)   Calculate the maximum permissible number (M) of simultaneous 16 kbit/s transmissions (with different code sequences) to static receivers from static transmitters.

(iv)   Calculate the maximum rate at which digits can be transmitted between two portable stations, assuming that M static stations are also transmitting. Use the value of M calculated in (iii).

(v)   Assuming that the transponder is to be used exclusively and simultaneously by 100 portable transmitters transmitting to portable receivers, calculate the maximum rate at which digits can be transmitted.

(vi)   For the application considered in (v), calculate the maximum digit rate which could be obtained if the transponder bandwidth were suitably modified.

Note on hard limiter: Assume that the transponder consists of a low-noise amplifier (bandwidth 20 MHz), a frequency changer, a hard-limiter and an output amplifier. At the input to the hard-limiter, a given signal of power P_1 from one terrestrial transmitter may be regarded as embedded in a gaussian process of bandwidth 20 MHz and power P_2, made up of up-link noise and signals from other transmitters using different code sequences. The fraction of the transponder EIRP useful to a terrestrial receiver receiving the given signal is then

$$\frac{\pi P_1}{4(P_1 + P_2)}.$$

For further information see: Schwartz, J. W., Aein, J. M., and Kaiser, J., 'Modulation techniques for multiple access to a hard-limiting satellite repeater', *Proc. IEEE*, **54**, No. 5, pp. 763–77, May 1966.

Solution outline

(i) $10 \log (G_R/T) = 13 \, \text{dB/K}$, where $G_R = 4\pi A/\lambda^2$ and $A = 0.7\pi(0.75)^2$ (Appendix 8). Hence $T = 555$ K.

(ii)   The area covered is approximately $4\pi(41\,000)^2/G_R$ (km)^2, where $10 \log (G_R/1000) = 7$, and is therefore approximately that of a circle of diameter 2300 km.

(iii)   For the required error rate, $\frac{1}{2} \exp(-E/N_0) \leqslant 10^{-6}$ (Section 2.2.3) and hence $E/N_0 > 13.12$ after de-spreading in a terrestrial receiver. This requires the ratio of noise spectral density to signal power to be less than $(13.12 \times 16\,000)^{-1}$, i.e. 4.76×10^{-6}.

Since the G_R/T ratio for the satellite is 5.012 and the combined absorption and free-space losses for the 41 000 km path are 3.766×10^{20},

$$\frac{P_1}{P_1 + P_2} = \frac{10^{5.5} \times 5.012}{(M \times 10^{5.5} \times 5.012) + (k \times 2 \times 10^7 \times 3.766 \times 10^{20})}$$

$$= (M + 0.065\,58)^{-1}.$$

Hence the satellite EIRP useful for each transmission is $\pi 10^5/4(M + 0.065\,58)$. Assuming that the remainder of the satellite EIRP contributes accompanying noise uniformly distributed over the 20MHz band, for the required error rate

$$\frac{\left\{10^5 - \dfrac{\pi 10^5}{4(M + 0.065\,58)}\right\} \dfrac{G_R}{(2 \times 10^7)(3.766 \times 10^{20})} + kT}{\left\{\dfrac{\pi 10^5}{4(M + 0.065\,58)}\right\} \dfrac{G_R}{(3.766 \times 10^{20})}} < 4.76 \times 10^{-6},$$

where G_R and T refer to the terrestrial receiver and $G_R/T = 19.95$. Hence $M = 71$.

(iv)   In this case

$$P_1 = 10^{3.3}/(3.766 \times 10^{20})$$

and

$$P_2 = \{71 \times 10^{5.5}/(3.766 \times 10^{20})\} + \{k(2 \times 10^7)/5.012\}.$$

Hence the useful satellite EIRP is 6.97 W.

Since the G_R/T ratio for the terrestrial receiver is 2.512, the ratio of received signal power to 'noise' power spectral density is

$$\frac{6.97}{\{(10^5 - 6.97)/(2 \times 10^7)\} + \{k(3.766 \times 10^{20})/2.512\}} = 986.$$

Hence for an E/N_0 ratio of 13.12 a digit rate of $986/13.12 = 75.2$ bit/s is possible.

(v)   Proceeding as in (iii) but with terrestrial EIRP $10^{3.3}$ W and $M = 100$, transmission at a digit rate R requires

$$\frac{\left\{10^5 - \dfrac{\pi 10^5}{441.6}\right\} \dfrac{G_R}{(2 \times 10^7)(3.766 \times 10^{20})} + kT}{\left\{\dfrac{\pi 10^5}{441.6}\right\} \dfrac{G_R}{(3.766 \times 10^{20})}} < \frac{1}{13.12R}.$$

Since $G_R/T = 2.512$, it follows that R cannot exceed about 7700 bit/s.

(vi)   Denote the transponder bandwidth by B MHz. Proceeding as in (v) gives

$$\frac{\left[10^5 - \dfrac{\pi 10^5}{4\{100 + (B \times 0.5197)\}}\right] \dfrac{G_R}{(B \times 10^6)(3.766 \times 10^{20})} + kT}{\left[\dfrac{\pi 10^5}{4\{100 + (B \times 0.5197)\}}\right] \dfrac{G_R}{(3.766 \times 10^{20})}} < \frac{1}{13.12R},$$

where $G_R/T = 2.512$ again. The upper limit of R may be shown to be greatest when B is approximately 96 and is then about 12 860 bit/s.

Appendix

A.1 The Fourier transform

A.1.1 Definition of Fourier transform

The definition used in this book is that the Fourier transform $V_1(f)$ [or $V_2(\omega)$, where $\omega = 2\pi f$] of the function $v(t)$ is given by:

$$V_1(f) = \int_{-\infty}^{\infty} v(t)\exp(-j2\pi ft)\,dt \qquad \left[V_2(\omega) = \int_{-\infty}^{\infty} v(t)\exp(-j\omega t)\,dt\right].$$

The conditions for the existence of its Fourier transform are usually satisfied when $v(t)$ represents a finite-energy electrical signal, an electrical signal for which $\int_{-\infty}^{\infty} |v(t)|^2\,dt$ is finite. Note that when $v(t)$ is real,

$$V_1(-f) = V_1^*(f) \qquad [V_2(-\omega) = V_2^*(\omega)],$$

where the symbol * denotes complex conjugate.

The inverse transform is

$$v(t) = \int_{-\infty}^{\infty} V_1(f)\exp(j2\pi ft)\,df \qquad \left[= \frac{1}{2\pi}\int_{-\infty}^{\infty} V_2(\omega)\exp(j\omega t)\,d\omega\right].$$

Note that if $V_1(f)$ $[V_2(\omega)]$ is an even function of $f[\omega]$,

$$v(t) = 2\int_{0}^{\infty} V_1(f)\cos(2\pi ft)\,df \qquad \left[= \frac{1}{\pi}\int_{0}^{\infty} V_2(\omega)\cos(\omega t)\,d\omega\right].$$

The transforms of two simple but important functions are given in Fig. A.1.

The transform of the gaussian pulse $\exp(-\pi t^2)$ is the gaussian function $\exp(-\pi f^2)[\exp(-\omega^2/4\pi)]$.

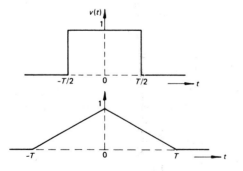

	$V_1(f)$	$V_2(\omega)$
	$T\dfrac{\sin(\pi fT)}{\pi fT}$	$T\dfrac{\sin(\omega T/2)}{(\omega T/2)}$
	$T\left\{\dfrac{\sin(\pi fT)}{\pi fT}\right\}^2$	$T\left\{\dfrac{\sin(\omega T/2)}{(\omega T/2)}\right\}^2$

Fig. A.1 Two important Fourier transforms

A.1.2   Properties of the Fourier transform

(i)   The transform of $v(at)$ is $1/a\, V_1\, (f/a)\, [1/a\, V_2\, (\omega/a)]$.

(ii)   The transform of $d/dt\, \{v(t)\}$ is $j2\pi f\, V_1\, (f)\, [j\omega V_2(\omega)]$.

(iii)   The transform of $v(t-t_0)$ is $V_1(f)\exp\left(-j2\pi f t_0\right)[V_2(\omega)\exp\left(-j\omega t_0\right)]$.

(iv)   The transform of $v(t)\exp\left(j2\pi f_0 t\right)[v(t)\exp\left(j\omega_0 t\right)]$ is $V_1(f-f_0)[V_2(\omega-\omega_0)]$.

(v)   If $U_1(f)\,[U_2(\omega)]$ is the transform of $u(t)$, the transform of $\int_{-\infty}^{\infty} u(\tau)v(t-\tau)\,d\tau$ is $U_1(f)V_1(f)[U_2(\omega)V_2(\omega)]$, i.e. the transform of the convolution of $u(t)$ and $v(t)$ is the product of the transforms of $u(t)$ and $v(t)$.

(vi)   The transform of $u(t)v(t)$ is $\int_{-\infty}^{\infty} U_1(x)\,V_1\,(f-x)\,dx\,[1/2\pi \int_{-\infty}^{\infty} U_2(x)\,V_2\,(\omega-x)\,dx]$.

(vii)   When $v(t)$ is real and finite-energy, its autocorrelation $R(\tau)$ is defined by:

$$R(\tau) = \int_{-\infty}^{\infty} v(t+\tau)v(t)\,dt.$$

$R(\tau)$ is an even function of τ and its transform is $|V_1(f)|^2\,[|V_2(\omega)|^2]$. Putting $\tau = 0$ in the inverse transformation,

$$R(0) = \int_{-\infty}^{\infty} v^2(t)\,dt = \int_{-\infty}^{\infty} |V_1(f)|^2\,df = \frac{1}{2\pi}\int_{-\infty}^{\infty} |V_2(\omega)|^2\,d\omega.$$

If $v(t)$ represents the voltage across (or the current through) a resistance of one ohm, $R(0)$ is the energy in $v(t)$ and therefore $|V_1(f)|^2\,[\,=|V_2(\omega)|^2\,]$ represents the energy per Hz of the spectrum.

A.1.3   Dirac's delta function

The Dirac delta function $\delta(x)$ is defined by

$$\delta(x) = 0 \qquad \text{if } x \neq 0$$

$$\int_{-\infty}^{\infty} \delta(x)\,dx = 1.$$

It has the useful 'sifting' property that

$$\int_{-\infty}^{\infty} g(x)\delta(x-x_0)\,dx = g(x_0).$$

The transform of $\delta(t-t_0)$ is $\exp\left(-j2\pi f t_0\right)[\exp\left(-j\omega t_0\right)]$ and the transform of $\exp\left(j2\pi f_0 t\right)[\exp\left(j\omega_0 t\right)]$ is $\delta(f-f_0)[2\pi\delta(\omega-\omega_0)]$.

A.1.4   Complex envelope and modulation theorem

When $v(t)$ is real and $V_1(f)\,[V_2(\omega)]$ is restricted to narrow bands around $-f_c$ and $+f_c$ $[-\omega_c$ and $+\omega_c]$, it is sometimes useful to write $v(t)$ in the form of a modulated sinewave thus:

$$v(t) = \mathrm{Re}\,[\tilde{v}(t)\exp\left(j2\pi f_c t\right)] \qquad [= \mathrm{Re}\,[\tilde{v}(t)\exp\left(j\omega_c t\right)]]$$

where $\tilde{v}(t)$ is called the complex envelope of $v(t)$. Then

$$V_1(f) = \tfrac{1}{2}\{\tilde{V}_1(f-f_c) + \tilde{V}_1^*(-f-f_c)\}$$

$$[V_2(\omega)] = \tfrac{1}{2}\{\tilde{V}_2(\omega-\omega_c) + \tilde{V}_2^*(-\omega-\omega_c)\}$$

where $\tilde{V}_1(f)[\tilde{V}_2(\omega)]$ is the transform of $\tilde{v}(t)$, a result sometimes called the 'modulation theorem'.

A.1.5 Periodic functions

The Fourier series representation of a periodic function $v(t)$ with period T is

$$v(t) = \sum_{n=-\infty}^{\infty} c_n \exp\left\{ j\frac{2\pi nt}{T} \right\},$$

where

$$c_n = \frac{1}{T} \int_{-T/2}^{T/2} v(t) \exp\left\{ -j\frac{2\pi nt}{T} \right\} dt.$$

Note that when $v(t)$ is real, $c_{-n} = c_n^*$. Two useful Fourier-series representations are shown in Fig. A.2.

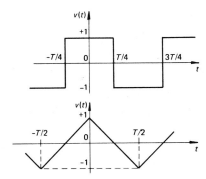

$$v(t) = \frac{4}{\pi}\left\{ \cos\left(2\pi t/\tau\right) - \frac{1}{3}\cos\left(6\pi t/T\right) \right.$$

$$\left. + \frac{1}{5}\cos\left(10\pi t/T\right) - \cdots \right\}$$

$$v(t) = \frac{8}{\pi^2}\left\{ \cos\left(2\pi t/T\right) + \left(\frac{1}{3}\right)^2 \cos\left(6\pi T/t\right) \right.$$

$$\left. + \left(\frac{1}{5}\right)^2 \cos\left(10\pi t/T\right) + \cdots \right\}$$

Fig. A.2 Examples of Fourier series

Although the definition of Fourier transform (Appendix 1.1) cannot be applied to a periodic function (it is not finite-energy), the periodic function $v(t)$ can be expressed as the inverse transform of a sum of delta functions thus:

$$v(t) = \int_{-\infty}^{\infty} V_1(f) \exp\left(j2\pi ft\right) df \qquad \left[= \frac{1}{2\pi} \int_{-\infty}^{\infty} V_2(\omega) \exp\left(j\omega t\right) d\omega \right],$$

where

$$V_1(f) = \sum_{n=-\infty}^{\infty} c_n \delta\left(f - \frac{n}{T}\right) \qquad \left[V_2(\omega) = \sum_{n=-\infty}^{\infty} c_n 2\pi\delta\left(\omega - \frac{2\pi n}{T}\right) \right].$$

When $v(t)$ is the train of equally spaced impulses

$$v(t) = \sum_{m=-\infty}^{\infty} \delta\left(t - mT\right),$$

$$V_1(f) = \frac{1}{T}\sum_{n=-\infty}^{\infty} \delta\left(f - \frac{n}{T}\right) \qquad \left[V_2(\omega) = \frac{2\pi}{T}\sum_{n=-\infty}^{\infty} \delta\left(\omega - \frac{2\pi n}{T}\right) \right].$$

The autocorrelation $R(\tau)$ of the real periodic function $v(t)$ is given by

$$R(\tau) = \frac{1}{T} \int_{-T/2}^{T/2} v(t+\tau)v(t)\,dt.$$

It may be shown that

$$R(\tau) = \sum_{n=-\infty}^{\infty} |c_n|^2 \exp\left\{j\frac{2\pi n\tau}{T}\right\}$$

and thus $R(\tau)$ may be regarded as the inverse transform of

$$\sum_{n=-\infty}^{\infty} |c_n|^2 \,\delta\left(f-\frac{n}{T}\right) \quad \left[\sum_{n=-\infty}^{\infty} |c_n|^2 \,2\pi\delta\left(\omega-\frac{2\pi n}{T}\right)\right],$$

where

$$|c_n|^2 = \frac{1}{T} \int_{-T/2}^{T/2} R(\tau) \exp\left\{-j\frac{2\pi n\tau}{T}\right\}d\tau,$$

or since $R(\tau)$ is real and even

$$|c_n|^2 = \frac{2}{T} \int_{0}^{T/2} R(\tau) \cos\left\{\frac{2\pi n\tau}{T}\right\} d\tau.$$

$|c_0|^2$ is the DC power in $v(t)$ and $2|c_n|^2$ is the n-th harmonic power in $v(t)$.

A periodic function $v(t)$ consisting of a finite-energy function $u(t)$ repeated at intervals of T may be written:

$$v(t) = \sum_{m=-\infty}^{\infty} u(t-mT).$$

By the sifting property of the delta function, $v(t)$ may be written as the convolution of $u(t)$ and a periodic train of impulses thus:

$$v(t) = \int_{-\infty}^{\infty} u(\tau)\left\{\sum_{m=-\infty}^{\infty} \delta(t-mT-\tau)\right\}d\tau.$$

Hence $v(t)$ is the inverse transform of the product

$$U_1(f)\frac{1}{T}\sum_{n=-\infty}^{\infty}\delta\left(f-\frac{n}{T}\right) \quad \left[U_2(\omega)\frac{2\pi}{T}\sum_{n=-\infty}^{\infty}\delta\left(\omega-\frac{2\pi n}{T}\right)\right],$$

where $U_1(f)$ $[U_2(\omega)]$ is the transform of $u(t)$. This expression may be interpreted as a line spectrum consisting of delta functions with envelope $U_1(f)/T$ $[U_2(\omega)2\pi/T]$. Thus the DC power in $v(t)$ is $|U_1(0)|^2/T^2$ $[|U_2(0)|^2/T^2]$ and the n-th harmonic power in $v(t)$ is

$$\frac{2}{T^2}\left|U_1\left(\frac{n}{T}\right)\right|^2 \quad \left[\frac{2}{T^2}\left|U_2\left(\frac{2\pi n}{T}\right)\right|^2\right].$$

The power density spectrum may be written

$$\frac{1}{T^2}|U_1(f)|^2 \sum_{n=-\infty}^{\infty}\delta\left(f-\frac{n}{T}\right) \quad \left[\frac{2\pi}{T^2}|U_2(\omega)|^2 \sum_{n=-\infty}^{\infty}\delta\left(\omega-\frac{2\pi n}{T}\right)\right],$$

where $|U_1(f)|^2$ $[|U_2(\omega)|^2]$ is the transform of the autocorrelation of $u(t)$.

A.1.6 Random processes

Useful electrical signals are functions of time which are to some degree unpredictable. Such functions are called random processes. We consider here random processes for which the associated probability distributions are invariant under time shifts and are therefore called stationary. We shall also assume ergodicity, which means that we need not distinguish between averages taken over an ensemble of different sample waveforms and averages taken over samples of one waveform taken at different times.

The autocorrelation $R(\tau)$ of such an ergodic random process may be defined by

$$R(\tau) = \lim_{T \to \infty} \left[\frac{1}{T} \int_{-T/2}^{T/2} v(t+\tau)v(t)\mathrm{d}t \right].$$

When $R(\tau)$ is a finite-energy function, its Fourier transform $S_1(f)\,[S_2(\omega)]$ is the power spectrum of the process and the relations:

$$S_1(f) = \int_{-\infty}^{\infty} R(\tau) \exp(-\mathrm{j}2\pi f\tau)\mathrm{d}\tau \qquad \left[S_2(\omega) = \int_{-\infty}^{\infty} R(\tau)\exp(-\mathrm{j}\omega\tau)\mathrm{d}\tau \right]$$

and

$$R(\tau) = \int_{-\infty}^{\infty} S_1(f) \exp(\mathrm{j}2\pi f\tau)\mathrm{d}f \qquad \left[= \frac{1}{2\pi} \int_{-\infty}^{\infty} S_2(\omega) \exp(\mathrm{j}\omega\tau)\,\mathrm{d}\omega \right]$$

are known as the Wiener-Khintchine relations. These relations are frequently used to determine the power spectra of electrical signals.

When a random process contains DC or periodic components, its autocorrelation is the sum of a finite-energy function and a DC or periodic function. In such a case the power spectrum is the sum of a continuous spectrum and a line spectrum. If the process is regarded as the sum of its ensemble average and a random process with a finite-energy autocorrelation function, the line contribution to its power spectrum is the power spectrum of its ensemble average and the continuous contribution is the power spectrum of the difference between the process and its ensemble average.

A.2 The response of linear channels

Denote the input to a linear channel by $u(t)$ with Fourier transform $U_1(f)\,[U_2(\omega)]$ and denote the output of that channel by $v(t)$ with Fourier transform $V_1(f)\,[V_2(\omega)]$. The input and output are related by the expression:

$$V_1(f) = U_1(f)\,H_1(f) \qquad [V_2(\omega) = U_2(\omega)\,H_2(\omega)],$$

where $H_1(f)\,[H_2(\omega)]$ is the transfer function of the channel. The transfer function is the ratio of the complex amplitudes of the output and the input when the input is a sinewave.

An alternative expression relating the output and the input is:

$$v(t) = \int_{-\infty}^{\infty} u(\tau)h(t-\tau)\mathrm{d}\tau = \int_{-\infty}^{\infty} h(\tau)u(t-\tau)\mathrm{d}\tau,$$

where $h(t)$ is the impulse response of the channel. $h(t)$ is the output when the input is $\delta(t)$. The Fourier transform of $h(t)$ is $H_1(f)\,[H_2(\omega)]$ and therefore for a realizable channel (for which $h(t)$ is real) $H_1(-f) = H_1^*(f)\,[H_2(-\omega) = H_2^*(\omega)]$.

Under certain conditions, the complex envelope of a signal passing through a band-pass channel behaves as if it were passing through a low-pass channel (the equivalent baseband channel) with transfer function $H_1(f+f_c)\,[H_2(\omega+\omega_c)]$. Those conditions obtain when f_c is greater than both the bandwidth of the signal and the bandwidth of the channel.

If the transfer function is written in modulus-argument form thus

$$H_2(\omega) = |H_2(\omega)| \exp\{-j\phi(\omega)\},$$

$\phi(\omega)$ represents the phase lag introduced by the channel. If in a band-pass channel $\phi(\omega)$ is approximately linear around $\omega = \omega_c$, the first-order Taylor expansion of $\phi(\omega)$ is

$$\phi(\omega) \simeq \phi(\omega_c) + (\omega - \omega_c)\frac{d\phi}{d\omega}\bigg|_{\omega = \omega_c}.$$

It follows that the channel delays the complex envelope of a bandpass signal by $|d\phi/d\omega|_{\omega = \omega_c}$, which is therefore the envelope (or group) delay of the channel.

A.3 rect (x) and sinc (x)

The rectangular function rect (x) is defined by

$$\text{rect}(x) = \begin{cases} 1 & |x| < \tfrac{1}{2} \\ 0 & |x| > \tfrac{1}{2} \end{cases}$$

and the function sinc (x) by

$$\text{sinc}(x) = \frac{\sin(\pi x)}{\pi x}.$$

The Fourier transform of rect (t/T) may be shown to be T sinc (fT). Hence by convolving rect (t/T) with itself the Fourier transform of a triangular pulse of unit height and width $2T$ may be shown to be T sinc^2 (fT). The shapes of sinc (x) and sinc^2 (x) are shown in Fig. A.3, which also illustrates the important property:

$$\int_{-\infty}^{\infty} \text{sinc}(x)dx = \int_{-\infty}^{\infty} \text{sinc}^2(x)dx = 1.$$

The inverse transform of rect (f/f_1) may be shown to be f_1 sinc $(f_1 t)$.

A.4 erfc (x)

The error function erf (x) is defined by

$$\text{erf}(x) = \frac{2}{\sqrt{\pi}} \int_0^x \exp(-y^2)dy$$

and the complementary error function erfc (x) by

$$\text{erfc}(x) = 1 - \text{erf}(x).$$

A graph of erfc (x) is given in Fig. A.4.

The probability density function $p(v)$ for a random variable v gaussian (or normally) distributed with zero mean and standard deviation σ is given by

$$p(v) = \frac{1}{\sigma\sqrt{2\pi}} \exp\left\{-\frac{v^2}{2\sigma^2}\right\}.$$

The probability $P(V)$ that v exceeds some value V is given by

$$P(V) = \int_v^{\infty} p(v)dv$$

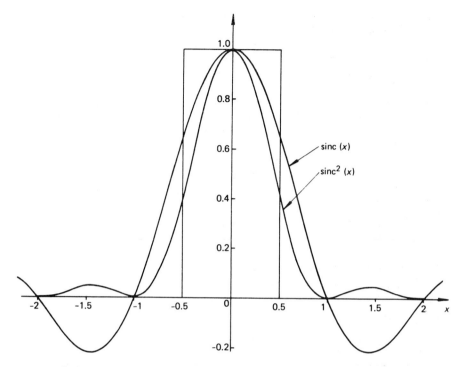

Fig. A.3 Graph of sinc(x) and sinc^2(x) showing rectangle of equal area

and is easily shown to equal $1/2$ erfc $\{V/\sigma\sqrt{2}\}$.

For large x,

$$\mathrm{erfc}\,(x) \sim (1/x\sqrt{\pi})\exp(-x^2).$$

A.5 Insertion gain, insertion loss and return loss

If the insertion of a two-port network between a sinusoidal source and a load (Fig. A.5) changes the power dissipated in the load from P_{20} to P_2, the network is said to have an insertion gain of $10\log(P_2/P_{20})$ dB and an insertion loss of $10\log(P_{20}/P_2)$ dB at the frequency of the source.

The return loss $S(\mathrm{j}\omega)$ of a one-port network at angular frequency ω is defined by

$$S(\mathrm{j}\omega) = -20\log|\rho(\mathrm{j}\omega)|\ \mathrm{dB}$$

where $\rho(\mathrm{j}\omega)$ is the reflexion coefficient, the ratio of the reflected voltage to the incident voltage. When connected to a source of resistance R, a one-port network of impedance $Z(\mathrm{j}\omega)$ has a reflexion coefficient

$$\rho(\mathrm{j}\omega) = \frac{Z(\mathrm{j}\omega) - R}{Z(\mathrm{j}\omega) + R}.$$

A.6 Noise temperature, noise bandwidth and noise factor

If the one-sided power spectral density of the noise available from a one-port network is $N(f)$ W/Hz at frequency f Hz, that network is said to have a noise temperature $N(f)/k$ K

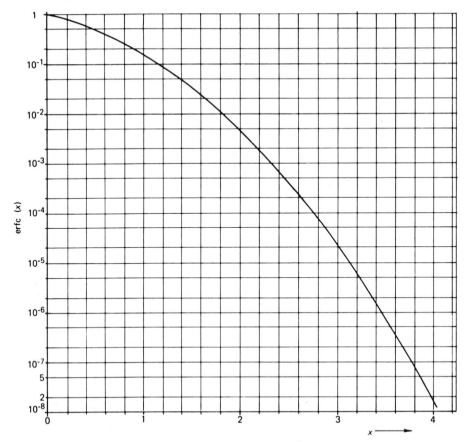

Fig. A.4 Graph of $\mathrm{erfc}(x) \equiv 1 - \dfrac{2}{\sqrt{\pi}} \displaystyle\int_0^x \exp(-y^2)\,\mathrm{d}y$

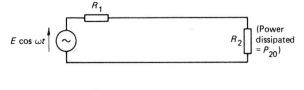

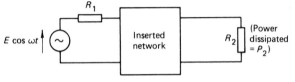

Fig. A.5

at frequency f, where k is Boltzmann's constant ($k \simeq 1.38 \times 10^{-23}$ J/K). The power spectral density of thermal noise from a resistor at a physical temperature T K is kT W/Hz and therefore the noise temperature of a resistor is equal to its absolute physical temperature in kelvins.

When the input to a two-port network with transfer function $H(f)$ is white noise with one-sided power spectral density N_0, the output noise power is

$$N_0 \int_0^\infty |H(f)|^2 \, \mathrm{d}f.$$

It is often convenient to regard this quantity as the product of $N_0|H(f_0)|^2$ (the output noise spectral density at a given frequency f_0) and B_n (the noise bandwidth of the network) defined by

$$B_n = \frac{\displaystyle\int_0^\infty |H(f)|^2 \, \mathrm{d}f}{|H(f_0)|^2}.$$

The noise factor F of a two-port network is defined as the ratio of the total noise power available at the output port to the noise power available at the output port due solely to thermal noise in the source connected to the input port. If the gain G of the network is defined as the power available at the output divided by the power available from the source, the noise power available at the output is kT_sGFB_n. Since F is dependent on the temperature (T_s K) of that source it is conventional to quote F for the case $T_s = 290$. Under certain conditions F is equal to the signal-to-noise ratio at the input divided by the signal-to-noise ratio at the output.

The noise temperature (T_N K) of a two-port network is defined as the noise temperature the source would have if at the network output the available noise power due solely to the source were equal to that due to noise originating within the network. T_N is independent of the noise actually produced by the source and is therefore more convenient than noise factor as a measure of network noise in situations in which the source noise temperature may change. When G is defined as above, the noise power available from the network may be written $k(T_s + T_N)GB_n$, from which it follows that $T_N = (F-1)T_s$.

In some applications network behaviour is frequency-dependent and interest is concentrated on a narrow band of frequencies. In such cases it is helpful to refer to the noise factor (or temperature) at a given frequency and in the above definitions of F and T_N 'noise power' should be replaced by 'noise power spectral density'.

A.7 FM advantage

Suppose that the waveform $v(t)$ at the input to an FM demodulator is the sum of the modulated wave

$$A\cos\left\{2\pi f_c t + \frac{\Delta f}{f_m}\sin 2\pi f_m t\right\}$$

and gaussian noise with one-sided power spectral density N_0 restricted to a frequency band of width B Hz centred on f_c. Writing $v(t)$ in the form

$$v(t) = A'(t)\cos\{2\pi f_c t + \phi(t)\}$$

and assuming the output of the demodulator to be $K\mathrm{d}\phi(t)/\mathrm{d}t$, the signal power at the demodulator output is $2\pi^2 K^2(\Delta f)^2$.

If the input signal power ($A^2/2$) is much greater than the input noise power (N_0B), the effects of modulation and noise may be regarded as independent and it may be shown (see also Section 5.4.3) that the one-sided noise power spectral density at the demodulator output is $8\pi^2 K^2 N_0 f^2/A^2$.

It follows that the effect of the demodulator followed by a low-pass filter cutting off at *b* Hz is to increase the signal-to-noise ratio by the factor

$$\frac{3}{2}\left(\frac{\Delta f}{b}\right)^2 \frac{B}{b},$$

which is sometimes called the FM improvement.

If the FM demodulator and low-pass filter are compared to an SSB demodulator with input noise restricted to a bandwidth *b*, for the same input signal power and noise power spectral density the output signal-to-noise ratio in the FM case is greater than that in the SSB case by the factor

$$\frac{3}{2}\left(\frac{\Delta f}{b}\right)^2,$$

which is usually called the FM advantage. In a form more appropriate for more general modulating waves this factor may be rewritten as $3\,(\Delta f_{\text{r.m.s.}}/b)^2$.

A.8 Systems aspects of antennas

The power gain $G\,(\theta, \phi)$ of a transmitting antenna in a given direction specified by polar co-ordinates θ and ϕ is defined as the ratio of the power per unit solid angle radiated in the given direction to the power which would be radiated per solid angle if all the power supplied to the antenna were radiated isotropically. When no direction is specified, power gain is understood to be the value of $G\,(\theta,\phi)$ in the direction in which it is greatest.

The effective area (or effective aperture) $A\,(\theta,\,\phi)$ of a receiving antenna is the ratio of the power available at its terminals to the power per unit area in a plane wave arriving from the direction specified by θ and ϕ, assuming that the polarisation of the plane wave is that which the antenna would radiate. For an aperture antenna, the ratio of the maximum value of $A\,(\theta,\,\phi)$ to the geometrical area is called the aperture efficiency.

For antennas that do not contain non-reciprocal devices, power gain and effective area vary in the same manner with direction and it may be shown[*] that

$$G\,(\theta,\phi) = \frac{4\pi}{\lambda^2}\,A\,(\theta,\phi),$$

where λ is the wavelength of the radiation.

If power P_R is available at the terminals of a receiving antenna when power P_T is supplied to a distant transmitting antenna, the ratio P_T/P_R is called the path loss. When propagation is line-of-sight and there are no losses due to absorption or scattering in the medium between the antennas or due to diffraction or reflexion by obstacles, 'free-space' conditions are said to obtain and

$$P_R = P_T G_T G_R \left(\frac{\lambda}{4\pi r}\right)^2,$$

where

    r   is the distance between the antennas (in the same units as λ),

    G_T is the power gain of the transmitting antenna in the direction of the receiving antenna,

    G_R is the power gain of the receiving antenna in the direction of the transmitting antenna.

---

* Panter, R. F., *Communication Systems Design: Line-of-Sight and Troposcatter Systems*, McGraw-Hill, 1972.

The quantity $(4\pi r/\lambda)^2$ is referred to as the 'free-space loss' or the 'basic transmission loss'. The product $P_T G_T$ is known as the 'equivalent isotropically radiated power' (EIRP).

For antennas transmitting or receiving plane-polarised waves, the concept of equivalent length is sometimes used. The equivalent length of an antenna is the length of a rod perpendicular to the direction of maximum radiation which, when carrying a uniform current equal to the current at the terminals of the antenna, would produce at a distant point in that direction the same field strength as that produced by the antenna. When an antenna is in the receiving mode, the open-circuit voltage appearing at its terminals is equal to the product of its effective length, and the component in the direction of its polarisation of the E-field which would exist in the absence of the antenna. For a plane-polarised plane wave, the root-mean-square value E of the E-field can be calculated from the power per unit area (power density) in W/m^2 by equating the latter to E^2/η, where η is approximately $120\pi \; \Omega$ for air. In other media η is approximately $120\pi \sqrt{\mu/\varepsilon}$, where μ is the relative permeability and ε the relative permittivity.

A.9 Information theory

A.9.1 Entropy

Consider a simple digital communication system in which a sequence x of symbols is transmitted over a noisy channel and received as the sequence y. The symbols constituting x and y are taken from an 'alphabet' of n symbols.

The probability $p(i)$ that a symbol selected for transmission is the i-th symbol of the alphabet is called the 'a priori' probability. If the symbols of x are chosen independently, the 'entropy' $H(x)$ of x is defined by:

$$H(x) = - \sum_{i=1}^{n} p(i) \log_2 p(i) \text{ bits.}$$

$H(x)$ may be interpreted as the average self-information per transmitted symbol. Choosing $p(i) = 1/n$ for all $1 \leqslant i \leqslant n$ gives the maximum possible value of $H(x)$ for a given value of n.

The conditional probability $p(i|j)$ that a signalling element received as the j-th symbol was transmitted as the i-th symbol is called the 'a posteriori' probability. The entropy $H(x|y)$ of the transmitted sequence when the received sequence is known is called the 'equivocation' and is defined by:

$$H(x|y) = - \sum_{i=1}^{n} \sum_{j=1}^{n} p(i,j) \log_2 p(i|j),$$

where $p(i,j)$ is the joint probability of the i-th symbol being transmitted and the j-th symbol received.

The information gained by the receiver when the i-th symbol is transmitted and the j-th symbol received is $\log_2 p(i|j) - \log_2 p(i)$. Averaging over all combinations of i and j gives for the information gained per symbol:

$$\sum_{i=1}^{n} \sum_{j=1}^{n} p(i,j) \{ \log_2 p(i|j) - \log_2 p(i) \},$$

which may be written $H(x) - H(x|y)$.

In analogue communication systems in which the transmitted signal is the continuous voltage waveform $v(t)$, the entropy H for each independent sample of $v(t)$ is defined by:

$$H = - \int_{-\infty}^{\infty} p(v) \log_2 p(v) \, dv \text{ bit/sample,}$$

where $p(v)$ is the probability density function of $v(t)$. The form of $p(v)$ which maximises H

for a given signal power is the gaussian distribution. When $p(v)$ is gaussian with mean square value N,

$$H = \ln \sqrt{(2\pi e N)}.$$

A.9.2 Shannon's mean power theorem

The capacity C of a channel of bandwidth W Hz in which a signal of average power S is accompanied by noise of power N is given by:

$$C = W\log_2\left(1 + \frac{S}{N}\right) \text{ bit/s.}$$

C is the maximum rate at which the channel can pass information with an arbitrarily small frequency of errors. The formula assumes that both signal and noise have gaussian statistical properties.

Bibliography

Background reading

The prior knowledge which the reader is assumed to have is available in a wide variety of books. Some of the more recently published relevant books are listed below.

1 Coates, R. W., *Modern Communications Systems*, Macmillan, Second Edition 1983
2 Connor, F. R., *Antennas*, Edward Arnold, 1972
3 Connor, F. R., *Modulation*, Edward Arnold, Second Edition 1982
4 Connor, F. R., *Networks*, Edward Arnold, 1972
5 Connor, F. R., *Noise*, Edward Arnold, Second Edition 1983
6 Connor, F. R., *Signals*, Edward Arnold, Second Edition 1983
7 Connor, F. R., *Wave Transmission*, Edward Arnold, 1972
8 Dunlop, J., and Smith, D. G., *Telecommunications Engineering*, Van Nostrand Reinhold (UK), 1984
9 Haykin, S., *Communication Systems*, Wiley, Second Edition 1983
10 Kennedy, G., *Electronic Communication Systems*, McGraw-Hill, Third Edition 1983
11 Lathi, B. P., *Modern Digital and Analog Communication Systems*, Holt-Saunders, 1983
12 Lynn, P. A., *An Introduction to the Analysis and Processing of Signals*, Macmillan, Second Edition 1982
13 Pettai, J. R., *Noise in Receiving Systems*, Wiley, 1984
14 Pierce, J. R., and Posner, E. C., *Introduction to Communication Science and Systems*, Plenum Press, 1980
15 Poularakis, A. D., and Seely, S., *Signals and Systems*, PWS Engineering, Boston, 1985
16 Roden, M. S., *Digital and Data Communication Systems*, Prentice-Hall, 1982
17 Shanmugan, K. S., *Digital and Analog Communication Systems*, Wiley, 1979
18 Stremler, F. G., *Introduction to Communication Systems*, Addison-Wesley, Second Edition, 1982
19 Usher, M. J., *Information Theory for Information Technologists*, Macmillan 1984

Further reading

Throughout the text of Chapters 1–8 there are references to publications in which the reader can find explanations or further information.

Journals

The attention of readers is drawn to the special issues of professional journals, which appear from time to time and which provide useful introductions to current trends and achievements in important areas of telecommunications. Some recent examples are the following.

1 *IEEE Trans.*, COM-30, No. 5, Part I, May 1982, Special Issue on Spread Spectrum Communications
2 *IEEE Trans.*, COM-30, No. 10, October 1982, Special Issue on Phase-locked Loops
3 *IEEE Trans.*, VT-33, No. 3, August 1984, Special Issue on Mobile Radio Communications
4 *Proc. IEEE*, **72**, No. 11, November 1984, Special Issue on Satellite Communication Networks
5 *IEEE Journal*, SAC-3, No. 1, January 1985, Special Issue on Broadcasting Satellites
6 *Proc. IEEE*, **73**, No. 4, April 1985, Special Issue on Visual Communication Systems

Index

A compression law, 10–11, 28–9
a posteriori probability, 40, 269
a priori probability, 39, 269
active channel, 114, 132
adjacent-channel crosstalk, *see* crosstalk
adjacent-channel rejection, 183, 196
aloha, 115
alternate mark inversion (AMI), 147, 149–50, 161–2, 176
amplitude shift keying (ASK), 40–41, 51–2, 57–8
amplitude-to-phase conversion, 149
angle diversity, 187
antenna gain, 268–9
antenna noise temperature, *see* noise
aperture efficiency, 268
aperture-to-medium coupling loss, 218–19
asynchronous digital transmission, 38
atmospheric absorption, 217–20
atmospheric noise, *see* noise
attenuation
    ionospheric, 222–3, 246–7
    line, 4, 20, 117, 136, 138
autocorrelation
    finite-energy signals, 260
    periodic functions, 262
    random processes, 145–6, 159–64, 263
automatic request repeat (ARQ), 42, 59, 65–6
Autospec code, 42, 59–60
availability, 94

Back porch, 75
bandwidth
    programme sound, 14
    telephony, 2
    video signal, 74, 81–2
Barker sequences, 151, 178–9
baseband signal, 124
basic transmission loss, *see* free-space loss
baud, the, 37–8, 51–2
BCH codes, 47–8, 64–6
Bennett's formula, 149, 171–2
bi-polar six-zero substitution (B6ZS), 150, 175–6
bi-polar violations, 150
blanking, 74–5
block codes, 42–8, 59–66
blocking, 94, 107

Boltzmann's constant (*k*), 266
boundary tracing, 78–9, 91
box-car waveform, 26–7
Bullington, estimate of diffraction loss, 232
busy channel, 114
busy hour, 96, 114–16

Calling rate, 106–7
carrier spacing, 113
Carson's rule, 124–6, 138–9
catastrophic error propagation, 68
central delta step, 12
cepstral analysis, 13
channel capacity, 270
channel, frequency hopping, 191
character interleaving, 150
characteristic polynomial, 190
chip-rate, 191
chirp, 188, 191–2, 212–13
chromaticity co-ordinates, 83–5
chrominance signals, 76–7
chrominance values, 76, 85
circuit switching, 95–6
coaxial cable, 117–18, 136
code division multiple access (CDMA), 191, 255–7
code polynomial, 45
code rate, 48
codes for error control, 59–70
coherence bandwidth, 223
coherent FSK, 41
coherent PSK, 41, 58
colorimeter, 75
colour burst, 76
colour triangle, 83–5
comb jammer, 210–11
companding
    advantage, 10
    instantaneous, 8–11, 27–9, 78
    near-instantaneous (NICAM), 16, 82
    syllabic, 7, 26
        in delta modulation, 12–13, 31
complementary colours, 83
complex envelope, 260–61
composite video signal, 74–5
concentrator, 93–4, 97–101
consecutive zeros, 48, 64–5

conventional loading, 115–16, 127, 136, 139–40
convolution, 260, 263
convolutional codes, 43, 48–51, 67–70
coset, 45, 60–61, 64
cosine spectrum pulse, 156–8
cosmic noise, *see* noise
critical angle, 222
critical frequency, 222
crosstalk
    adjacent channel, 3, 113, 121
    far end (FEXT), 3, 148, 165–7
    near end (NEXT), 3, 148, 165–7
    recommended levels, 3, 15
    stereophonic sound, 34
cut-off frequency of loaded cable, 5, 20
cyclic code, 45–8, 61–6

D layer, 221–2, 246–7
data transmission over carrier telephony systems, 127, 140–41
dBm0, 114
dBm0p, 114
dBr, 114
DC restoration, 127
de Bruijn diagram, 48–51, 67–70
decimation, 208–9
decision feedback, 42, 66–7
decision feedback equalisation, 145
dejitteriser, 152
delay distortion, 2, 51, 73, 127–8
delay probability, 100–101, 108–10
delta function, 260–62
delta modulation
    adaptive, 12–13, 31, 78
    exponential, 12, 30–32
    linear, 11–13, 29–30, 78
deltasigma modulation, 12, 30, 78
Deygout, estimate of diffraction loss, 217, 232–3
di-bit, 41, 52
differential PCM (DPCM), 77–8
differential phase distortion (television), 77, 86
differentially coherent demodulation, 41
differentially encoded PSK (DPSK), 41, 52, 250–51, 255–7
diffraction, 215–17, 221, 231–4
digit interleaving, 150, 180–81
digital encoding
    programme sound, 16, 33
    telephony, 8–14, 27–32
    two-tone images, 78–9, 87–91
    video signals, 77–8
dipulse code, 160–61
direct-sequence spread spectrum, 189–91
distortion due to errors
    delta modulation, 31–2

PCM, 27, 33
distributor, 94
dither, 16, 77
diversity branches, 187
diversity combining, 187–8, 202–5
Dolby noise-reduction systems, 15
double superhet, 196
duobinary transmission, 157–8

E layer, 221–2
echo suppressor, 3–4
echoes, 3–4, 6–7, 51, 124, 128
effective area, aperture, 268
effective height of obstacle, 231
effective radius of earth, 215, 227
emphasis
    programme sound, 17, 34–5, 128, 244
    SSB/FDM on cables, 121
    SSB/FDM/FM, 125, 138–9, 252–5
encoders, cyclic codes, 47, 63
energy spectrum, 260
ensemble average, 146, 263
entropy, 269–70
envelope delay, *see* group delay
Epstein and Peterson, estimate of diffraction loss, 232–3
equal-gain combining, 188, 205
equalisable distortion (FM), 126
equalisation, adaptive, 144–5
equivalent baseband channel, 263–4
equivalent four-wire repeaters, 120
equivalent isotropically radiated power (EIRP), 269
equivalent length
    antenna, 269
    slant paths, 217–19
equivalent peak power, 115
equivocation, 269
erf( ), erfc( ), 55, 264–6
ergodicity, 263
erlang, the, 96
Erlang's formula, 98–100, 104–5
error concealment, 16
error-correction coding, 42–51, 59–65
error-detection coding, 37, 42, 59
error extension, 150, 157–8
error probability, 38, 41
error rate, 38
expanders
    companding, *see* companding
    switching, 94
exponential delta modulation, *see* delta modulation
eye opening, 144, 158–9
eye pattern, 143–4, 158–9

F-layer, 221–2
facsimile, analogue, 73, 81, 127

fades
  duration of, 186, 201–2
  frequency of, *see* fading rate
fading
  fast and slow, 186, 219, 223
  mobile radio, 231, 251–2
  sky-wave, 223, 246–8
  troposcatter, 219, 250–51
  use of FSK, 40–41, 52
fading rate, 201, 219, 223, 251–2
far-end crosstalk (FEXT), *see* crosstalk
field frequency, 74
field synchronisation, 74
figure of merit (*G/T*), 252–5
finite-energy signal, 259
five-unit code, 37
flicker, 74
FM
  advantage, 267–8
  feedback (FMFB), 184–5, 200–201
  improvement, 197–8, 200–201, 268
  threshold, 183–5, 197, 201
forward scattering loss, 218–20
forward error correction, 42–51, 59–70
four-wire operation, 6–7
Fourier series, 261–2
Fourier transform, 259–64
frame, in TDM, 150
frame-alignment word (FAW), 150–51,
  176–80
free-space loss, 268–9
Fresnel zones, 216–17, 230–34
frequency diversity, 187, 246–7
frequency-division multiplex (FDM), 52, 113,
  252–3
frequency hopping, 188, 191–2, 210–12
frequency modulation in carrier telephony,
  122–6
frequency offset, 51, 121, 126
frequency-selective fading, 223, 246–7
frequency-shift keying (FSK), 40–41, 52, 58
front porch, 75
full availability, 94

Galactic noise, *see* noise
Galois field, 47, 64–5
gamma correction, 76
gaussian probability distribution, 55–8,
  264–6
gaussian random process, 186, 201–2
generating function, convolutional code,
  50–51, 69–70
generator matrix, 44–6, 60–65
generator polynomial, 46–8, 61–5
grading, 94–5, 105–6
granularity, 16
ground-wave prediction, 221, 235–6
group, carrier telephony, 121–3

group code, 44
group delay, 2, 127, 264
group pilot, 121, 127–8
group velocity, 237–8
grouping, 93–5
gyro frequency, 222

Hagelbarger code, 43
Hamming bound, 60–61
Hamming code, 42–3, 59–60
Hamming distance, 43–5, 48–51, 59–60
Hamming weight, 45
hard limiting, 255–7
harmonic distortion, 14–15
harmonic powers, 262
header, 96
Heaviside's distortionless condition, 4, 20
HF radio systems, 235–9, 246–50
hierarchies,
  PCM/TDM, 151–2, 180–81
  SSB/FDM, 121–2
high-density binary (HDB) encoding, 150,
  174–6
Holbrook and Dixon, 114–15
holding time, 96
hue, 77, 83
Huffman code, 78, 87–8
hybrid transformer, 3, 5–7, 21–5, 119–20,
  135
hypergroup, 122
hypothetical reference circuit, 116

I, Q chrominance signals, 76–7, 85–6
idling pattern, 12
image signals, 183
image frequency, 195–6
impulse response, 263
incoherent FSK, 41, 202–3
information feedback, 42
insertion gain, 265
insertion loss, 265
instantaneous companding, *see* companding
integrate-and-dump receiver, 38–9, 56–7,
  248–50
interference in FM systems, 125–6, 139, 244
interlaced scanning, 74
interleaved AMI, 147, 164
intermediate frequency (IF), 183, 196
intermodulation distortion, programme
  sound, 15
intermodulation interference, 116–19, 122–3,
  132
intermodulation noise, *see* intermodulation
  interference
intermodulation product accumulation,
  118–19, 132–4, 136–7,
intersymbol interference (ISI), 51, 143–5,
  154–9, 223, 250–51

inverse Fourier transform, 259
ionosphere
  propagation in, 222–3, 236–9
  sounding, 238–9
  structure, 221–2

Jitter, 127, 147–9, 152, 169–74
jitter accumulation, 148–9, 169–70, 172–4
justification, 151–2, 180–81

Kell factor, 74, 81–2
knife-edge diffraction, 216–17, 231–4

Laemmel code, 88–9
Lender's duobinary scheme, 158
lightning, interference due to, 225
likelihood functions, 40
limited availability, 94–5, 105–6
line build-out network, 119–20, 136
line codes, 149–50, 174–6
line synchronisation, 74–5
linear code, *see* group code
linear delta modulation, *see* delta modulation
linear shift registers, 190–91, 205–7
linked compressor-expander, 15, 128, 247–8
listening tests, 1, 14
loading, *see* lumped loading, *and* noise
  loading
local oscillator, 183, 195–6
locking range (PLL), 199
log-normal distribution, 3, 115, 131–2
loop filter (PLL), 184–5
lost-call probability, 97–100, 104–8
lost-calls-cleared, 98–100
lost-calls-held, 107–8
luminance, 73, 75–6
luminosity coefficients, 75–6, 83
lumped loading, 4, 20–21

Manchester code, 147, 160–61
mastergroup, 121–2
matched filter, 39, 56–8, 147, 212–13
maximal length sequences, 190–91, 206–9
maximal ratio combining, 187–8, 203–5, 250–51
maximum a-posteriori probability decision, 40
maximum likelihood decision, 40
maximum usable frequency, 222
message switching, 95–6
metric, 48–50
Millington, 235–6
minimum (Hamming) distance, 43–5, 48, 61–3
mistuning jitter, 149, 171–3
modem, 52, 127
modified refractive index, 227–8
modulation theorem, 260–61
μ-compression law, 10, 27–8

multi-channel peak factor, 115
multi-hop sky-wave transmission, 222–3, 236–7, 246–7
multitone signalling, 248–50

Narrow-band gaussian process, *see* gaussian
  random process
near-end crosstalk (NEXT), *see* crosstalk
near-instantaneously companded audio
  multiplex (NICAM), *see* companding
negative-exponential distribution, 96–7, 103
negative-impedance repeater, 7, 24
negative modulation, television, 74–5
noise,
  antenna, 223–5, 240
  atmospheric, 223–5
  cosmic, 224, 241–2
  galactic, 224, 241–2
  phase-thermal, 125, 138
  thermal, 115–17, 132, 136–7, 148, 167–9, 265–6
  sky, 223–4, 241
  solar, 224, 241
noise bandwidth, 198–9, 267
noise factor
  attenuator, 240
  cascaded networks, 239–40
  definition, 267
noise loading, 116–17, 139–40
noise margin, 40, 158–9
noise-power ratio, 117, 139–40
noise reduction, 15, 32
noise requirements
  programme sound, 15–16
  telephony, 2, 115–16
noise temperature
  definition, 265–6
  system, 223–41
non-coherent FSK, *see* incoherent FSK
non-return-to-zero (NRZ) pulse train, 159–60
non-systematic jitter, 148–9, 169
NTSC system, 76–7, 85–6
Nyquist transmission rate, 143
Nyquist's theorem on vestigial symmetry, 143, 154–5

Obstacle gain, 217
O'Dell's grading rule, 94
off-line operation, 37
Ohm's law of hearing, 2
on-line operation, 38
on–off keying, 40, 55, 57
outage rate, 186, 201–5, 251–2
overlap group, 178–80
overloading by FDM signal, 114–15, 136–7

Packet switching, 96
pair-selected ternary (PST) encoding, 176
PAL system, 76–7, 86–7

parity-check codes, 16, 42–8, 59–66
parity-check matrix, 44, 60, 62–3
parity-check polynomial, 46
partial-response signalling, 143, 157–8
path loss, 268
pattern-induced jitter, *see* systematic jitter
pels, picture elements, 78
phase-locked loop (PLL), 148–9, 184–5,
   197–200
phase plane, 199
phase-shift keying (PSK), 40–41, 188–9
phase-thermal noise, *see* noise
phonemes, 13
Piccolo, *see* multitone signalling
pilot frequencies in carrier telephony, 119–21,
   127–8
pilot-tone stereophonic system, 17, 33–5
pitch extraction, 13
plane-earth transmission formula, 230–31
plesiosynchronism, 151–2
Poisson distribution, 103–4
Poisson process, 96–7
polar modulation stereophonic system, 17
positive modulation, television, 74
power density, radiation, 269
power-separating filters, 119–20, 147–8
power spectrum, 145–6, 159–64, 263
pre-emphasis, *see* emphasis
primary PCM multiplex equipment, 150
primitive BCH codes, 48
primary colours, 75
process gain, 188–9, 191, 209–10
programme sound, channel requirements,
   14–15, 127–8
progressive grading, 94–5
propagation constant (γ) of line, 4
pseudonoise modulation, 188–91, 209–10,
   245
pseudoternary encoding, 147, 149–50, 161–4
psophometer, 2, 15, 114, 136–7
pulse-code modulation (PCM), 8–11, 16,
   27–9, 33, 77, 82, 140
pulse compression, 191–2
pulse stretcher, 26
pulse stuffing, *see* justification
pure-chance traffic, 96–7, 103–4

Quadrature VSB transmission, 140–41
quantisation distortion
   delta modulation 12, 30–31
   PCM, 8–11, 16, 27–8, 33, 140
quaternary PSK (QPSK), 41, 52, 58

Radio-frequency (RF) amplifier, 183, 195–6
radio horizon, 227
raised-cosine-spectrum pulse, 143, 155–6,
   158–9
rake, 223

random access, 110–11
random process, 263
Rayleigh fading, 186, 201–3
Rayleigh probability distribution, 57, 186–7,
   201
Rayleigh's law of scattering, 217
recovery effect, 235–6
rect( ), 264
rectangular-spectrum pulse, 154
redundant ternary encoding, 147–8, 150,
   167–9, 174–6
reference atmosphere, 215, 227
reference white, 75, 83–4
reflexion at atmospheric layer, 228–30
reflexion at earth's surface, 215, 230–31
reflexion coefficient, 215, 228, 265
refractive index
   ionosphere, 222
   modified, 227–8
   troposphere, 215
regulation, 119–20
repeater
   coaxial cable, 119–21, 135–7
   regenerative, 147–8, 164–5
   telephone circuit, 6–7, 22, 24–5
resonant circuit, 171, 195, 248–50
return loss, 265
return loss measurement, 23–4
return-to-zero (RZ) pulse train, 145–6
rician distribution, 186–7, 223
run-length encoding, 78, 87–9

Satellite systems, 217–20, 241, 252–7
saturation, colour television, 77, 83
scanning, television and facsimile, 73–4
scatter volume, 217
scattering angle, 218, 250
scattering by rain, 217–19
scrambling, 52, 70–71
SECAM system, 77
section, of repeatered link, 118, 147
selection combining, 187, 202–5, 250–52
Shannon's mean-power theorem, 270
shift-and-add property, 191
sidetone, 3, 5
signalling speed, 37–8, 51–2
sinc ( ), 264–5
skip distance, 222, 236–9
sky noise, *see* noise
sky wave, 222
slope clipping in delta modulation, 12–13
slotted aloha, 111
solar noise, *see* noise
space diversity, 187, 250–52
space switching, 93
spectra of digital signals, 145–7
spectrum shaping, 146–7
speech interpolation, 108–10

speech voltmeter, 2
spread-spectrum transmission, 188–92, 209–13
spurious responses, 195–6
SSB/FDM signal statistics, 114–15
SSB/FDM/FM signal spectrum, 124–5, 138–9
standard array, 45, 60, 64
start-stop telegraphy, 38
state-transition diagram, 176–8
stationarity, 263
stereophonic broadcasting, 17, 33–5
store-and-forward telegraphy, 109–10
Strowger exchanges, 93
submarine carrier telephony, 113, 120–21
supergroup, 121–2, 138, 140
superhet receiver, 183, 195–6
supermaster group, 121–2
surface wave, 221, 235–6
survivor, 48–50
syllabic compandor, *see* companding
syllabically companded delta modulation, *see* companding
synchronous digital transmission, 38
syndrome, 44–6, 60, 62–4
system noise temperature, *see* noise temperature
system value, 243–4
systematic convolutional code, 69
systematic jitter, 149, 169–74

Telegraph codes, 37
telegraph distortion, 38–9
telephony, channel requirements, 2–4
television
    colour, 75–7, 83–7
    monochrome, 73–5, 81–2
television over carrier-telephony systems, 128
telex, 37
ternary signalling, 55–6
test tone, 2, 113–14
thermal noise, *see* noise
threshold-misalignment jitter, 149, 173–4
time-assigned speech interpolation (TASI), *see* speech interpolation

time-division multiplex (TDM), 150–52
timing extraction, 147–9, 169–74
torn-tape, 95, 109
traffic, 96–7
transfer function, 263–4
transmission deviations, SSB/FDM/FM, 124, 126
transparency, line codes, 149
transversal filter, 51, 144–5, 159
trellis diagram, 48–50, 67
tributary, 151–2, 180–81
tri-stimulus values, 75
troposcatter, 217–20, 234–5, 250–51
tuned circuit, *see* resonant circuit
two-tone telegraphy, 246–7
two-wire operation, 6–7

U, V chrominance signals, 76, 85–6
urban UHF mobile radio, 251–2

Van Duuren code, 37, 42, 59, 65–6
vestigial-sideband (VSB) transmission, 40, 74, 127–8, 140–41
vestigial symmetry,
    frequency domain, 140–41, 154–5
    time domain, 143
video signal, 73–5
virtual height, 237–9
virtual path, 237–9
visual acuity, 73, 81
Viterbi decoding, 48–50, 67
vocoder, 13–14
voice-frequency telegraphy (VFT), 52, 126–7
voiced/unvoiced phonemes, 13
volume, 3, 114, 131–2
volume unit (vu), 3

Waiting time, 100–101, 109–10
white-block-skipping (WBS) encoding, 78–9, 89–90
white-noise loading, *see* noise loading
white-noise test set, 116–17, 139–40
Wiener-Khintchine relations, 263

Zero-relative-level point, 113–14